THE ANCIENT LIFE-HISTORY OF THE EARTH

International Science Library

THE ANCIENT LIFE-HISTORY OF THE EARTH

A COMPREHENSIVE OUTLINE OF THE PRINCIPLES AND LEADING FACTS OF PALAEONTOLOGICAL SCIENCE

BY

H. ALLEYNE NICHOLSON

M. D., D. Sc., M. A., Ph. D. (Gött.), F. R. S. E., F. L. S.

PROFESSOR OF NATURAL HISTORY IN THE UNIVERSITY OF ST. ANDREWS

The Werner Company
Book Manufacturers
Akron, Ohio

WERNER EDITION

PREFACE.

THE study of Palæontology, or the science which is concerned with the living beings which flourished upon the globe during past periods of its history, may be pursued by two parallel but essentially distinct paths. By the one method of inquiry, we may study the anatomical characters and structure of the innumerable extinct forms of life which lie buried in the rocks simply as so many organisms, with but a slight and secondary reference to the *time* at which they lived. By the other method, fossil animals are regarded principally as so many landmarks in the ancient records of the world, and are studied *historically* and as regards their relations to the chronological succession of the strata in which they are entombed. In so doing, it is of course impossible to wholly ignore their structural characters, and their relationships with animals now living upon the earth; but these points are held to occupy a subordinate place, and to require nothing more than a comparatively general attention.

In a former work, the Author has endeavored to furnish a summary of the more important facts of Palæontology regarded in its strictly scientific aspect, as a mere department of the great science of Biology. The present work, on the other hand, is an attempt to treat Palæontology more especially from its historical side, and in its more intimate relations with Geology. In accordance with this object, the introductory portion of the work is devoted to a consideration of the general principles of Palæontology, and the bearings of this science upon various geological problems—such as the mode of formation of the sedimentary rocks, the reactions of living beings upon the crust of the earth, and the sequence in time of the fossiliferous formations. The second portion of the work deals exclusively with Historical Palæontology, each formation being considered separately, as regards its lithological nature and subdivisions, its relations to other formations, its geographical distribution, its mode of origin, and its characteristic life-forms.

In the consideration of the characteristic fossils of each successive period, a general account is given of their more important zoological characters and their relations to living forms; but the technical language of Zoology has been avoided, and the aid of illustrations has been freely called into use. It may therefore be hoped that the work may be found to be available for the purposes of both the Geological and the Zoological student; since it is essentially an outline of Historical Palæontology, and the student of either of the above-mentioned sciences must perforce possess some knowledge of the last. Whilst primarily intended for students, it may be added that the method of treatment adopted has been so far untechnical as not to render the work useless to the general reader who may desire to acquire some knowledge of the subject of such vast and universal interest.

In carrying out the object which he has held before him, the Author can hardly expect, from the nature of the materials with which he has to deal, that he has kept himself absolutely clear of errors, both of omission and commission. The subject, however, is one to which he has devoted the labor of many years, both in studying the researches of others and in personal investigations of his own; and he can only trust that such errors as may exist will be found to belong chiefly to the former class, and to be neither serious nor numerous. It need only be added that the work is necessarily very limited in its scope, and that the necessity of not assuming a thorough previous acquaintance with Natural History in the reader has inexorably restricted its range still further. The Author does not, therefore, profess to have given more than a merely general outline of the subject; and those who desire to obtain a more minute and detailed knowledge of Palæontology, must have recourse to other and more elaborate treatises.

UNITED COLLEGE, ST. ANDREWS.
October 2, 1876.

CONTENTS.

PART I.

PRINCIPLES OF PALÆONTOLOGY.

INTRODUCTION.

CHAPTER III.

CHAPTER IV.

CHAPTER V.

CHAPTER VI.

PART II.

HISTORICAL PALÆONTOLOGY.

CHAPTER VII.

CHAPTER VIII.

CHAPTER IX.

CHAPTER X.

CHAPTER XI.

CHAPTER XII.

CHAPTER XIII.

CHAPTER XIV.

CHAPTER XV.

CHAPTER XVI.

CHAPTER XVII.

CHAPTER XVIII.

CHAPTER XIX.

CHAPTER XX.

CHAPTER XXI.

CHAPTER XXII.

CHAPTER XXIII.

LIST OF ILLUSTRATIONS.

PART I.

PRINCIPLES OF PALÆONTOLOGY.

THE
ANCIENT LIFE-HISTORY
OF
THE EARTH.

INTRODUCTION.

The Laws of Geological Action.

Under the general title of "Geology" are usually included at least two distinct branches of inquiry, allied to one another in the closest manner, and yet so distinct as to be largely capable of separate study. *Geology,** in its strictest sense, is the science which is concerned with the investigation of the materials which compose the earth, the methods in which those materials have been arranged, and the causes and modes of origin of these arrangements. In this limited aspect, Geology is nothing more than the Physical Geography of the past, just as Physical Geography is the Geology of to-day; and though it has to call in the aid of Physics, Astronomy, Mineralogy, Chemistry, and other allies more remote, it is in itself a perfectly distinct and individual study. One has, however, only to cross the threshold of Geology to discover that the field and scope of the science cannot be thus rigidly limited to purely physical problems. The study of the physical development of the earth throughout past ages brings us at once in contact with the forms of animal and vegetable life which peopled its surface in bygone epochs, and it is found impossible adequately to com-

* Gr. *gē*, the earth; *logos*, a discourse.

prehend the former, unless we possess some knowledge of the latter. However great its physical advantages may be, Geology remains imperfect till it is wedded with Palæontology,* a study which essentially belongs to the vast complex of the Biological Sciences, but at the same time has its strictly geological side. Dealing, as it does, wholly with the consideration of such living beings as do not belong exclusively to the present order of things, Palæontology is, in reality, a branch of Natural History, and may be regarded as substantially the Zoology and Botany of the past. It is the ancient life-history of the earth, as revealed to us by the labors of palæontologists, with which we have mainly to do here; but before entering upon this, there are some general questions, affecting Geology and Palæontology alike, which may be very briefly discussed.

The working geologist, dealing in the main with purely physical problems, has for his object to determine the material structure of the earth, and to investigate, as far as may be, the long chain of causes of which that structure is the ultimate result. No wider or more extended field of inquiry could be found; but philosophical geology is not content with this. At all the confines of his science, the transcendental geologist finds himself confronted with some of the most stupendous problems which have ever engaged the restless intellect of humanity. The origin and primæval constitution of the terrestrial globe, the laws and geologic action through long ages of vicissitude and development, the origin of life, the nature and source of the myriad complexities of living beings, the advent of man, possibly even the future history of the earth, are amongst the questions with which the geologist has to grapple in his higher capacity.

These are problems which have occupied the attention of philosophers in every age of the world, and in periods long antecedent to the existence of a science of geology. The mere existence of cosmogonies in the religion of almost every nation, both ancient and modern, is a sufficient proof of the eager desire of the human mind to know something of the origin of the earth on which we tread. Every human being who has gazed on the vast panorama of the universe, though it may have been but with the eyes of a child, has felt the longing to solve, however imperfectly, "the riddle of the painful earth," and has, consciously or unconsciously, elaborated some sort of a theory as to the why and wherefore of what he sees. Apart from the

* Gr. *palaios*, ancient; *onta*, beings; *logos*, discourse.

profound and perhaps inscrutable problems which lie at the bottom of human existence, men have in all ages invented theories to explain the common phenomena of the material universe; and most of these theories, however varied in their details, turn out on examination to have a common root, and to be based on the same elements. Modern geology has its own theories on the same subject, and it will be well to glance for a moment at the principles underlying the old and the new views.

It has been maintained, as a metaphysical hypothesis, that there exists in the mind of man an inherent principle, in virtue of which he believes and expects that what has been, will be; and that the course of nature will be a continuous and uninterrupted one. So far, however, from any such belief existing as a necessary consequence of the constitution of the human mind, the real fact seems to be that the contrary belief has been almost universally prevalent. In all old religions, and in the philosophical systems of almost all ancient nations, the order of the universe has been regarded as distinctly unstable, mutable, and temporary. A beginning and an end have always been assumed, and the course of terrestrial events between these two indefinite points has been regarded as liable to constant interruption by revolutions and catastrophes of different kinds, in many cases emanating from supernatural sources. Few of the more ancient theological creeds, and still fewer of the ancient philosophies, attained body and shape without containing, in some form or other, the belief in the existence of periodical convulsions, and of alternating cycles of destruction and repair.

That geology, in its early infancy, should have become imbued with the spirit of this belief, is no more than might have been expected; and hence arose the at one time powerful and generally-accepted doctrine of "Catastrophism." That the succession of phenomena upon the globe, whereby the earth's crust had assumed the configuration and composition which we find it to possess, had been a discontinuous and broken succession, was the almost inevitable conclusion of the older geologists. Everywhere in their study of the rocks they met with apparently impassable gaps, and breaches of continuity that could not be bridged over. Everywhere they found themselves conducted abruptly from one system of deposits to others totally different in mineral character or in stratigraphical position. Everywhere they discovered that well-marked and easily recognizable groups of animals and plants were succeeded, without the intermediation of any obvious lapse of time, by other assemblages of organic beings of a different character. Everywhere they found evidence that the earth's crust had undergone changes of such magnitude as to render it seemingly irrational to suppose that they could have been produced by any process now in existence. If we add to the above the prevalent belief of the time as to the comparative brevity of the period which had elapsed since the birth of the globe, we can readily understand the general acceptance of some form of catastrophism amongst the earlier geologists.

As regards its general sense and substance, the doctrine of catastrophism held that the history of the earth, since first it emerged from the primitive chaos, had been one of periods of repose, alternating with catastrophes and cataclysms of a more or less violent character. The periods of tranquillity were supposed to have been long and protracted; and during each of them it was thought that one of the great geological "formations" was deposited. In each of these periods, therefore, the condition of the earth was supposed to be much the same as it is now—sediment was quietly accumulated at the bottom of the sea, and animals and plants flourished uninterruptedly and in successive generations. Each period of tranquillity, however, was believed to have been, sooner or later, put an end to by a sudden and awful convulsion of nature, ushering in a brief and paroxysmal period, in which the great physical forces were unchained and permitted to spring into a portentous activity. The forces of subterranean fire, with their concomitant phenomena of earthquake and volcano, were chiefly relied upon as the efficient cause of these periods of spasm and revolution. Enormous elevations of portions of the earth's crust were thus believed to be produced, accompanied by corresponding and equally gigantic depressions of other portions. In this way new ranges of mountains were produced, and previously existing ranges levelled with the ground, seas were converted into dry land, and continents buried beneath the ocean—catastrophe following catastrophe, till the earth was rendered uninhabitable, and its races of animals and plants were extinguished, never to reappear in the same form. Finally, it was believed that this feverish activity ultimately died out, and that the ancient peace once more came to reign upon the earth. As the abnormal throes and convulsions began to be relieved, the dry land and sea once more resumed their relations of stability, the conditions of life were once more established, and new races of animals and plants sprang into existence, to last until the supervention of another fever-fit.

Such is the past history of the globe, as sketched for us, in alternating scenes of fruitful peace and revolutionary destruction, by the earlier geologists. As before said, we cannot wonder at the former general acceptance of Catastrophistic doctrines. Even in the light of our present widely-increased knowledge, the series of geological monuments remains a broken and imperfect one; nor can we ever hope to fill up completely the numerous gaps with which the geological record is defaced. Catastrophism was the natural method of accounting for these gaps, and, as we shall see, it possesses a basis of truth. At present, however, catastrophism may be said to be nearly extinct, and its place is taken by the modern doctrine of "Continuity" or "Uniformity"—a doctrine with which the name of Lyell must ever remain imperishably associated.

The fundamental thesis of the doctrine of Uniformity is, that, in spite of all apparent violations of continuity, the sequence of geological phenomena has in reality been a regular and uninterrupted one; and that the vast changes which can be shown to have passed over the earth in former periods have been the result of the slow and ceaseless working of the ordinary physical forces—acting with no greater intensity than they do now, but acting through enormously prolonged periods. The essential element in the theory of Continuity is to be found in the allotment of indefinite time for the accomplishment of the known series of geological changes. It is obviously the case, namely, that there are two possible explanations of all phenomena which lie so far concealed in "the dark backward and abysm of time," that we can have no direct knowledge of the manner in which they were produced. We may, on the one hand, suppose them to be the result of some very powerful cause, acting through a short period of time. That is Catastrophism. Or, we may suppose them to be caused by a much weaker force operating through a proportionately prolonged period. This is the view of the Uniformitarians. It is a question of *energy* versus *time;* and it is *time* which is the true element of the case. An earthquake may remove a mountain in the course of a few seconds; but the dropping of the gentle rain will do the same, if we extend its operations over a millennium. And this is true of all agencies which are now at work, or ever have been at work, upon our planet. The Catastrophists, believing that the globe is but, as it were, the birth of yesterday, were driven of necessity to the conclusion that its history had been checkered by the intermittent action of paroxysmal and almost inconceivably potent forces. The Uniformitarians, on the other hand, maintaining the "adequacy of existing causes," and denying that the known physical forces ever acted in past time with greater intensity than they do at present, are, equally of necessity, driven to the conclusion that the world is truly in its "hoary eld," and that its present state is really the result of the tranquil and regulated action of known forces through unnumbered and innumerable centuries.

The most important point for us, in the present connection, is the bearing of these opposing doctrines upon the question as to the origin of the existing terrestrial order. On any doctrine of uniformity that order has been evolved slowly, and, according to law, from a pre-existing order. Any doctrine of catastrophism, on the other hand, carries with it, by implication, the belief that the present order of things was brought about suddenly and irrespective of any pre-existent order; and it is important to hold clear ideas as to which of these beliefs is the true one. In the first place, we may postulate that the world had a beginning, and, equally, that the existing terrestrial order had a beginning. However far back we may go, geology does not, and cannot, reach the actual beginning of the world; and we are, therefore, left simply to our own speculations on this point. With regard, however, to the existing terrestrial order, a great deal can be discovered, and to do so is one of the principal tasks of geological science. The first steps in the production of that order lie buried in the profound and unsearchable depths of a past so prolonged as to present itself to our finite minds as almost an eternity. The last steps are in the prophetic future, and can be but dimly guessed at. Between the remote past and the distant future, we have, however, a long period which is fairly open to inspection; and in saying a "long" period, it is to be borne in mind that this term is used in its *geological* sense. Within this period, enormously long as it is when measured by human standards, we can trace with reasonable certainty the progressive march of events, and can determine the laws of geological action, by which the present order of things has been brought about.

The natural belief on this subject doubtless is, that the world, such as we now see it, possessed its present form and configuration from the beginning. Nothing can be more natural than the belief that the present continents and oceans have always been where they are now; that we have always had the same mountains and plains; that our rivers have always had their present courses, and our lakes their present positions, that our climate has always been the same; and that our animals and plants have always been identical with those now familiar to us. Nothing could be more natural than such a belief, and nothing could be further removed from the actual truth. On the contrary, a very slight acquaintance with geology shows us, in the words of Sir John Herschel, that "the actual configuration of our continents and islands, the coast-lines of our maps, the direction and elevation of our mountain-chains, the courses of our rivers, and the soundings of our oceans, are not things primordially arranged in the construction of our globe, but results of successive and complex actions on a former state of things; *that,* again, of similar actions on another still more remote; and so on, till the original and really permanent state is pushed altogether out of sight and beyond the reach even of imagination; while on the other hand, a similar, and, as far as we can see, interminable vista is opened out for the future, by which the habitability of our planet is secured amid the total abolition on it of the present theatres of terrestrial life."

Geology, then, teaches us that the physical features which now distinguish the earth's surface have been produced as the ultimate result of an almost endless succession of precedent changes. Palæontology teaches us, though not yet in such assured accents, the same lesson. Our present animals and plants have not been produced, in their innumerable forms, each as we now know it, as the sudden, collective, and simultaneous birth of a renovated world. On the contrary, we have the clearest evidence that some of our existing animals and plants made their appearance upon the earth at a much earlier period than others. In the confederation of animated nature some races can boast of an immemorial antiquity, whilst others are comparative *parvenus.* We have also the clearest evidence that the animals and plants which now inhabit the globe have been preceded, over and over again, by other different assemblages of animals and plants, which have flourished in successive periods of the earth's history, have reached their culmination, and then have given way to a fresh series of living beings. We have, finally, the clearest evidence that these successive groups of animals and plants (faunæ and floræ) are to a greater or less extent directly connected with one another. Each group is, to a greater or less extent, the lineal descendant of the group which immediately preceded it in point of time, and is more or less fully concerned with giving origin to the group which immediately follows it. That this law of "evolution" has prevailed to a great extent is quite certain; but it does not meet all the exigencies of the case, and it is probable that its action has been supplemented by some still unknown law of a different character.

We shall have to consider the question of geological "continuity" again. In the meanwhile, it is sufficient to state that this doctrine is now almost universally accepted as the basis of all inquiries, both in the domain of geology and that of palæontology. The advocates of continuity possess one of immense advantage over those who believe in violent and revolutionary convulsions, that they call into play only agencies of which we have actual knowledge. We *know* that certain forces are now at work, producing certain modifications in the present condition of the globe; and we *know* that these forces are capable of producing the vastest of the changes which geology brings under our consideration, provided we assign a time proportionately vast for their operation. On the other hand, the advocates of catastrophism, to make good their views, are compelled to invoke forces and actions, both destructive and restorative, of which we have, and can have, no direct knowledge. They endow the whirlwind and the earthquake, the central fire and the rain from heaven, with powers as mighty as ever imagined in fable, and they build up the fragments of a repeatedly shattered world by the intervention of an intermittently active creative power.

It should not be forgotten, however, that from one point of view there is a truth in catastrophism which is sometimes overlooked by the advocates of continuity and uniformity. Catastrophism has, as its essential feature, the proposition that the known and existing forces of the earth at one time acted with much greater intensity and violence than they do at present, and they carry down the period of this excessive action to the commencement of the present terrestrial order. The Uniformitarians, in effect, deny this proposition, at any rate as regards any period of the earth's history of which we have actual cognizance. If, however, the "nebular hypothesis" of the origin of the universe be well founded—as is generally admitted—then, beyond question, the earth is a gradually cooling body, which has at one time been very much hotter than it is at present. There has been a time, therefore, in which the igneous forces of the earth, to which we owe the phenomena of earthquakes and volcanoes, must have been far more intensely active than we can conceive of from anything that we can see at the present day. By the same hypothesis, the sun is a cooling body, and must have at one time possessed a much higher temperature than it has at present. But increased heat of the sun would seriously alter the existing conditions affecting the evaporation and precipitation of moisture on our earth; and hence the aqueous forces may also have acted at one time more powerfully than they do now. The fundamental principle of catastrophism is, therefore, not wholly vicious; and we have reason to think that there must have been periods—very remote, it is true, and perhaps unrecorded in the history of the earth—in which the known physical forces may have acted with an intensity much greater than direct observation would lead us to imagine. And this may be believed, altogether irrespective of those great secular changes by which hot or cold epochs are produced, and which can hardly be called "catastrophistic," as they are produced gradually, and are liable to recur at definite intervals.

Admitting, then, that there *is* a truth at the bottom of the once current doctrines of catastrophism, still it remains certain that the history of the earth has been one of *law* in all past time, as it is now. Nor need we shrink back affrighted at the vastness of the conception—the vaster for its very vagueness—that we are thus compelled to form as to the duration of *geological time.* As we grope our way backward through the dark labyrinth of the ages, epoch succeeds to epoch, and period to period, each looming more gigantic in its outlines and more shadowy in its features, as it rises, dimly revealed, from the mist and vapor of an older and ever-older past. It is useless to add century to century or millennium to millennium. When we pass a certain boundary-line, which, after all, is reached very soon, figures cease to convey to our finite faculties any real notion of the periods with which we have to deal. The astronomer can employ material illustrations to give form and substance to our conceptions of celestial space; but such a resource is unavailable to the geologist. The few thousand years of which we have historical evidence sink into absolute insignificance besides the unnnumbered æons which unroll themselves one by one as we penetrate the dim recesses of the past, and decipher with feeble vision the ponderous volumes in which the record of the earth is written. Vainly does the strained intellect seek to overtake an ever-receding commencement, and toil to gain some adequate grasp of an apparently endless succession. A beginning there must have been, though we can never hope to fix its point. Even speculation droops her wings in the attenuated atmosphere of a past so remote, and the light of imagination is quenched in the darkness of a history so ancient. In *time,* as in *space,* the confines of the universe must ever remain concealed from us; and of the end we know no more than of the beginning. Inconceivable as is to us the lapse of "geological time," it is no more than "a mere moment of the past, a mere infinitesimal portion of eternity." Well may "the human heart, that weeps and trembles," say, with Richter's pilgrim through celestial space, "I will go no farther; for the spirit of man acheth with this infinity. Insufferable is the glory of God. Let me lie down in the grave, and hide me from the persecution of the Infinite, for end, I see, there is none."

CHAPTER I.

THE SCOPE AND MATERIALS OF PALÆONTOLOGY.

The study of the rock-masses which constitute the crust of the earth, if carried out in the methodical and scientific manner of the geologist, at once brings us, as has been before remarked, in contact with the remains or traces of living beings which formerly dwelt upon the globe. Such remains are found, in greater or less abundance, in the great majority of rocks; and they are not only of great interest in themselves, but they have proved of the greatest importance as throwing light upon various difficult problems in geology, in natural history, in botany, and in philosophy. Their study constitutes the science of palæontology; and though it is possible to proceed to a certain length in geology and zoology without much palæontological knowledge, it is hardly possible to attain to a satisfactory general acquaintance with either of these subjects without having mastered the leading facts of the first. Similarly, it is not possible to study palæontology without some acquaintance with both geology and natural history.

PALÆONTOLOGY, then, is the science which treats of the living beings, whether animal or vegetable, which have inhabited the earth during past periods of its history. Its object is to eludicate, as far as may be, the structure, mode of existence, and habits of all such ancient forms of life; to determine their position in the scale of organized beings; to lay down the geographical limits within which they flourished; and to fix the period of their advent and disappearance. It is the ancient life-history of the earth; and were its record complete, it would furnish us with a detailed knowledge of the form and relations of all the animals and plants which have at any period flourished upon the land-surfaces of the globe or inhabited its waters; it would enable us to determine precisely their succession in time; and it would place in our hands an unfailing key to the problems of evolution. Unfortunately, from causes which will be subsequently discussed, the palæontological record is extremely imperfect, and our knowledge is interrupted by gaps, which not only bear a large proportion to our solid information, but which in many cases are of such a nature that we can never hope to fill them up

Fossils.—The remains of animals or vegetables which we now find entombed in the solid rock, and which constitute the working material of the palæontologist, are termed "fossils,"* or "petrifactions." In most cases, as can be readily understood, fossils are the actual hard parts of animals and plants which were in existence when the rock in which they are now found was being deposited. Most fossils, therefore, are of the nature of the shells of shell-fish, the skeletons of coral-zoophytes, the bones of vertebrate animals, or the wood, bark, or leaves of plants. All such bodies are more or less of a hard consistence to begin with, and are capable of resisting decay for a longer or shorter time—hence the frequency with which they occur in the fossil condition. Strictly speaking, however, by the term "fossil" must be understood "any body, *or the traces of the existence of any body,* whether animal or vegetable, which has been buried in the earth by natural causes" (Lyell). We shall find, in fact, that many of the objects which we have to study as "fossils" have never themselves actually formed parts of any animal or vegetable, though they are due to the former existence of such organisms, and indicate what was the nature of these. Thus the footprints left by birds, or reptiles, or quadrupeds upon sand or mud, are just as much proofs of the former existence of these animals as would be bones, feathers, or scales, though in

* Lat. *fossus*, dug up.

themselves they are inorganic. Under the head of fossils, therefore, come the footprints of air-breathing vertebrate animals; the tracks, trails, and burrows of sea-worms, crustaceans, or molluscs; the impressions left on the sand by stranded jelly-fishes; the burrows in stone or wood of certain shell-fish; the "moulds" or "casts" of shells, corals, and other organic remains; and various other bodies of a more or less similar nature.

FOSSILIZATION.—The term "fossilization" is applied to all those processes through which the remains of organized beings may pass in being converted into fossils. These processes are numerous and varied; but there are three principal modes of fossilization which alone need be considered here. In the first instance, the fossil is to all intents and purposes an actual portion of the original organized being—such as a bone, a shell, or a piece of wood. In some rare instances, as in the case of the body of the Mammoth discovered embedded in ice at the mouth of the Lena in Siberia, the fossil may be preserved almost precisely in its original condition, and even with its soft parts uninjured. More commonly, certain changes have taken place in the fossil, the principal being the more or less removal of the organic matter originally present. Thus bones become light and porous by the removal of their gelantine, so as to cleave to the tongue on being applied to that organ; whilst shells become fragile, and lose their primitive colors. In other cases, though practically the real body it represents, all the cavities of the fossil, down to its minutest recesses, may have become infiltrated with mineral matter. It need hardly be added, that it is in the more modern rocks that we find the fossils, as a rule, least changed from their former condition; but the original structure is often more or less completely retained in some of the fossils from even the most ancient formations.

In the second place, we very frequently meet with fossils in the state of "casts" or moulds of the original organic body. What occurs in this case will be readily understood if we imagine any common bivalve shell, as an Oyster, or Mussel, or Cockle, embedded in clay or mud. If the clay were sufficiently soft and fluid, the first thing would be that it would gain access to the interior of the shell, and would completely fill up the space between the valves. The pressure, also, of the surrounding matter would insure that the clay would everywhere adhere closely to the exterior of the shell. If now we suppose the clay to be in any way hardened so as to be converted into stone, and if we were to break up the stone, we should obviously have the following state of parts. The clay which filled the shell would form an accurate cast of the *interior* of the shell, and the clay outside would give us an exact impression or cast of the *exterior* of the shell (fig. 1). We should have, then, two casts, an interior and an exterior, and the two would be very different to one another, since the inside of a shell is very unlike the outside. In the case, in fact, of many univalve shells, the interior cast or "mould" is so unlike the exterior cast, or unlike the shell itself, that it may be difficult to determine the true origin of the former.

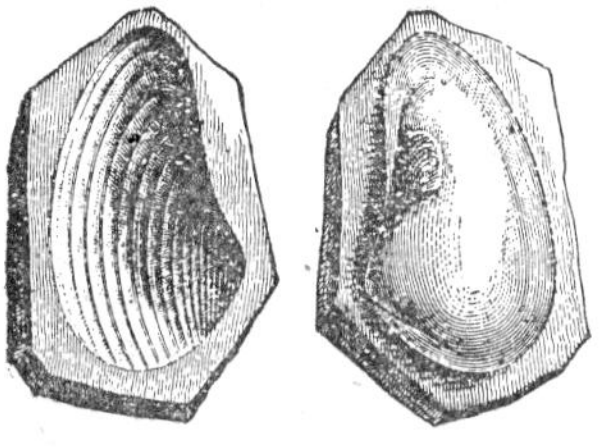

Fig. 1.—*Trigonia longa*, showing casts of the exterior and interior of the shell. —Cretaceous (Neocomian.)

It only remains to add that there is sometimes a further complication. If the rock be very porous and permeable by water, it may happen that the original shell is entirely dissolved away, leaving the interior cast loose, like the kernel of a nut, within the case formed by the exterior cast. Or it may happen that subsequent to the attainment of this state of things, the space thus left vacant between the interior and exterior cast—the space, that is, formerly occupied by the shell itself—may be filled up by some foreign mineral deposited there by the infiltration of water. In this last case the splitting open of the rock would reveal an interior cast, an exterior cast, and finally a body which would have the exact form of the original shell, but which would be really a much later formation, and which would not exhibit under the microscope the minute structure of shell.

In the third class of cases we have fossils which present with the greatest accuracy the external form, and even sometimes the internal minute structure, of the original organic body, but which, nevertheless, are not themselves truly organic, but have been formed by a "replacement" of the particles of the primitive organism by some material substance. The most elegant example of this is afforded by fossil wood which has been "silicified" or converted into flint (*silex*). In such cases we have fossil wood which presents the rings of growth and fibrous structure of recent wood, and which under the microscope exhibits the minutest vessels which characterize ligneous tissue, together with the even more minute markings of the vessels (fig. 2.) The whole, however, instead of being composed of the original carbonaceous matter of wood, is now converted into flint. The only explanation that can be given of this by no means rare phenomenon, is that the wood must have undergone a slow process of decay in water charged with silica or flint in solution. As each successive particle of wood was removed

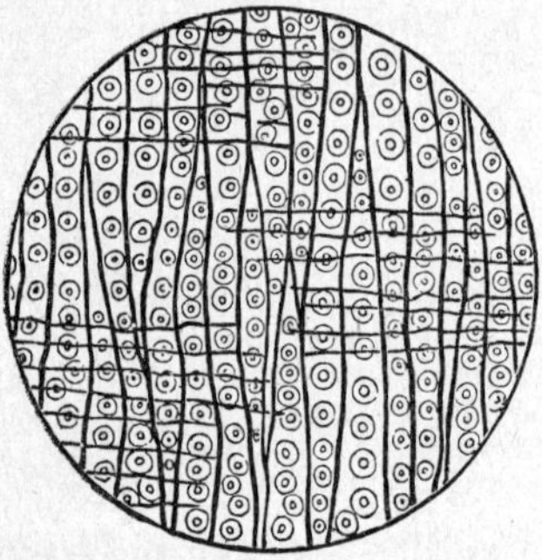

Fig, 2. —Microscopic section of the silicified wood of a Conifer (*Sequoia*) cut in the long direction of the fibres. Post-tertiary? Colorado. (Original.)

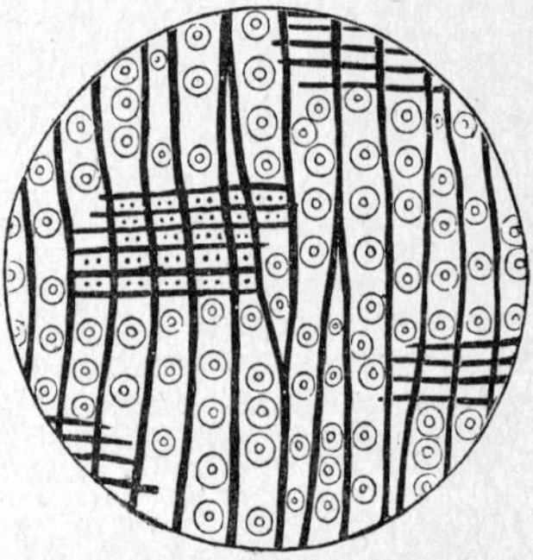

Fig. 3.—Microscopic section of the wood of the common Larch (*Abies larix*), cut in the long direction of the fibres. In both the fresh and the fossil wood (fig. 2) are seen the discs characteristic of coniferous wood. (Original.)

by decay, its place was taken by a particle of flint deposited from the surrounding water, till ultimately the entire wood was silicified. The process, therefore, resembles what would take place if we were to pull down a house built of brick by successive bricks, replacing each brick as removed by a piece of stone of precisely the same size and form. The result of this would be that the house would retain its primitive size, shape, and outline, but it would finally have been converted from a house of brick into a house of stone. Many other fossils besides wood—such as shells, corals, sponges, &c.—are often found silicified; and this may be regarded as the commonest form of fossilization by replacement. In other cases, however, though the principle of the process is the same, the replacing substance may be iron pyrites, oxide of iron, sulphur, malachite, magnesite, talc, &c.; but it is rarely that the replacement with these minerals is so perfect as to preserve the more delicate details of internal structure.

CHAPTER II.

THE FOSSILIFEROUS ROCKS.

Fossils are found in rocks, though not universally or promiscuously; and it is therefore necessary that the palæontologist should possess some acquaintance with, at any rate, those rocks which yield organic remains, and which are therefore said to be "*fossiliferous.*" In geological language, all the materials which enter into the composition of the solid crust of the earth, be their texture what it may—from the most impalpable mud to the hardest granite—are termed "rocks;" and for our present purpose we may divide these into two great groups. In the first division are the *Igneous Rocks*—such as the lavas and ashes of volcanoes—which are formed within the body of the earth itself, and which owe their structure and origin to the action of heat. The Igneous Rocks are formed primarily below the surface of the earth, which they only reach as the result of volcanic action; they are generally destitute of distinct "stratification," or arrangement in successive layers; and they do not contain fossils, except in the comparatively rare instances where volcanic ashes have enveloped animals or plants which were living in the sea or on the land in the immediate vicinity of the volcanic focus. The second great division of rocks is that of the *Fossiliferous, Aqueous,* or *Sedimentary* Rocks. These are formed at the surface of the earth, and, as implied by one of their names, are invariably deposited in water. They are produced by vital or chemical action, or are formed from the "sediment" produced by the disintegration and reconstruction of previously existing rocks, without previous solution; they mostly contain fossils; and they are arranged in distinct layers or "strata." The so-called "aerial" rocks which, like the beds of blown sand, have been formed by the action of the atmosphere, may also contain fossils; but they are not of such importance as to require special notice here.

For all practical purposes, we may consider that the Aqueous Rocks are the natural cemetery of the animals and plants of bygone ages; and it is therefore essential that the palæonto-

Fig. 4.—Sketch of Carboniferous strata at Kinghorn, in Fife, showing stratified beds (limestone and shales) surmounted by an unstratified mass of trap. (Original).

logical student should be acquainted with some of the principal facts as to their physical characters, their minute structure and mode of origin, their chief varieties, and their historical succession.

The Sedimentary or Fossiliferous Rocks form the greater portion of that part of the earth's crust which is open to our examination, and are distinguished by the fact that they are regularly "stratified" or arranged in distinct and definite layers or "strata." These layers may consist of a single material, as in a block of sandstone, or they may consist of different materials. When examined on a large scale, they are always found to consist of alternations of layers of different mineral composition. We may examine any given area, and find in it nothing but one kind of rock—sandstone, perhaps, or limestone.

In all cases, however, if we extend our examination sufficiently far, we shall ultimately come upon different rocks; and, as a general rule, the thickness of any particular set of beds is comparatively small, so that different kinds of rock alternate with one another in comparatively small spaces.

As regards the origin of the Sedimentary Rocks, they are for the most part "derivative" rocks, being derived from the wear and tear of pre-existent rocks. Sometimes, however, they owe their origin to chemical or vital action, when they would more properly be spoken of simply as Aqueous Rocks. As to their mode of deposition, we are enabled to infer that the materials which compose them have formerly been spread out by the action of water, from what we see going on every day at the mouths of our great rivers, and on a smaller scale wherever there is running water. Every stream, where it runs into a lake or into the sea, carries with it a burden of mud, sand, and rounded pebbles, derived from the waste of the rocks which

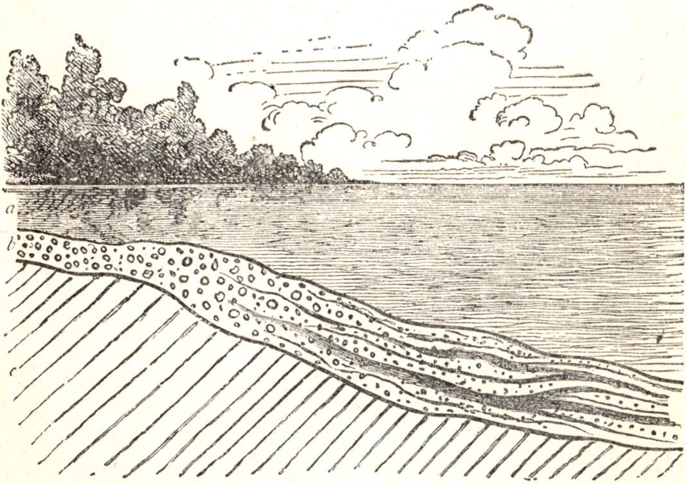

Fig. 5.—Diagram to illustrate the formation of sedimentary deposits at the point where a river debouches into the sea.

form its bed and banks. When these materials cease to be impelled by the force of the moving water, they sink to the bottom, the heaviest pebbles, of course, sinking first, the smaller pebbles and sand next, and the finest mud last. Ultimately, therefore, as might have been inferred upon theoretical grounds and as is proved by practical experience, every lake becomes a receptacle for a series of stratified rocks produced by the streams flowing into it. These deposits may vary in different parts of the lake, according as one stream brought down one kind of material and another stream contributed another material; but in all cases the materials will bear ample evidence that they were produced, sorted, and deposited by running water. The finer beds of clay or sand will all be arranged in thicker or thinner layers or laminæ; and if there are any beds of pebbles these will all be rounded or smooth, just like the water-worn pebbles of any brook-course. In all probability, also, we should find in some of the beds the remains of fresh-water shells or plants or other organisms which inhabited the lake at the time these beds were being deposited.

In the same way large rivers—such as the Ganges or Mississippi—deposit all the materials which they bring down at their mouths, forming in this way their "deltas." Whenever such a delta is cut through, either by man or by some channel of the river altering its course, we find that it is composed of a succession of horizontal layers or strata of sand or mud, varying in mineral composition, in structure, or in grain, according to the nature of the material brought down by the river at different periods. Such deltas, also, will contain the remains of animals which inhabit the river, with fragments of the plants which grew on its banks, or bones of the animals which lived in its basin.

Nor is this action confined, of course, to large rivers only, though naturally most conspicuous in the greatest bodies of water. On the contrary, all streams, of whatever size are engaged in the work of wearing down the dry land, and of transporting the materials thus derived from higher to lower levels, never resting in this work till they reach the sea.

Lastly, the sea itself—irrespective of the materials delivered into it by rivers—is constantly preparing fresh stratified deposits by its own action. Upon every coast-line the sea is constantly eating back into the land and reducing its component rocks to form the shingle and sand which we see upon every shore. The materials thus produced are not, however, lost, but are ultimately deposited elsewhere in the form of new stratified accumulations, in which are buried the remains of animals inhabiting the sea at the time.

Whenever, then, we find anywhere in the interior of the land any series of beds having these characters—composed, that is, of distinct layers, the particles of which, both large and small, show distinct traces of the wearing action of water—whenever and wherever we find such rocks, we are justified in assuming that they have been deposited by water in the manner above mentioned. Either they were laid down in some former lake by the combined action of the streams which flowed into it; or they were deposited at the mouth of some ancient river, forming its delta; or they were laid down at the bottom of the ocean. In the first two cases, any fossils which the beds might contain would be the remains of fresh-water or terrestrial organisms. In the last case, the majority, at any rate, of the fossils would be the remains of marine animals.

The term "formation" is employed by geologists to express "any group of rocks which have some character in common, whether of origin, age, or composition" (Lyell); so that we may speak of stratified and unstratified formations, aqueous or igneous formations, fresh-water or marine formations, and so on.

Chief Divisions of the Aqueous Rocks.

The Aqueous Rocks may be divided into two great sections, the Mechanically-formed and the Chemically-formed, including under the last head all rocks which owe their origin to vital action, as well as those produced by ordinary chemical agencies.

A. Mechanically-formed Rocks.—These are all those Aqueous Rocks of which we can obtain proofs that their particles have been mechanically transported to their present situation. Thus, if we examine a piece of *conglomerate* or puddingstone, we find it to be composed of a number of rounded pebbles embedded in an enveloping matrix or paste, which is usually of a sandy nature, but may be composed of carbonate of lime (when the rock is said to be "Calcareous conglomerate"). The pebbles in all conglomerates are worn and rounded by the action of water in motion, and thus show that they have been subjected to much mechanical attrition whilst they have been mechanically transported for a greater or less distance from the rock of which they originally formed part. The analogue of the old conglomerates at the present day is to be found in the great beds of shingle and gravel which are formed by the action of the sea on every coast-line, and which are composed of water-worn and well-rounded pebbles of different sizes. A *breccia* is a mechanically-formed rock, very similar to a conglomerate, and consisting of larger or smaller fragments of rock embedded in a common matrix. The fragments, however, are in this case all more or less angular, and are not worn or rounded. The fragments in breccias may be of large size, or they may be comparatively small (fig. 6); and the matrix may be composed of sand (arenaceous) or of carbonate of lime (calcareous). In the case of an ordinary sandstone, again, we have a rock which may be regarded as simply a very fine-grained conglomerate or breccia, being composed of small grains of sand (silica), sometimes rounded, sometimes more or less angular, cemented together by some such substance as oxide of iron, silicate of iron, or carbonate of lime. A sandstone, therefore, like a conglomerate, is a mechanically-formed rock, its component grains being equally the result of mechanical attrition and having equally been transported from a distance; and the same is true of the ordinary sand of the sea-shore, which is nothing more than an unconsolidated sandstone. Other so-called sands and sandstones, though equally mechanical in their origin, are truly calcareous in their nature, and are more or less entirely composed of carbonate of lime. Of this kind are the shell-sand so common on our coasts, and the coral-sand which is so largely formed in the neighborhood of coral-reefs. In these cases the rock is composed of fragments of the skeletons of shell-fish and numerous other marine animals, together, in many instances, with the remains of certain sea-weeds (*Corallines, Nullipores,* &c.) which are endowed with the power of secreting carbonate of lime from the sea-water. Lastly, in certain rocks still finer in their texture than sandstones, such as the various mud-rocks and shales, we can still recognize a mechanical source and origin. If slices of

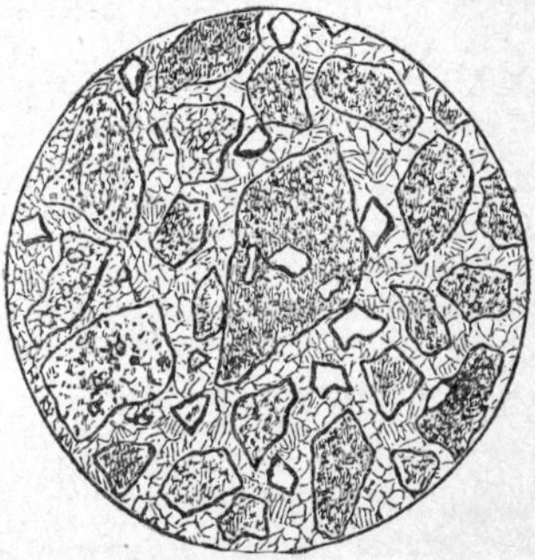

Fig. 6.—Microscopic section of a calcareous breccia in the Lower Silurian (Coniston Limestone) of Shap Wells, Westmoreland. The fragments are all of small size, and consist of angular pieces of transparent quartz, volcanic ashes, and limestone embedded in a matrix of crystalline limestone. (Original.)

any of these rocks sufficiently thin to be transparent are examined under the microscope, it will be found that they are composed of minute grains of different sizes, which are all more or less worn and rounded, and which clearly show, therefore, that they have been subjected to mechanical attrition.

All the above-mentioned rocks, then, are *mechanically-formed* rocks; and they are often spoken of as "Derivative Rocks," in consequence of the fact that their particles can be shown to have been mechanically *derived* from other pre-existent rocks. It follows from this that every bed of any mechanically-formed rock is the measure and equivalent of a corresponding amount of destruction of some older rock. It is not necessary to enter here into a minute account of the subdivisions of these rocks, but it may be mentioned that they may be divided into two principal groups, according to their chemical composition. In the one group we have the so-called *Arenaceous* (Lat. *arena,* sand) or *Siliceous* Rocks, which are essentially composed of larger or smaller grains of flint or silica. In this group are comprised ordinary sand, the varieties of sandstone and grit, and most conglomerates and breccias. We shall, however, afterwards see that some siliceous rocks are of organic origin. In the second group are the so-called *Argillaceous* (Lat. *argilla,* clay) Rocks, which contain a larger or smaller amount of clay or hydrated silicate of alumina in their composition. Under this head come clays, shales, marls, marl-slate, clay-slates, and most flags and flagstones.

B. Chemically-formed Rocks.—In this section are comprised all those Aqueous or Sedimentary Rocks which have been formed by chemical agencies. As many of these chemical agencies, however, are exerted through the medium of living beings, whether animals or plants, we get into this section a number of what may be called "*organically-formed rocks.*" These are of the greatest possible importance to the palæontologist, as being to a greater or less extent composed of the actual remains of animals or vegetables, and it will therefore be necessary to consider their character and structure in some detail.

By far the most important of the chemically-formed rocks are the so-called *Calcareous Rocks* (Lat. *calx,* lime), comprising all those which contain a large proportion of carbonate of lime, or are wholly composed of this substance. Carbonate of lime is soluble in water holding a certain amount of carbonic acid gas in solution; and it is, therefore, found in larger or smaller quantity dissolved in all natural waters, both fresh and salt, since these waters are always to some extent charged with the above-mentioned solvent gas. A great number of aquatic animals, however, together with some aquatic plants, are endowed with the power of separating the lime thus held in solution in the water, and of reducing it again to its solid condition. In this way shell-fish, crustaceans, sea-urchins, corals, and an immense number of other animals, are enabled to construct their skeletons; whilst some plants form hard structures within their tissues in a precisely similar manner. We do meet with some calcareous deposits, such as the "stalactites" and "stalagmites" of caves, the "calcareous tufa" and "travertine" of some hot springs, and the spongy calcareous deposits of so-called "petrifying springs," which are purely chemical in their origin, and owe nothing to the operation of living beings. Such deposits are formed simply by the precipitation of carbonate of lime from water, in consequence of the evaporation from the water of the carbonic acid gas which formerly held the lime in solution; but, though sometimes forming masses of considerable thickness and of geological importance, they do not concern us here. Almost all the limestones which occur in the series of the stratified rocks are, primarily at any rate, of *organic* origin, and have been, directly or indirectly, produced by the action of certain lime-making animals or plants, or both combined. The presumption as to all the calcareous rocks, which cannot be clearly shown to have been otherwise produced, is that they are thus organically formed; and in many cases this presumption can be readily reduced to a certainty. There are many varieties of the calcareous rocks, but the following are those which are of the greatest importance:—

Chalk is a calcareous rock of a generally soft and pulverulent texture, and with an earthy fracture. It varies in its purity, being sometimes almost wholly composed of carbonate of lime, and at other times more or less intermixed with foreign matter. Though usually soft and readily reducible to powder, chalk is occasionally, as in the north of Ireland, tolerably hard and compact; but it never assumes the crystalline aspect and stony density of limestone, except it be in immediate contact with some mass of igneous rock. By means of the microscope, the true nature and mode of formation of chalk can be determined with the greatest ease. In the case of the harder varieties, the examination can be conducted by means of slices ground down to a thinness sufficient to render them transparent; but in the softer kinds the rock must be disintegrated under water, and the *débris* examined microscopically. When investigated by either of these methods, chalk is found to be a genuine organic rock, being composed of the shells or hard parts of innumerable marine animals of different kinds, some entire, some fragmentary, cemented together by a matrix of very finely granular carbonate of lime. Foremost amongst the animal remains which so largely compose chalk are the shells of the minute creatures which will be subsequently spoken of under the name of *Foraminifera* (fig. 7), and which, in spite of their microscopic dimensions, play a more important part in the process of lime-making than perhaps any other of the larger inhabitants of the ocean.

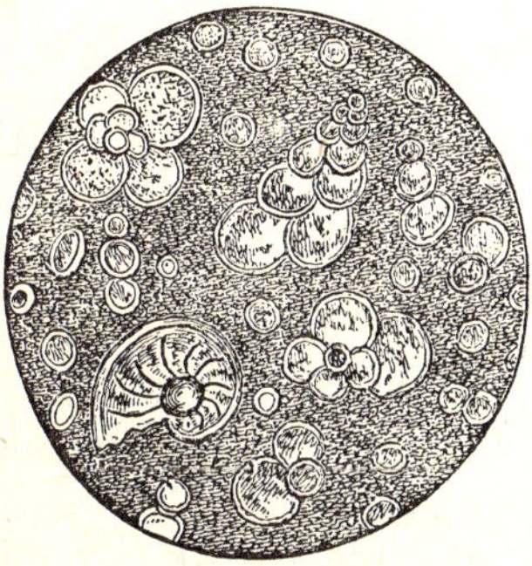

Fig. 7.—Section of Gravesend Chalk, examined by transmitted light and highly magnified. Besides the entire shells of *Globigerina*, *Rotalia*, and *Textularia* numerous detached chambers of *Globigerina* are seen. (Original.)

As chalk is found in beds of hundreds of feet in thickness, and of great purity, there was long felt much difficulty in satisfactorily accounting for its mode of formation and origin. By the researches of Carpenter, Wyville Thomson, Huxley, Wallich, and others, it has, however, been shown that there is now forming, in the profound depths of our great oceans, a deposit which is in all essential respects identical with chalk, and which is generally known as the "Atlantic ooz," from its having been first discovered in that sea. This ooze is found at great depths (5000 to over 15,000 feet) in both the Atlantic and Pacific, covering enormously large areas of the sea-bottom, and it presents itself as a whitish-brown, sticky, impalpable mud, very like greyish chalk when dried. Chemical examination shows that the ooze is composed almost wholly of carbonate of lime, and microscopical examination proves it to be of organic origin, and to be made up of the remains of living beings. The principal forms of these belong to the *Foraminifera*, and the commonest of these are the irregularly-chambered shells of *Globigerina*, absolutely indistinguishable from the *Globigerinæ* which are so largely present in the chalk (fig. 8). Along with these occur fragments of the skeletons of other larger creatures, and a certain proportion of the flinty cases of minute animal and vegetable organisms (*Polycystina* and *Diatoms*). Though many of the minute animals, the hard parts of which form the ooze, undoubtedly live at or near the surface of the sea, others, probably, really live near the bottom; and the ooze itself forms a congenial home for numerous sponges, sea-lilies, and other marine animals which flourish at great depths in the sea. There is thus established an intimate and most interesting parallelism between the chalk and the ooze of modern oceans. Both are formed essentially in the same way, and the latter only requires consolidation to become actually converted into chalk. Both are fundamentally organic deposits, apparently requiring a great depth of water for their accumulation, and mainly composed of the remains of *Foraminifera,* together with the entire or broken skeletons of other marine animals of greater dimensions. It is to be remembered, however, that the ooze, though strictly representative of the chalk, cannot be said in any proper sense to be actually *identical* with the formation so called by geologists. A great lapse of time separates the two, and though composed of the remains of representative classes or groups of animals, it is only in the case of the lowly-organized *Globigerinæ,* and of some other organisms of little higher grade, that we find absolutely the same kinds or *species* of animals in both.

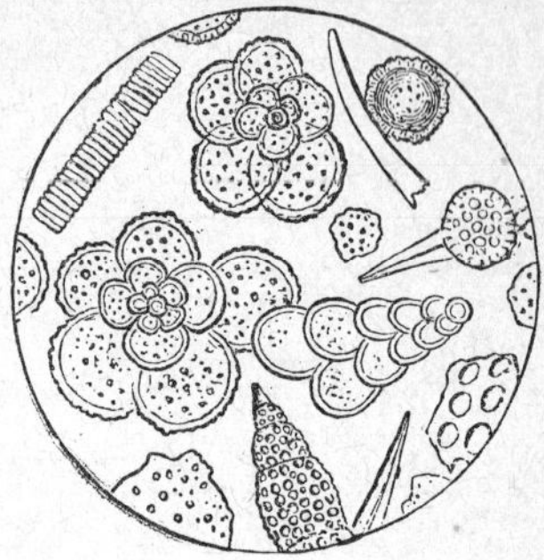

Fig. 8.—Organisms in the Atlantic Ooze, chiefly *Foraminifera* (*Globigerina* and *Textularia*), with *Polycystina* and sponge-spicules; highly magnified. (Original.)

Limestone, like chalk, is composed of carbonate of lime, sometimes almost pure, but more commonly with a greater or less intermixture of some foreign material, such as alumina or silica. The varieties of limestone are almost innumerable, but the great majority can be clearly proved to agree with chalk in being essentially of organic origin, and in being more or less largely composed of the remains of living beings. In many instances the organic remains which compose limestone are so large as to be readily visible to the naked eye, and the rock is at once seen to be nothing more than an agglomeration of the skeletons, generally fragmentary, of certain marine animals, cemented together by matrix of carbonate of lime. This is the case, for example, with the so-called "Crinoidal Limestones" and "Encrinital Marbles" with which the geologist is so familiar, especially as occurring in great beds amongst the older formations of the earth's crust. These are seen, on weathered or broken surfaces, or still better in polished slabs (fig. 9), to be com-

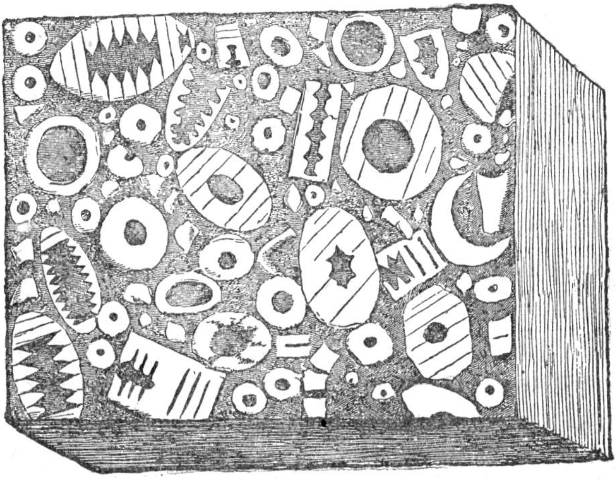

Fig. 9.—Slab of Crinoidal marble, from the Carboniferous limestone of Dent, in Yorkshire, of the natural size. The polished surface intersects the columns of the Crinoids at different angles, and thus gives rise to varying appearances. (Original.)

posed more or less exclusively of the broken stems and detached plates of sea-lilies (*Crinoids*). Similarly, other limestones are composed almost entirely of the skeletons of corals; and such old coralline limestones can readily be paralleled by formations which we can find in actual course of production at the present day. We only need to transport ourselves to the islands of the Pacific, to the West Indies, or to the Indian Ocean, to find great masses of lime formed similarly by living corals, and well known to every one under the name of "coral-reefs." Such reefs are often of vast extent, both superficially and in vertical thickness, and they fully equal in this respect any of the coralline limestones of bygone ages. Again, we find other limestones—such as the celebrated "Nummulitic Limestone" (fig. 10), which sometimes attains a thickness of some thousands of feet—which are almost entirely made up of the shells of *Foraminifera*. In the case of the "Nummulitic Limestone," just mentioned, these shells are of a large size, varying from the size of a split pea up to that of a florin. There are, however, as we shall see, many other limestones, which are likewise largely made up of

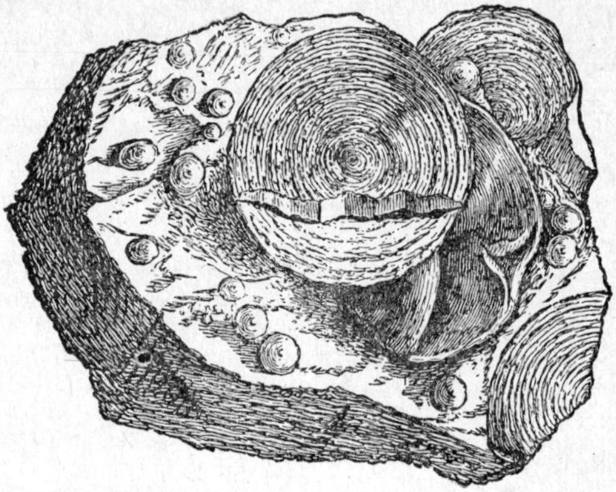

Fig. 10.—Piece of Nummulitic Limestone from the Great Pyramid. Of the natural size. (Original.)

Foraminifera, but in which the shells are very much more minute, and would hardly be seen at all without the microscope.

We may, in fact, consider that the great agents in the production of limestones in past ages have been animals belonging to the *Crinoids,* the *Corals,* and the *Foraminifera.* At the present day, the Crinoids have been nearly extinguished, and the few known survivors seem to have retired to great depths in the ocean; but the two latter still actively carry on the work of lime-making, the former being very largely helped in their operations by certain lime-producing marine plants (*Nullipores* and *Corallines*). We have to remember, however, that though the limestones, both ancient and modern, that we have just spoken of, are truly organic, they are not necessarily formed out of the remains of animals which actually lived on the precise spot where we now find the limestone itself. We may find a crinoidal limestone, which we can show to have been actually formed by the successive growth of generations of sea-lilies *in place;* but we shall find many others in which the rock is made up of innumerable fragments of the skeletons of these creatures, which have been clearly worn and rubbed by the sea-waves, and which have been mechanically transported to their present site.

In the same way, a limestone may be shown to have been an actual coral-reef, by the fact that we find in it great masses of coral, growing in their natural position, and exhibiting plain proofs that they were simply quietly buried by the calcareous sediment as they grew; but other limestones may contain only numerous rolled and water-worn fragments of corals. This is precisely paralleled by what we can observe in our existing coral-reefs. Parts of the modern coral-islands and coral-reefs are really made up of corals, dead or alive, which actually grew on the spot where we now find them; but other parts are composed of limestone-rock ("coral-rock"), or of a loose and ("coral-sand"), which is organic in the sense that it is composed of lime formed by living beings, but which, in truth, is composed of fragments of the skeletons of these living beings, mechanically transported and heaped together by the sea. To take another example nearer home, we may find great accumulations of calcareous matter formed *in place*, by the growth of shell-fish, such as oysters or mussels; but we can also find equally great accumulations on many of our shores in the form of "shell-sand" which is equally composed of the shells of molluscs, but which is formed by the trituration of these shells by the mechanical power of the sea-waves. We thus see that though all these limestones are primarily organic, they not uncommonly become "mechanically-formed" rocks in a secondary sense, the materials of which they are composed being formed by living beings, but having been mechanically transported to the place where we now find them.

Many limestones, as we have seen, are composed of large and conspicuous organic remains, such as strike the eye at once. Many others, however, which at first sight appear compact, more or less crystalline, and nearly devoid of traces of life, are found, when properly examined, to be also composed of the remains of various organisms. All the commoner limestones, in fact from the Lower Silurian period onwards, can be easily proved to be thus *organic* rocks, if we investigate weathered or polished surfaces with a lens, or, still better, if we cut thin slices of the rock and grind these down till they are transparent. When thus examined, the rock is usually found to be composed of innumerable entire or fragmentary fossils, cemented together by a granular or crystalline matrix of carbonate of lime (figs. 11 and 12). When the matrix is granular, the rock is precisely similar to chalk, except that it is harder and less earthy in texture, whilst the fossils are only occasionally referable to the *Foraminifera*.

In other cases, the matrix is more or less crystalline, and when this crystallization has been carried to a great extent, the original organic nature of the rock may be greatly or completely obscured thereby. Thus, in limestones which have been greatly altered or "metamorphosed" by the combined action of heat and pres-

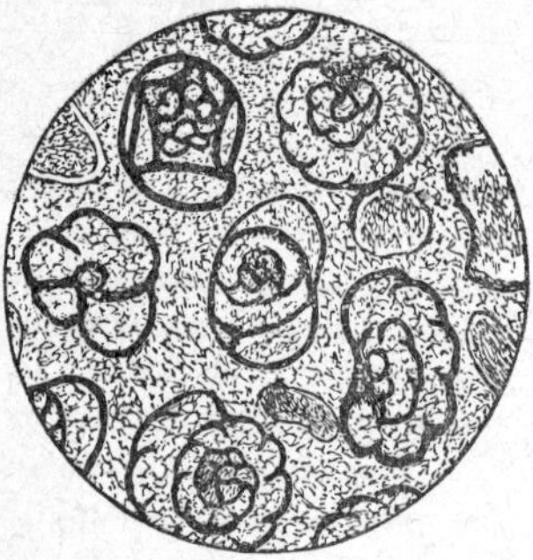

Fig. 11.—Section of Carboniferous Limestone from Spergen Hill, Indiana. U. S., showing numerous large-sized *Foraminifera* (*Endothyra*) and a few oolitic grains; magnified. (Original.)

Fig.12.—Section of Ccniston Limestone (Lower Silurian) from Keisley, Westmoreland; magnified. The matrix is very coarsely crystalline, and the included organic remains are chiefly stems of Crinoids. (Original.)

sure, all traces of organic remains become annihilated, and the rock becomes completely crystalline throughout. This, for example, is the case with the ordinary white "statuary marble," slices of which exhibit under the microscope nothing but an aggregate of beautifully transparent crystals of carbonate of lime, without the smallest traces of fossils. There are also other cases, where the limestone is not necessarily high crystalline, and where no metamorphic action in the strict sense has taken place, in which, nevertheless, the microscope fails to reveal any evidence that the rock is organic. Such cases are somewhat obscure, and doubtless depend on different causes in different instances; but they do not affect the important generalization that limestones are fundamentally the product of the operation of living beings. This fact remains certain; and when we consider the vast superficial extent occupied by calcareous deposits, and the enormous collective thickness of these, the mind cannot fail to be impressed with the immensity of the period demanded for the formation of these by the agency of such humble and often microscopic creatures as Corals, Sea-lilies, Foraminifers, and Shell-fish.

Amongst the numerous varieties of limestone, a few are of such interest as to deserve a brief notice. *Magnesia limestone* or *dolomite*, differs from ordinary limestone in containing a certain proportion of carbonate of magnesia along with the carbonate of lime. The typical dolomites contain a large proportion of carbonate of magnesia, and are highly crystalline. The ordinary magnesian limestones (such as those of Durham in the Permian series, and the Guelph Limestones of North America in the Silurian series) are generally of a yellowish, buff, or brown color, with a crystalline or pearly aspect, effervescing with acid much less freely than ordinary limestone, exhibiting numerous cavities from which fossils have been dissolved out, and often assuming the most varied and singular forms in consequence of what is called "concretionary action." Examination with the microscope shows that these limestones are composed of an aggregate of minute but perfectly distinct crystals, but that minute organisms of different kinds, or fragments of larger fossils, are often present as well. Other magnesian limestones, again, exhibit no striking external peculiarities by which the presence of magnesia would be readily recognized, and though the base of the rock is crystalline, they are replete with the remains of organized beings. Thus many of the magnesian limestones of the Carboniferous series of the North of England are very like ordinary limestone to look at, through effervescing less freely with acids, and the microscope proves them to be charged with remains of *Foraminifera* and other minute organisms.

Marbles are of various kinds, all limestones which are sufficiently hard and compact to take a high polish going by this name. Statuary marble, and most of the celebrated foreign marbles, are "metamorphic" rocks, of a highly crystalline nature, and having all traces of their primitive organic structure obliterated. Many other marbles, however, differ from ordinary limestone simply in the matter of density. Thus, many marbles (such as Derbyshire marble) are simply "crinoidal limestones" (fig. 9); whilst various other British marbles exhibit innumerable organic remains under the microscope. Black marbles owe their color to the presence of very minute particles of carbonaceous matter, in some cases at any rate; and they may either be metamorphic, or they may be charged with minute fossils such as *Foraminifera* (*e.g.*, the black limestones of Ireland, and the black marble of Dent, in Yorkshire).

"*Oolitic*" *limestones*, or "*oolites*," as they are often called, are of interest both to the palæontologist and geologist. The peculiar structure to which they owe their name is that the rock is more or less entirely composed of spheroidal or oval grains, which vary in size from the head of a small pin or less up to the size of a pea, and which may be in almost immediate contact with one another, or may be cemented together by a more or less abundant calcareous matrix. When the grains are pretty nearly spherical and are in tolerably close contact, the rock looks very like the roe of a fish, and the name of "oolite" or 'egg-stone" is in allusion to this. When the grains are of the size of peas or upwards, the rock is often called a "pisolite" (Lat. *pisum*, a pea). Limestones having this peculiar structure are especially abundant in the Jurassic formation, which is often called the "Oolitic series" for this reason; but essentially similar limestones occur not uncommonly in the Silurian, Devonian, and Carboniferous formations, and, indeed, in almost all rock groups in which limestones are largely developed. Whatever may be the age of the formation in which they occur, and whatever may be the size of their component "eggs," the structure of oolitic limestones is fundamentally the same. All the ordinary oolitic limestones, namely, consist of little spherical or ovoid "concretions," as they are termed, cemented together by a larger or smaller amount of crystalline carbonate of lime, together, in many instances, with numerous organic remains of different kinds (fig. 13). When examined in polished slabs, or in thin sections prepared for the microscope, each of these little concretions is seen to consist of numerous concentric coats of carbonate of lime, which sometimes simply surround an imaginary center, but which, more commonly, have been successively deposited round some foreign body, such as a little crystal of quartz, a cluster of sand-grains, or a minute shell. In other cases, as in some of the beds of the Carboniferous limestone in the North of England, where the limestone is highly "arenaceous," there is a modification of the

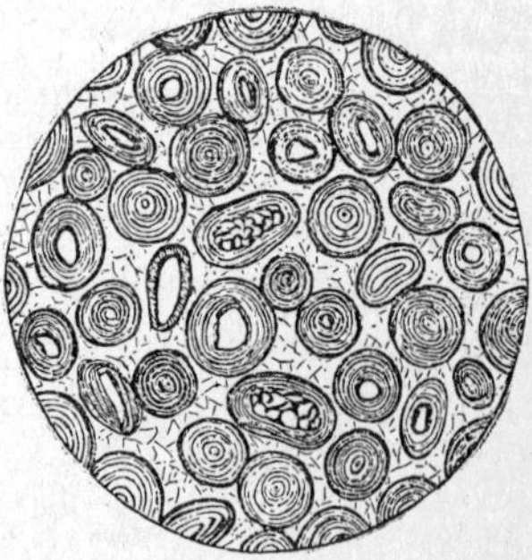

Fig. 13.—Slice of oolitic limestone from the Jurassic series (Coral Rag) of Weymouth; magnified. (Original.)

oolitic structure. Microscopic sections of these sandy limestones (fig. 14) show numerous generally angular or oval grains of silica or flint, each of which is commonly surrounded by a thin coating of carbonate of lime, or sometimes by several such coats, the whole being cemented together along with the shells of *Foraminifera* and other minute fossils by a matrix of crystalline calcite. As compared with typical oolites, the concretions in these limestones are usually much more irregular in shape, often lengthened out and almost cylindrical, at other times angular, the central nucleus being of large size, and the surrounding envelope of lime being very thin, and often exhibiting no concentric structure. In both these and the ordinary oolites, the structure is fundamentally the same. Both have been formed in a sea, probably of no great depth, the waters of which were charged with carbonate of lime in solution, whilst the bottom was formed of sand intermixed with minute shells and fragments of the skeletons of larger marine animals. The excess of lime in the sea-water was precipitated round the sand-grains, or round the smaller shells, as so many nuclei, and this precipitation must often have taken place time after time, so as to give rise to the concentric structure so characteristic of oolitic concretions. Finally, the oolitic grains thus produced were cemented together by a further precipitation of crystalline carbonate of lime from the waters of the ocean.

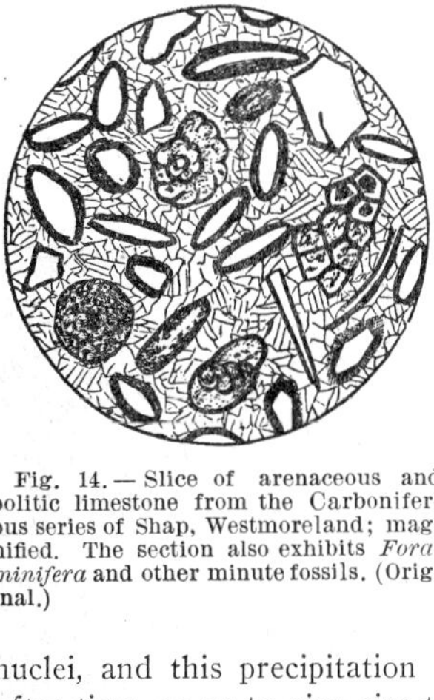

Fig. 14.—Slice of arenaceous and oolitic limestone from the Carboniferous series of Shap, Westmoreland; magnified. The section also exhibits *Foraminifera* and other minute fossils. (Original.)

Phosphate of Lime is another lime-salt, which is of interest to the palæontologist. It does not occur largely in the stratified series, but it is found in considerable beds * in the Laurentian formation, and less abundantly in some later rock-

* Apart from the occurrence of phosphate of lime in actual beds in the stratified rocks, as in the Laurentian and Silurian series, this salt may also occur disseminated through the rock, when it can only be detected by chemical analysis. It is interesting to note that Dr. Hicks has recently proved the occurrence of phosphate of lime in this disseminated form in rocks as old as the Cambrian, and that in quantity quite equal to what is generally found to be present in the later fossiliferous rocks. This affords a chemical proof that animal life flourished abundantly in the Cambrian seas.

groups, whilst it occurs abundantly in the form of nodules in parts of the Cretaceous (Upper Greensand) and Tertiary deposits. Phosphate of lime forms the larger proportion of the earthy matters of the bones of Vertebrate animals, and also occurs in less amount in the skeletons of certain of the Invertebrates (*e.g., Crustacæ*). It is, indeed, perhaps more distinctively than carbonate of lime, an organic compound; and though the formation of many known deposits of phosphate of lime cannot be positively shown to be connected with the previous operation of living beings, there is room for doubt whether this salt is not in reality always primarily a product of vital action. The phosphatic nodules of the Upper Greensand are erroneously called "coprolites," from the belief originally entertained that they were the droppings or fossilized excrements of extinct animals; and though this is not the case, there can be little doubt but that the phosphate of lime which they contain is in this instance of organic origin.* It appears, in fact, that decaying animal matter has a singular power of determining the precipitation around it of mineral salts dissolved in water. Thus, when any animal bodies are undergoing decay at the bottom of the sea, they have a tendency to cause the precipitation from the surrounding water of any mineral matter which may be dissolved in it; and the organic body thus becomes a center round which the mineral matters in question are deposited in the form of a "concretion" or "nodule.." The phosphatic nodules in question were formed in a sea in which phosphate of lime, derived from the destruction of animal skeletons, was held largely in solution; and a precipitation of it took place round any body, such as a decaying animal substance, which happened to be lying on the sea-bottom, and which offered itself as a favorable nucleus. In the same way we may explain the formation of the calcareous nodules, known as "septaria" or "cement stones," which occur so commonly in the London Clay and Kimmeridge Clay, and in which the principal ingredient is carbonate of lime. A similar origin is to be ascribed to the nodules of clay iron-stone (impure carbonate of iron) which occur so abundantly in the shales of the Carboniferous series and in other argillaceous deposits; and a parallel modern example is to be found in the nodules of

* It has been maintained, indeed, that the phosphatic nodules so largely worked for agricultural purposes, are in themselves actual organic bodies or true fossils. In a few cases this admits of demonstration, as it can be shown that the nodule is simply an organism (such as a sponge) infiltrated with phosphate of lime (Sollas); but there are many other cases in which no actual structure has yet been shown to exist, and as to the true origin of which it would be hazardous to offer a positive opinion.

manganese, which were found by Sir Wyville Thomson, in the Challenger, to be so numerously scattered over the floor of the Pacific at great depths. In accordance with this mode of origin, it is exceedingly common to find in the center of all these nodules, both old and new, some organic body, such as a bone, a shell, or a tooth, which acted as the original nucleus of precipitation, and was thus preserved in a shroud of mineral matter. Many nodules, it is true, show no such nucleus; but it has been affirmed that all of them can be shown, by appropriate microscopical investigation, to have been formed round an original organic body to begin with (Hawkins Johnson).

The last lime-salt which need be mentioned is *gypsum*, or *sulphate of lime*. This substance, apart from the other modes of occurrence, is not uncommonly found interstratified with the ordinary sedimentary rocks, in the form of more or less irregular beds; and in these cases it has a palæontological importance, as occasionally yielding well-preserved fossils. Whilst its exact mode of origin is uncertain, it cannot be regarded as in itself an organic rock, though clearly the product of chemical action. To look at, it is usually a whitish or yellowish-white rock, as coarsely crystalline as loaf-sugar, or more so; and the microscope shows it to be composed entirely of crystals of sulphate of lime.

We have seen that the *calcareous* or lime-containing rocks are the most important of the group of organic deposits; whilst the siliceous or flint-containing rocks may be regarded as the most important, most typical and most generally distributed of the mechanically-formed rocks. We have, however, now briefly to consider certain deposits which are more or less completely formed of flint; which, nevertheless, are essentially organic in their origin.

Flint or silex, hard and intractable as it is, is nevertheless capable of solution in water to a certain extent, and even of assuming, under certain circumstances, a gelatinous or viscous condition. Hence, some hot-springs are impregnated with silica to a considerable extent; it is present in small quantity in sea-water; and there is reason to believe that a minute proportion must very generally be present in all bodies of fresh water as well. It is from this silica dissolved in the water that many animals and some plants are enabled to construct for themselves flinty skeletons; and we find that these animals and plants are and have been sufficiently numerous to give rise to very considerable deposits of siliceous matter by the mere accumulation of their skeletons. Amongst the animals which require special mention in this connection are the microscopic organisms which are known to the naturalist as *Polycystina.* These little creatures are the lowest possible grade of organization, very closely related to the animals which we have previously spoken of as *Foraminifera,* but differing in the fact that they secrete a shell or skeleton composed of flint instead of lime. The Polycystina occur abundantly in our present seas; and their shells are present in some numbers in the ooze which is found at great depths in the Atlantic and Pacific oceans, being easily recognized by their exquisite shape, their glassy transparency, the general presence of longer or shorter spines, and the sieve-like perforations in the walls. Both in Barbadoes and in the Nicobar islands occur geological formations which are composed of the flinty skeletons of these microscopic animals; the deposit in the former locality attaining a great thickness, and having been long known to workers with the microscope under the name of "Barbadoes earth" (fig. 15).

Fig. 15. — Shells of *Polycystina* from "'Barbadoes earth;" greatly magnified. (Original.)

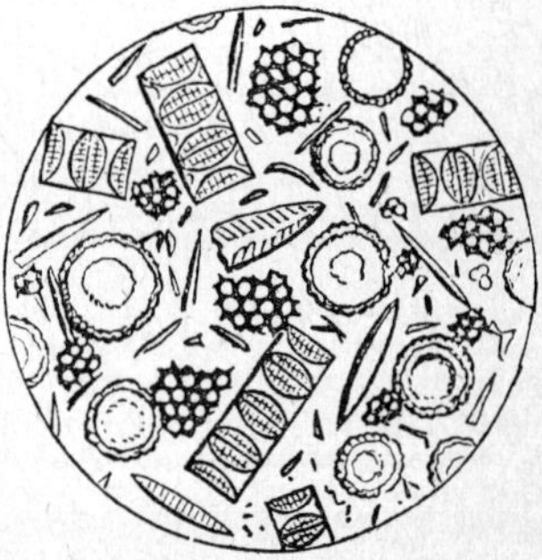

Fig. 16.—Cases of Diatoms in the Richmond "Infusorial earth;" highly magnified. (Original.)

In addition to flint-producing animals, we have also the great group of fresh-water and marine microscopic plants known as *Diatoms,* which likewise secrete a siliceous skeleton, often of great beauty. The skeletons of Diatoms are found abundantly at the present day in lake-deposits, guano, the silt of estuaries, and in mud which covers many parts of the sea-bottom; they have been detected in strata of great age; and in spite of their microscopic dimensions, they have not uncommonly accumulated to form deposits of great thickness, and of considerable superficial extent. Thus the celebrated deposit of "tripoli" ("Polirschiefer") of Bohemia, largely worked as polishing-powder, is composed wholly, or almost wholly, of the flinty cases of Diatoms, of which it is calculated that no less than forty-one thousand millions go to make up a single cubic inch of the stone. Another celebrated deposit is the so-called "Infusorial earth" of Richmond in Virginia, where there is a stratum in places thirty feet thick, composed almost entirely of the microscopic shells of Diatoms.

Nodules or layers of *flint,* or the impure variety of flint known as *chert,* are found in limestones of almost all ages from the Silurian upwards; but they are especially abundant in the chalk. When these flints are examined in thin and transparent slices under the microscope, or in polished sections, they are found to contain an abundance of minute organic bodies—such as *Foraminifera,* sponge-spicules, &c.—embedded in a siliceous basis. In many instances the flint contains larger organisms—such as a Sponge or Sea-urchin. As the flint has completely surrounded and infiltrated the fossils which it contains, it is obvious that it must have been deposited from sea-water in a gelatinous condition, and subsequently have hardened. That silica is capable of assuming this viscous and soluble condition is known; and the formation of flint may therefore be regarded as due to the separation of silica from the sea-water and its deposition round some organic body in a state of chemical change or decay, just as nodules of phosphate of lime or carbonate of iron are produced. The existence of numerous organic bodies in flint has long been known; but it should be added that a recent observer (Mr. Hawkins Johnson) asserts that the existence of an organic structure can be demonstrated by suitable methods of treatment, even in the actual matrix or basis of the flint. *

In addition to the deposits formed of flint itself, there are other siliceous deposits formed by certain *silicates,* and also of organic origin. It has been shown, namely—by observations carried out in our present seas—that the shells of *Foraminifera* are liable to become completely infiltrated by silicates (such as "glauconite," or silicate of iron and potash). Should the actual calcareous shell become dissolved away subsequent to this infiltration—as is also liable to occur—then, in place of the shells of

* It has been asserted that the flints of the chalk are merely fossil sponges. No explanation of the origin of flint, however, can be satisfactory, unless it embraces the origin of chert in almost all great limestones from the Silurian upwards, as well as the common phenomenon of the silification of organic bodies (such as corals and shells) which are known with certainty to have been originally calcareous.

the *Foraminifera,* we get a corresponding number of green sandy grains of glauconite, each grain being the *cast* of a single shell. It has thus been shown that the green sand found covering the sea-bottom in certain localities (as found by the Challenger expedition along the line of the Agulhas current) is really organic, and is composed of casts of the shells of *Foraminifera.* Long before these observations had been made, it had been shown by Professor Ehrenberg that the green sands of various geological formations are composed mainly of the internal casts of the shells of *Foraminifera;* and we have thus another and a very interesting example how rock-deposits of considerable extent and of geological importance can be built up by the operation of the minutest living beings.

As regards *argillaceous* deposits, containing *alumina* or *clay* as their essential ingredient, it cannot be said that any of these have been actually shown to be of organic origin. A recent observation by Sir Wyville Thomson would, however, render it not improbable that some of the great argillaceous accumulations of past geological periods may be really organic. This distinguished observer, during the cruise of the Challenger, showed that the calcareous ooze which has been already spoken of as covering large areas of the floor of the Atlantic and Pacific at great depths, and which consists almost wholly of the shells of *Foraminifera,* gave place at still greater depths to a red ooze consisting of impalpable clayey mud, colored by oxide of iron, and devoid of traces of organic bodies. As the existence of this widely-diffused red ooze, in mid-ocean, and at such great depths, cannot be explained on the supposition that it is a sediment brought down into the sea by rivers, Sir Wyville Thomson came to the conclusion that it was probably formed by the action of the sea-water upon the shells of *Foraminifera.* These shells, though mainly consisting of lime, also contain a certain proportion of alumina, the former being soluble in the carbonic acid dissolved in the sea-water, whilst the latter is insoluble. There would further appear to be grounds for believing that the solvent power of the sea-water over lime is considerable increased at great depths. If, therefore, we suppose the shells of *Foraminifera* to be in course of deposition over the floor of the Pacific, at certain depths they would remain unchanged, and would accumulate to form a calcareous ooze; but at greater depths they would be acted upon by the water, their lime would be dissolved out, their form would disappear, and we should simply have left the small amount of alumina which they previously contained.

In process of time this alumina would accumulate to form a bed of clay; and as this clay has been directly derived from the decomposition of the shells of animals, it would be fairly entitled to be considered an organic deposit. Though not finally established, the hypothesis of Sir Wyville Thomson on this subject is of the greatest interest to the palæontologist, as possibly serving to explain the occurrence, especially in the older formations, of great deposits of argillaceous matter which are entirely destitute of traces of life.

It only remains, in this connection, to shortly consider the rock-deposits in which *carbon* is found to be present in greater or less quantity. In the great majority of cases where rocks are found to contain carbon or carbonaceous matter, it can be stated with certainty that this substance is of organic origin, though it is not necessarily derived from vegetables. Carbon derived from the decomposition of animal bodies is not uncommon; though it never occurs in such quantity from this source as it may do when it is derived from plants. Thus, many limestones are more or less highly bituminous; the celebrated siliceous flags or so-called "bituminous schists" of Caithness are impregnated with oily matter apparently derived from the decomposition of the numerous fishes embedded in them; Silurian shales containing Graptolites, but destitute of plants, are not uncommonly "anthracitic," and contain a small percentage of carbon derived from the decay of these zoophytes; whilst the petroleum so largely worked in North America has not improbably an animal origin. That the fatty compounds present in animal bodies should more or less extensively impregnate fossiliferous rock-masses, is only what might be expected; but the great bulk of the carbon which exists stored up in the earth's crust is derived from plants; and the form in which it principally presents itself is that of *coal*. We shall have to speak again, and at greater length, of coal, and it is sufficient to say here that all the true coals, anthracites, and lignites, are of organic origin, and consist principally of the remains of plants in a more or less altered condition. The bituminous shales which are found so commonly associated with beds of coal also derive their carbon primarily from plants; and the same is certainly, or probably, the case with similar shales which are known to occur in formations younger than the Carboniferous. Lastly, carbon may occur as a conspicuous constituent of rock-masses in the form of *graphite* or *black-lead*. In this form, it occurs in the shape of detached scales, of veins or strings, or sometimes of regular layers;* and there can be little doubt that in many instances it has an organic origin, though this is not capable of direct proof. When present, at any rate, in quantity, and in the form of layers associated with stratified rocks, as is often the case in the Laurentian formation, there can be little hesitation in regarding it as of vegetable origin, and as an altered coal.

CHAPTER III.

CHRONOLOGICAL SUCCESSION OF THE FOSSILIFEROUS ROCKS.

The physical geologist, who deals with rocks simply as rocks, and who does not necessarily trouble himself about what fossils they may contain, finds that the stratified deposits which form so large a portion of the visible part of the earth's crust are not promiscuously heaped together, but that they have a certain definite arrangement. In each country that he examines, he finds that certain groups of strata lie above certain other groups; and in comparing different countries with one another, he finds that, in the main, the same groups of rocks are always found in the same relative position to each other. It is possible, therefore, for the physical geologist to arrange the known stratified rocks into a successive series of groups, or "formations," having a certain definite order. The establishment of this physical order amongst the rocks introduces, however, at once the element of *time,* and the physical succession of the strata can be converted directly into a historical or *chronological* succession. This is obvious, when we reflect that any bed or set of beds of sedimentary origin is clearly and necessarily

* In the Huronian formation of Steel River, on the north shore of Lake Superior, there exists a bed of carbonaceous matter which is regularly interstratified with the surrounding rocks, and has a thickness of from 30 to 40 feet. This bed is shown by chemical analysis to contain about 50 per cent of carbon, partly in the form of graphite, partly in the form of anthracite; and there can be little doubt but that it is really a stratum of "metamorphic" coal.

younger than all the strata upon which it rests, and older than all those by which it is surmounted.

It is possible, then, by an appeal to the rocks alone, to determine in each country the general physical succession of the strata, and this "stratigraphical" arrangement, when once determined, gives us the *relative* ages of the successive groups. The task, however, of the physical geologist in this matter is immensely lightened when he calls in palæontology to his aid, and studies the evidence of the fossils embedded in the rocks. Not only is it thus much easier to determine the order of succession of the strata in any given region, but it becomes now for the first time possible to compare, with certainty and precision, the order of succession in one region with that which exists in other regions far distant. The value of fossils as tests of the relative ages of the sedimentary rocks depends on the fact that they are not indefinitely or promiscuously scattered through the crust of the earth,—as it is conceivable that they might be. On the contrary, the first and most firmly established law of Palæontology is, that *particular kinds of fossils are confined to particular rocks,* and *particular groups of fossils are confined to particular groups of rocks.* Fossils, then, are distinctive of the rocks in which they are found—much more distinctive, in fact, than the mere mineral character of the rock can be, for *that* commonly changes as a formation is traced from one region to another, whilst the fossils remain unaltered. It would therefore be quite possible for the palæontologist, by an appeal to the fossils alone, to arrange the series of sedimentary deposits into a pile of strata having a certain definite order. Not only would this be possible, but it would be found —if sufficient knowledge had been brought to bear on both sides—that the palæontological arrangement of the strata would coincide in its details with the stratigraphical or physical arrangement.

Happily for science, there is no such division between the palæontologist and the physical geologist as here supposed; but by the combined researches of the two, it has been found possible to divide the entire series of stratified deposits into a number of definite *rock-groups* or *formations,* which have a recognized order of succession, and each of which is characterized by possessing an assemblage of organic remains which do not occur in association in any other formation. Such an *assemblage of fossils,* characteristic of any given formation, represents the *life* of the particular *period* in which the formation was deposited. In this way the past history of the earth becomes divided into a series of successive *life-periods,* each of which corresponds with the deposition of a particular *formation* or group of strata.

Whilst particular *assemblages* of organic forms characterize particular *groups* of rocks, it may be further said that, in a general way, each subdivision of each formation has its own peculiar fossils, by which it may be recognized by a skilled worker in Palæontology. Whenever, for instance, we meet with examples of the fossils which are known as *Graptolites,* we may be sure that we are dealing with Silurian rocks (leaving out of sight one or two forms doubtfully referred to this family). We may, however, go much further than this with perfect safety. If the Graptolites belong to certain genera, we may be quite certain that we are dealing with *Lower* Silurian rocks. Furthermore, if certain special forms are present, we may be even able to say to what exact subdivision of the Lower Silurian series they belong.

As regards particular fossils, however, or even particular classes of fossils, conclusions of this nature require to be accompanied by a tacit but well-understood reservation. So far as our present observation goes, none of the undoubted Graptolites have ever been discovered in rocks later than those known upon other grounds to be Silurian; but it is possible that they might at any time be detected in younger deposits. Similarly, the species and genera which we now regard as characteristic of the Lower Silurian, may at some future time be found to have survived into the Upper Silurian period. We should not forget, therefore, in determining the age of strata by palæontological evidence, that we are always reasoning upon generalizations which are the result of experience alone, and which are liable to be vitiated by further and additional discoveries.

When the palæontological evidence as to the age of any given set of strata is corroborated by the physical evidence, our conclusions may be regarded as almost certain; but there are certain limitations and fallacies in the palæontological method of inquiry which deserve a passing mention. In the first place, fossils are not always present in the stratified rocks; many aqueous rocks are unfossiliferous, through a thickness of hundreds or even thousands of feet of little-altered sediments; and even amongst beds which do contain fossils, we often meet with strata of many feet or yards in thickness which are wholly destitute of any traces of fossils. There are, therefore, to begin with, many cases in which there is no palæontological evidence extant or available as to the age of a given group of strata. In the second place, palæontological observers in different parts of the world are liable to give different names to the same fossils, and in all parts of the world they are occasionally liable to group together different fossils under the same title. Both these sources of fallacy require to be guarded against in reasoning as to the age of strata from their fossil remains. Thirdly, the mere fact of fossils being found in beds which are known by physical evidence to be of different ages, has commonly led palæontologists to describe them as different species. Thus the same fossil, occurring in successive groups of strata, and with the merely trivial and varietal differences due to the gradual change in its environment, has been repeatedly described as a distinct species, with a distinct name, in every bed in which it was found. We know, however, that many fossils range vertically through many groups of strata, and there are some which even pass through several formations. The mere fact of a difference of physical position ought never to be taken into account at all in considering and determining the true affinities of a fossil. Fourthly, the results of experience, instead of being an assistance, are sometimes liable to operate as a source of error. When once, namely, a generalization has been established that certain fossils occur in strata of a certain age, palæontologists are apt to infer that *all* beds containing similar fossils must be of the same age. There is a presumption, of course, that this inference would be correct; but it is not a conclusion resting upon absolute necessity, and there might be physical evidence to disprove it. Fifthly, the physical geologist may lead the palæontologist astray by asserting that the physical evidence as to the age and position of a given group of beds is clear and unequivocal, when such evidence may be, in reality, very slight and doubtful. In this way, the observer may be readily led into wrong conclusions as to the nature of the organic remains—often obscure and fragmentary—which it is his business to examine, or he may be led erroneously to think that previous generalizations as to the age of certain kinds of fossils are premature and incorrect. Lastly, there are cases in which, owing to the limited exposure of the beds, to their being merely of local development, or to other causes, the physical evidence as to the age of a given group of strata may be entirely uncertain and unreliable, and in which, therefore, the observer has to rely wholly upon the fossils which he may meet with.

In spite of the above limitations and fallacies, there can be no doubt as to the enormous value of palæontology in enabling us to work out the historical succession of the sedimentary rocks. It may even be said that in any case where there should appear to be a clear and decisive discordance between the physical and the palæontological evidence as to the age of a given series of beds, it is the former that is to be distrusted rather than the latter. The records of geological science contain not a few cases in which apparently clear physical evidence of superposition has been demonstrated to have been wrongly interpreted; but the evidence of palæontology, when in any way sufficient, has rarely been upset by subsequent investigations. Should we find strata containing plants of the Coal-measures apparently resting upon other strata with Ammonites and Belemnites, we may be sure that the physical evidence is delusive; and though the above is an extreme case, the presumption in all such instances is rather that the physical succession has been misunderstood or misconstrued, than that there has been a subversion of the recognized succession of life-forms.

We have seen, then, that as the collective result of observations made upon the superposition of rocks in different localities, from their mineral characters, and from their included fossils, geologists have been able to divide the entire stratified series into a number of different divisions or formations, each characterized by a *general* uniformity of mineral composition, and by a special and peculiar *assemblage* of organic forms. Each of these primary groups is in turn divided into a series of smaller divisions, characterized and distinguished in the same way. It is not pretended for a moment that all these primary rock-groups can anywhere be seen surmounting one another regularly.* There is no region upon the earth where all the stratified formations can be seen together; and, even when most of them occur in the same country, they can nowhere be seen all succeeding each other in their regular and uninterrupted succession. The reason of this is obvious. There are many places—to take a

* As we have every reason to believe that dry land and sea have existed, at any rate from the commencement of the Laurentian period to the present day, it is quite obvious that no one of the great formations can ever, under any circumstances, have extended over the entire globe. In other words, no one of the formations can ever have had a greater geographical extent than that of the seas of the period in which the formation was deposited. Nor is there any reason for thinking that the proportion of dry land to ocean has ever been materially different to what it is at present, however greatly the areas of sea and land may have changed as regards their place. It follows from the above, that there is no sufficient basis for the view that the crust of the earth is composed of a succession of concentric layers, like the coats of an onion, each layer representing one formation.

single example—where one may see the Silurian rocks, the Devonian, and the Carboniferous rocks succeeding one another regularly, and in their proper order. This is because the particular region where this occurs was always submerged beneath the sea while these formations were being deposited. There are, however, many more localities in which one would find the Carboniferous rocks resting unconformably upon the Silurians without the intervention of any strata which could be referred to the Devonian period. This might arise from one of two causes: 1. The Silurians might have been elevated above the sea immediately after their deposition, so as to form dry land during the whole of the Devonian period, in which case, of course, no strata of the later age could possibly be deposited in that area. 2. The Devonian might have been deposited upon the Silurian, and then the whole might have been elevated above the sea, and subjected to an amount of denudation sufficient to remove the Devonian strata entirely. In this case, when the land was again submerged, the Carboniferous rocks, or any younger formation, might be deposited directly upon Silurian strata. From one or other of these causes, then, or from subsequent disturbances and denudations, it happens that we can rarely find many of the primary formations following one another consecutively and in their regular order.

In no case, however, do we ever find the Devonian resting upon the Carboniferous, or the Silurian rocks reposing on the Devonian. We have therefore, by a comparison of many different areas, an established order of succession of the stratified formations, as shown in the subjoined ideal section of the crust of the earth (fig. 17).

The main subdivisions of the stratified rocks are known by the following names:—

1. Laurentian.
2. Cambrian (with Huronian?).
3. Silurian.
4. Devonian or Old Red Sandstone.
5. Carboniferous.
6. Permian } New Red Sandstone.
7. Triassic }
8. Jurassic or Oolitic.
9. Cretaceous.
10. Eocene.
11. Miocene.
12. Pliocene.
13. Post-tertiary.

Ideal Section of the Crust of the Earth.

Fig. 17.

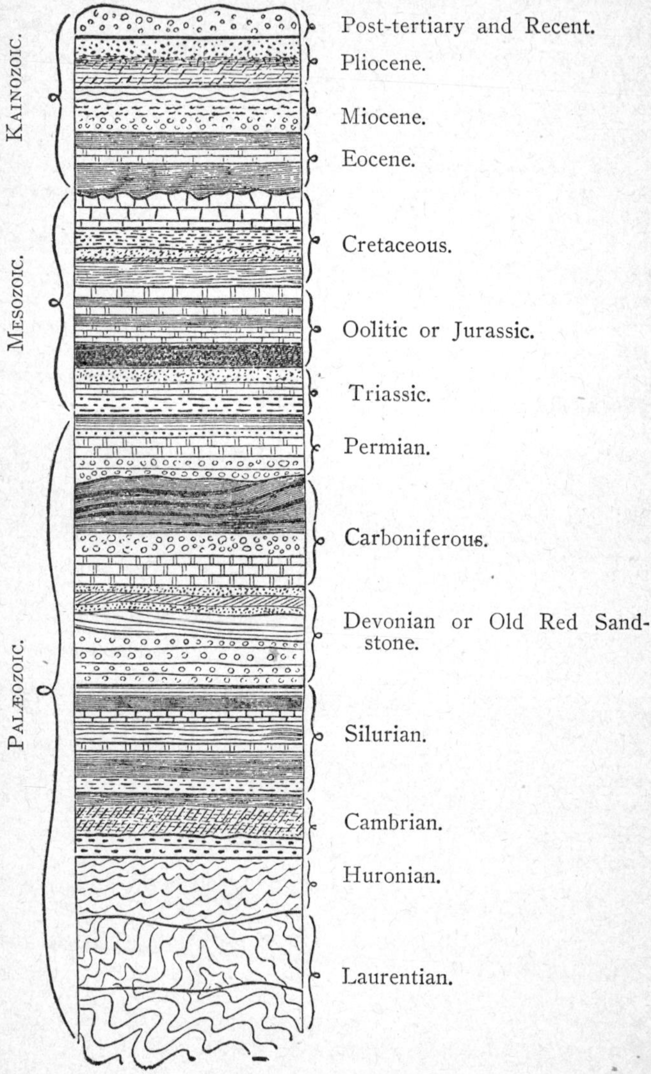

Of these primary rock divisions, the Laurentian, Cambrian, Silurian, Devonian, Carboniferous, and Permian are collectively grouped together under the name of the *Primary* or *Palæozoic* rocks (G. *palaios,* ancient; *zoe,* life). Not only do they constitute the oldest stratified accumulations, but from the extreme divergence between their animals and plants and those now in existence, they may appropriately be considered as belonging to an "Old-Life" period of the world's history. The Triassic, Jurassic, and Cretaceous systems are grouped together as the *Secondary* or *Mesozoic* formations (Gr. *mesos,* intermediate; *zoe,* life); the organic remains of this "Middle-Life" period being, on the whole, intermediate in their characters between those of the palæozoic epoch and those of more modern strata. Lastly, the Eocene, Miocene, and Pliocene formations are grouped together as the *Tertiary* or *Kainozoic rocks* (Gr. *kainos,* new; *zoe,* life); because they constitute a "New-Life" period, in which the organic remains approximate in character to those now existing upon the globe. The so-called *Post-Tertiary* deposits are placed with the Kainozoic, or may be considered as forming a separate *Quaternary* system.

CHAPTER IV.

THE BREAKS IN THE GEOLOGICAL AND PALÆONTOLOGICAL RECORD.

The term "contemporaneous" is usually applied by geologists to groups of strata in different regions which contain the same fossils, or an assemblage of fossils in which many identical forms are present. That is to say, beds which contain identical, or nearly identical, fossils, however widely separated they may be from one another in point of actual distance, are ordinarily believed to have been deposited during the same period of the earth's history. This belief, indeed, constitutes the keystone of the entire system of determining the age of strata by their fossil contents; and if we take the word "contemporaneous" in a general and strictly geological sense, this belief can be accepted as proved beyond denial. We must, however, guard ourselves against too literal an interpretation of the word "contemporaneous," and we must bear in mind the enormously-prolonged periods of time with which the geologist has to deal. When we say that two groups of strata in different regions are "contemporaneous," we simply mean that they were formed during the same geological period, and perhaps at different stages of that period, and we do not mean to imply that they were formed at precisely the same instant of time.

A moment's consideration will show us that it is only in the former sense that we can properly speak of strata being "contemporaneous;" and that, in points of fact, beds containing the same fossils, if occurring in widely distant areas, can hardly be "contemporaneous" in any literal sense; but that the very identity of their fossils is proof that they were deposited one after the other. If we find strata containing identical fossils within the limits of a single geographical region—say in Europe—then there is a reasonable probability that these beds are strictly contemporaneous, in the sense that they were deposited at the same time. There is a reasonable probability of this, because there is no improbability involved in the idea of an ocean occupying the whole area of Europe, and peopled throughout by many of the same species of marine animals. At the present day, for example, many identical species of animals are found living on the western coasts of Britain and the eastern coasts of North America, and beds now in course of deposition off the shores of Ireland and the seaboard of the state of New York would necessarily contain many of the same fossils. Such beds would be both literally and geologically contemporaneous; but the case is different if the distance between the areas where the strata occur be greatly increased. We find, for example, beds containing identical fossils (the Quebec or Skiddaw beds) in Sweden, in the north of England, in Canada, and in Australia. Now, if all these beds were contemporaneous, in the literal sense of the term, we should have to suppose that the ocean at one time extended uninterruptedly between all these points, and was peopled throughout the vast area thus indicated by many of the same animals. Nothing, however, that we see at the present day would justify us in imagining an ocean of such enormous extent, and at the same time so uniform in its depth, temperature, and other conditions of marine life, as to allow the same animals to flourish in it from end to end; and the example chosen is only one of a long and ever-recurring series. It is therefore much more reasonable to explain this, and all similar cases, as owing to the *migration* of the fauna, in whole or in part, from one marine area to another. Thus, we may suppose an ocean to cover what is now the European area, and to be peopled by certain species of animals. Beds of sediment—clay, sands, and limestones—will be deposited over the sea-bottom, and will entomb the remains of the animals as fossils. After this has lasted for a certain length of time, the European area may undergo elevation, or may become otherwise unsuitable for the perpetuation of its fauna; the result of which would be that some or all of the marine animals of the area would *migrate* to some more suitable region. Sediments would then be accumulated in the new area to which they had betaken themselves, and they would then appear, for the second time, as fossils in a set of beds widely separated from Europe. The second set of beds would, however, obviously not be strictly or literally contemporaneous with the first, but would be separated from them by the period of time required for the migration of the animals from the one area into the other. It is only in a wide and comprehensive sense that such strata can be said to be contemporaneous.

It is impossible to enter further into this subject here; but it may be taken as certain that beds in widely remote geographical areas can only come to contain the same fossils by reason of a migration having taken place of the animals of the one area to the other. That such migrations can and do take place is quite certain, and this is a much more reasonable explanation of the observed facts than the hypothesis that in former periods the conditions of life were much more uniform than they are at present, and that, consequently, the same organisms were able to range over the entire globe at the same time. It need only be added, that taking the evidence of the present as explaining the phenomena of the past—the only safe method of reasoning in geological matters—we have abundant proof that deposits which *are* actually contemporaneous, in the strict sense of the term, *do not contain the same fossils, if far removed from one another in point of distance.* Thus, deposits of various kinds are now in process of formation in our existing seas, as for example, in the Arctic Ocean, the Atlantic, and the Pacific, and many of these deposits are known to us by actual examination and observation with the sounding-lead and dredge. But it is hardly necessary to add that the animal remains contained in these deposits—the fossils of some future period—instead of being identical, are widely different from one another in their characters.

We have seen, then, that the entire stratified series is

capable of subdivision into a number of definite rock-groups or "formations," each possessing a peculiar and characteristic assemblage of fossils, representing the "life" of the "period" in which the formation was deposited. We have still to inquire shortly how it came to pass that two successive formations *should* thus be broadly distinguished by their life-forms, and why they should not rather possess at any rate a majority of identical fossils. It was originally supposed that this could be explained by the hypothesis that the close of each formation was accompanied by a general destruction of all the living beings of the period, and that the commencement of each new formation was signalized by the creation of a number of brand-new organisms, destined to figure as the characteristic fossils of the same. This theory, however, ignores the fact that each formation—as to which we have any sufficient evidence—contains a few, at least, of the life-forms which existed in the preceding period; and it invokes forces and processes of which we know nothing, and for the supposed action of which we cannot account. The problem is an undeniably difficult one, and it will not be possible here to give more than a mere outline of the modern views upon the subject. Without entering into the at present inscrutable question as to the manner in which new life-forms are introduced upon the earth, it may be stated that almost all modern geologists hold that the living beings of any given formation are in the main modified forms of others which have preceded them. It is not believed that any general or universal destruction of life took place at the termination of each geological period, or that a general introduction of new forms took place at the commencement of a new period. It is, on the contrary, believed that the animals and plants of any given period are for the most part (or exclusively) the lineal but modified descendants of the animals and plants of the immediately preceding period, and that some of them, at any rate, are continued into the next succeeding period, either unchanged, or so far altered as to appear as new species. To discuss these views in detail would lead us altogether too far, but there is one very obvious consideration which may advantageously receive some attention. It is obvious, namely, that the great discordance which is found to subsist between the animal life of any given formation and that of the next succeeding formation, and which no one denies, would be a fatal blow to the views just alluded to, unless admitting of some satisfactory explanation. Nor is this discordance one purely of life-forms, for there is often a physical break in the successions of strata as well. Let us therefore briefly consider how far these interruptions and breaks in the geological and palæontological record can be accounted for, and still allow us to believe in some theory of continuity as opposed to the doctrine of intermittent and occasional action.

In the first place, it is perfectly clear that if we admit the conception above mentioned of a continuity of life from the Laurentian period to the present day, we could never *prove* our view to be correct, unless we could produce in evidence fossil examples of *all* the kinds of animals and plants that have lived and died during that period. In order to do this, we should require, to begin with, to have access to an absolutely unbroken and perfect succession of all the deposits which have ever been laid down since the beginning. If, however, we ask the physical geologist if he is in possession of any such uninterrupted series, he will at once answer in the negative. So far from the geological series being a perfect one, it is interrupted by numerous gaps of unknown length, many of which we can never expect to fill up. Nor are the proofs of this far to seek. Apart from the facts that we have hitherto examined only a limited portion of the dry land, that nearly two-thirds of the entire area of the globe is inaccessible to geological investigation in consequence of its being covered by the sea, that many deposits can be shown to have been more or less completely destroyed subsequent to their deposition, and that there may be many areas in which living beings exist where no rock is in process of formation, we have the broad fact that rock-deposition only goes on to any extent in water, and that the earth must have always consisted partly of dry land and partly of water—at any rate, so far as any period of which we have geological knowledge is concerned. There *must*, therefore, always have existed, at some part or another of the earth's surface, areas where no deposition of rock was going on, and the proof of this is to be found in the well-known phenomenon of "*unconformability.*" Whenever, namely, deposition of sediment is continuously going on within the limits of a single ocean, the beds which are laid down succeed one another in uninterrupted and regular sequence. Such beds are said to be "conformable," and there are many rock-groups known where one may pass through fifteen or twenty thousand feet of strata without a break—indicating that the beds had been deposited in an area which remained continuously covered by the sea. On the other hand, we commonly find that there is no such regular succession when we pass from one great formation to another, but that, on the contrary, the younger formation rests "unconformably," as it is called, either upon the formation immediately preceding it in point of time, or upon some still older one. The essential physical feature of this unconformability is that the beds of the younger formation rest upon a worn and eroded surface formed by the beds of the older series (fig. 18); and a moment's consideration will show us what this indicates. It indicates, beyond the

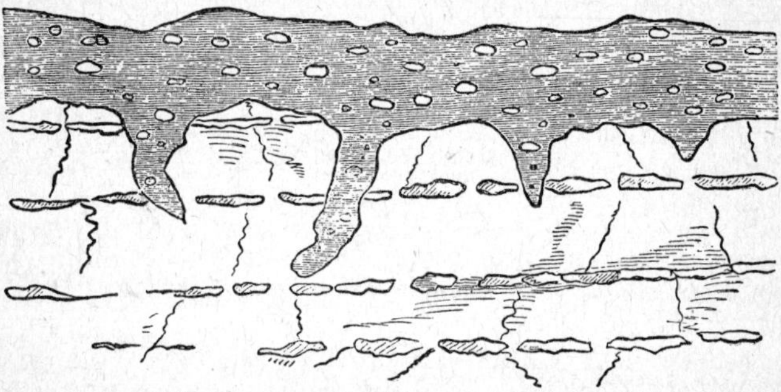

Fig. 18.—Section showing strata of Tertiary age (*a*) resting upon a worn and eroded surface of White Chalk (*b*), the stratification of which is marked by lines of flint.

possibility of misconception, that there was an interval between the deposition of the older series and that of the newer series of strata; and that during this interval the older beds were raised above the sea-level, so as to form dry land, and were subsequently depressed again beneath the waters, to receive upon their worn and wasted upper surface the sediments of the later group. During the interval thus indicated, the deposition of rock must of necessity have been proceeding more or less actively in other areas. Every unconformity, therefore, indicates that at the spot where it occurs, a more or less extensive series of beds must be actually *missing;* and though we may sometimes be able to point to these missing strata in other areas, there yet remains a number of unconformities for which we cannot at present supply the deficiency even in a partial manner.

It follows from the above that the series of stratified deposits is to a greater or less extent irremediably imperfect; and in this imperfection we have one great cause why we can never obtain a perfect series of all the animals and plants that have lived upon the globe. Wherever one of these great physical gaps occurs, we find, as we might expect, a corresponding break in the series of life-forms. In other words, whenever we find two formations to be unconformable, we shall always find at the same time that there is a great difference in their fossils, and that many of the fossils of the older formation do not survive into the newer, whilst many of those in the newer are not known to occur in the older. The cause of this is, obviously, that the lapse of time, indicated by the unconformability, has been sufficiently great to allow of the dying out or modification of many of the older forms of life, and the introduction of new ones by immigration.

Apart, however, altogether, from these great physical breaks and their corresponding breaks in life, there are other reasons why we can never become more than partially acquainted with the former denizens of the globe. Foremost amongst these is the fact that an enormous number of animals possess no hard parts of the nature of a skeleton, and are therefore incapable, under any ordinary circumstances, of leaving behind them any traces of their existence. It is true that there are cases in which animals in themselves completely soft-bodied are nevertheless able to leave marks by which their former presence can be detected. Thus every geologist is familiar with the winding and twisting "trails" formed on the surface of the strata by sea-worms; and the impressions left by the stranded carcases of Jelly-fishes on the fine-grained lithographic slates of Solenhofen supply us with an example of how a creature which is little more than "organized sea-water" may still make an abiding mark upon the sands of time. As a general rule, however, animals which have no skeletons are incapable of being preserved as fossils, and hence there must always have been a vast number of different kinds of marine animals of which we have absolutely no record whatever. Again, almost all the fossiliferous rocks have been laid down in water; and it is a necessary result of this that the great majority of fossils are the remains of aquatic animals. The remains of air-breathing animals, whether of the inhabitants of the land or of the air itself, are comparatively rare as fossils, and the record of the past existence of these is much more imperfect than is the case with animals living in water. Moreover, the fossiliferous deposits are not only almost exclusively aqueous formations, but the great majority are marine, and only a comparatively small number have been formed by lakes and rivers. It follows from the foregoing that the palæontological record is fullest and most complete so far as sea-animals are concerned, though even here we find enormous gaps, owing to the absence of hard structures in many great groups; of animals inhabiting fresh water our knowledge is rendered still further incomplete by the small proportion that fluviatile and lacustrine deposits bear to marine; whilst we have only a fragmentary acquaintance with the air-breathing animals which inhabited the earth during past ages.

Lastly, the imperfection of the palæontological record, due to the causes above enumerated, is greatly aggravated, especially as regards the earlier portion of the earth's history, by the fact that many rocks which contained fossils when deposited have since been rendered barren of organic remains. The principal cause of this common phenomenon is what is known as "metamorphism"—that is, the subjection of the rock to a sufficient amount of heat to cause a rearrangement of its particles. When at all of a pronounced character, the result of metamorphic action is invariably the obliteration of any fossils which might have been originally present in the rock. Metamorphism may effect rocks of any age, though naturally more prevalent in the older rocks, and to this cause must be set down an irreparable loss of much fossil evidence. The most striking example which is to be found of this is the great Laurentian series, which comprises some 30,000 feet of highly-metamorphosed sediments, but which, with one not wholly undisputed exception, has as yet yielded no remains of living beings, though there is strong evidence of the former existence in it of fossils.

Upon the whole, then, we cannot doubt that the earth's crust, so far as yet deciphered by us, presents us with but a very imperfect record of the past. Whether the known and admitted imperfections of the geological and palæontological records are sufficiently serious to account satisfactorily for the deficiency of direct evidence recognizable in some modern hypotheses, may be a matter of individual opinion. There can, however, be little doubt that they are sufficiently extensive to throw the balance of evidence decisively in favor of some theory of continuity, as opposed to any theory of intermittent and occasional action. The apparent breaks which divide the great series of the stratified rocks into a number of isolated formations, are not marks of mighty and general convulsions of nature, but are simply indications of the imperfection of our knowledge. Never, in all probability, shall we be able to point to a complete series of deposits, or a complete succession of life linking one great geological period to another. Nevertheless, we may well feel sure that such deposits and such an unbroken succession must have existed at one time. We are compelled to believe that nowhere in the long series of the fossiliferous rocks has there been a total break, but that there must have been a complete continuity of life, and a more or less complete continuity of sedimentation, from the Laurentian period to the present day. One generation hands on the lamp of life to the next, and each system of rocks is the direct offspring of those which preceded it in time. Though there has not been continuity in any given area, still the geological chain could never have been snapped at one point, and taken up again at a totally different one. Thus we arrive at the conviction that *continuity* is the fundamental law of geology, as it is of the other sciences, and that the lines of demarcation between the great formations are but gaps in our own knowledge.

CHAPTER V.

CONCLUSIONS TO BE DRAWN FROM FOSSILS.

We have already seen that geologists have been led by the study of fossils to the all-important generalization that the vast series of the Fossiliferous or Sedimentary Rocks may be divided into a number of definite groups or "formations," each of which is characterized by its organic remains. It may simply be repeated here that these formations are not properly and strictly characterized by the occurrence in them of any one particular fossil. It may be that a formation contains some particular fossil or fossils not occurring out of that formation, and that in this way an observer may identify a given group with tolerable certainty. It very often happens, indeed, that some particular stratum, or sub-group of series, contains peculiar fossils, by which its existence may be determined in various localities. As before remarked, however, the great formations are characterized properly by the association of certain fossils, by the predominance of certain families or orders, or by an *assemblage* of fossil remains representing the "life" of the period in which the formation was deposited.

Fossils, then, enable us to determine the *age* of the deposits in which they occur. Fossils further enable us to come to very important conclusions as to the mode in which the fossiliferous bed was deposited, and thus as to the condition of the particular district or region occupied by the fossiliferous bed at the time of the formation of the latter. If, in the first place, the bed contains the remains of animals such as now inhabit rivers, we know that it is "fluviatile" in its origin, and that it must at one time have either formed an actual riverbed, or been deposited by the overflowing of an ancient stream. Secondly, if the bed contain the remains of shell-fish, minute crustaceans, or fish, such as now inhabit lakes, we know that it is "lacustrine," and was deposited beneath the waters of a former lake. Thirdly, if the bed contain the remains of animals such as now people the ocean, we know that it is "marine" in its origin, and that it is a fragment of an old sea-bottom.

We can, however, often determine the conditions under which a bed was deposited with greater accuracy than this. If, for example, the fossils are of kinds resembling the marine animals now inhabiting shallow waters, if they are accompanied by the detached relics of terrestrial organisms, or if they are partially rolled and broken, we may conclude that the fossiliferous deposit was laid down in a shallow sea, in the immediate vicinity of a coast-line, or as an actual shore-deposit. If, again, the remains are those of animals such as now live in the deeper parts of the ocean, and there is a very sparing intermixture of extraneous fossils (such as the bones of birds or quadrupeds, or the remains of plants), we may presume that the deposit is one of deep water. In other cases, we may find, scattered through the rock, and still in their natural position, the valves of shells such as we know at the present day as living buried in the sand or mud of the sea-shore or of estuaries. In other cases, the bed may obviously have been an ancient coral-reef, or an accumulation of social shells, like Oysters. Lastly, if we find the deposit to contain the remains of marine shells, but that these are dwarfed of their fair proportions and distorted in figure, we may conclude that it was laid down in a brackish sea, such as the Baltic, in which the proper saltness was wanting, owing to its receiving an excessive supply of fresh water.

In the preceding, we have been dealing simply with the remains of aquatic animals, and we have seen that certain conclusions can be accurately reached by an examination of these. As regards the determination of the conditions of deposition from the remains of aerial and terrestrial animals, or from plants, there is not such an absolute certainty. The remains of land-animals would, of course, occur in "sub-aerial" deposits —that is, in beds, like blown sand, accumulated upon the land. Most of the remains of land-animals, however, are found in deposits which have been laid down in water, and they owe their present position to the fact that their former owners were drowned in rivers or lakes, or carried out to sea by streams. Birds, Flying Reptiles, and Flying Mammals might also similarly find their way into aqueous deposits; but it is to be remembered that many birds and mammals habitually spend a great part of their time in the water, and that these might therefore be naturally expected to present themselves as fossils in Sedimentary Rocks. Plants, again, even when undoubtedly such as must have grown on land, do not prove that the bed in which they occur was formed on land. Many of the remains of plants known to us are extraneous to the bed in which they are now found, having reached their present site by falling into lakes or rivers, or being carried out to sea by floods or gales of wind. There are, however, many cases in which plants have undoubtedly grown on the very spot where we now find them. Thus it is now generally admitted that the great coal-fields of the Carboniferous age are the result of the growth *in situ* of the plants which compose coal, and that these grew on vast marshy or partially submerged tracts of level alluvial land. We have, however, distinct evidence of old land-surfaces, both in the Coal-measures and in other cases (as, for instance, in the well-known "dirt-bed" of the Purbeck series). When, for example, we find the erect stumps of trees standing at right angles to the surrounding strata, we know that the surface through which these send their roots was at one time the surface of the dry land, or in other words, was an ancient soil (fig. 19).

Fig. 19.—Erect Tree containing Reptilian remains. Coal-measures, Nova Scotia. (After Dawson.)

In many cases fossils enable us to come to important conclusions as to the climate of the period in which they lived, but only a few instances of this can be here adduced. As fossils in the majority of instances are the remains of marine animals, it is mostly the temperature of the sea which can alone be determined in this way; and it is important to remember that, owing to the existence of heated currents, the marine climate of a given area does not necessarily imply a correspondingly warm climate in the neighboring land. Land-climates can only be determined by the remains of land-animals or land-plants, and these are comparatively rare as fossils. It is also important to remember that all conclusions on this head are really based upon the present distribution of animal and vegetable life on the globe, and are therefore liable to be vitiated by the following considerations:—

a. Most fossils are extinct, and it is not certain that the habits and requirements of any extinct animal were exactly similar to those of its nearest living relative.

b. When we get very far back in time, we meet with groups of organisms so unlike anything we know at the present day as to render all conjectures as to climate found upon their supposed habits more or less uncertain and unsafe.

c. In the case of marine animals, we are as yet very far from knowing the exact limits of distribution of many species within our present seas; so that conclusions drawn from living forms as to extinct species are apt to prove incorrect. For instance, it has recently been shown that many shells formerly believed to be confined to the Arctic Seas have, by reason of the extension of Polar currents, a wide range to the south; and this has thrown doubt upon the conclusions drawn from fossil shells as to the Arctic conditions under which certain beds were supposed to have been deposited.

d. The distribution of animals at the present day is certainly dependent upon other conditions beside climate alone; and the causes which now limit the range of given animals are certainly such as belong to the existing order of things. But the establishment of the present order of things does not date back in many cases to the introduction of the present species of animals. Even in the case therefore, of existing species of animals, it can often be shown that the past distribution of the species was different formerly to what it is now, not necessarily because the climate has changed, but because of the alteration of other conditions essential to the life of the species of conducing to its extension.

Still, we are in many cases able to draw completely reliable conclusions as to the climate of a given geological period, by an examination of the fossils belonging to that period. Among the more striking examples of how the past climate of a region may be deduced from the study of the organic remains contained in its rocks, the following may be mentioned: It has been shown that in Eocene times, or at the commencement of the Tertiary period, the climate of what is now Western Europe was of a tropical or sub-tropical character. Thus the Eocene beds are found to contain the remains of shells such as now inhabit tropical seas, as, for example, Cowries and Volutes; and with these are the fruits of palms, and the remains of other tropical plants. It has been shown, again, that in Miocene times, or about the middle of the Tertiary period, Central Europe was peopled with a luxuriant flora resembling that of the warmer parts of the United States, and leading to the conclusion that the mean annual temperature must have been at least 30° hotter than it is at present. It has been shown that, at the same time, Greenland, now buried beneath a vast ice-shroud, was warm enough to support a large number of trees, shrubs, and other plants, such as inhabit the temperate regions of the globe. Lastly, it has been shown, upon physical as well as palæontological evidence, that the greater part of the North Temperate Zone, at a comparatively recent geological period, has been visited with all the rigors of an Arctic climate, resembling that of Greenland at the present day. This is indicated by the occurrence of Arctic shells in the superficial deposits of this period, whilst the Musk-ox and the Reindeer roamed far south of their present limits.

Lastly, it was from the study of fossils that geologists learnt originally to comprehend a fact which may be regarded as of cardinal importance in all modern geological theories and speculations—namely, that the crust of earth is liable to local elevations and subsidences. For long after the remains of shells and other marine animals were for the first time observed in the solid rocks forming the dry land, and at great heights above the sea-level, attempts were made to explain this almost unintelligible phenomenon upon the hypothesis that the fossils in question were not really the objects they represented, but were in truth mere *lusus naturæ*, due to some "plastic virtue latent in the earth." The common-sense of scientific men, however, soon rejected this idea, and it was agreed by universal consent that these bodies really were the remains of animals which formerly lived in the sea. When once this was admitted, the further steps were comparatively easy, and at the present day no geological doctrine stands on a firmer basis than that which teaches us that our present continents and islands, fixed and immovable as they appear, have been repeatedly sunk beneath the ocean.

CHAPTER VI

THE BIOLOGICAL RELATIONS OF FOSSILS.

Not only have fossils, as we have seen, a most important bearing upon the sciences of Geology and Physical Geography, but they have relations of the most complicated and weighty character with the numerous problems connected with the study of living beings, or in other words, with the science of Biology. To such an extent is this the case, that no adequate comprehension of Zoology and Botany, in their modern form, is so much as possible without some acquaintance with the types of animals and plants which have passed away. There are also numerous speculative questions in the domain of vital science, which, if soluble at all, can only hope to find their key in researches carried out on extinct organisms. To discuss fully the biological relations of fossils would, therefore, afford matter for a separate treatise; and all that can be done here is to indicate very cursorily the principal points to which the attention of the palæontological student ought to be directed.

In the first place, the great majority of fossil animals and plants are "extinct"—that is to say, they belong to species which are no longer in existence at the present day. So far, however, from there being any truth in the old view that there were periodic destructions of all the living beings in existence upon the earth, followed by a corresponding number of new creations of animals and plants, the actual facts of the case show that the extinction of old forms and the introduction of new forms have been processes constantly going on throughout the whole of geological time. Every species seems to come into being at a certain definite point of time, and to finally disappear at another definite point; though there are few instances indeed, if there are any, in which our present knowledge would permit us safely to fix with precision the times of entrance and exit. There are, moreover, marked differences in the actual time during which different species remained in existence, and therefore corresponding differences in their "vertical range" or, in other words, in the actual amount and thickness of strata through which they present themselves as fossils. Some species are found to range through two or even three formations, and a few have an even more extended life. More commonly the species which begin in the commencement of a great formation die out at or before its close, whilst those which are introduced for the first time near the middle or end of the formation may either become extinct, or may pass on into the next succeeding formation. As a general rule, it is the animals which have the lowest and simplest organization that have the longest range in time, and the additional possession of microscopic or minute dimensions seems also to favor longevity. Thus some of the *Foraminifera* appear to have survived, with little or no perceptible alteration, from the Silurian period to the present day; whereas large and highly-organized animals, though long-lived as *individuals,* rarely seem to live long *specifically,* and have, therefore, usually a restricted vertical range. Exceptions to this, however are occasionally to be found in some "persistent types," which extend through a succession of geological periods with very little modification. Thus the existing Lampshells of the genus *Lingula* are little changed from the *Lingulæ* which swarmed in the Lower Silurian seas; and the existing Pearly Nautilus is the last descendant of a clan nearly as ancient. On the other hand, some forms are singularly restricted in their limits, and seem to have enjoyed a comparatively brief lease of life. An example of this is to be found in many of the *Ammonites*—close allies of the Nautilus—which are often confined strictly to certain zones of strata, in some cases of very insignificant thickness.

Of the *causes* of extinction amongst fossil animals and plants, we know little or nothing. All we can say is, that the attributes which constitute a *species* do not seem to be intrinsically endowed with permanence, any more than the attributes which constitute an *individual,* though the former may endure whilst many successive generations of the latter have disappeared. Each species appears to have its own life-period, its commencement, its culmination, and its gradual decay; and the life-periods of different species may be of very different duration.

From what has been said above, it may be gathered that our existing species of animals and plants are, for the most part, quite of modern origin, using the term "modern" in its geological acceptation. Measured by human standards, the majority of existing animals (which are capable of being preserved as fossils) are known to have a high antiquity; and some of them can boast of a pedigree which even the geologist may regard with respect. Not a few of our shell-fish are known to have commenced their existence at some point of the Tertiary period; one Lampshell (*Terebratulina caput-serpentis*) is believed to have survived since the Chalk; and some of the *Foraminifera* date, at any rate, from the Carboniferous period. We learn from this the additional fact that our existing animals and plants do not constitute an assemblage of organic forms which were introduced into the world collectively and simultaneously, but that they commenced their existence at very different periods, some being extremely old, whilst others may be regarded as comparatively recent animals. And this introduction of the existing fauna and flora was a slow and *gradual* process, as shown admirably by the study of the fossil shells of the Tertiary period. Thus, in the earlier Tertiary period, we find about 95 per cent of the known fossil shells to be species that are no longer in existence, the remaining 5 per cent being forms which are known to live in our present seas. In the middle of the Tertiary period we find many more recent and still existing species of shells, and the extinct types are much fewer in number; and this gradual introduction of forms now living goes on steadily, till, at the close of the Tertiary period, the proportions with which we started may be reversed, as many as 90 or 95 per cent of the fossil shells being forms still alive, while not more than 5 per cent may have disappeared.

All known animals at the present day may be divided into some five or six primary divisions, which are known technically as "*sub-kingdoms.*" Each of these sub-kingdoms* may be regarded as representing a certain type or plan of structure, and all the animals comprised in each are merely modified forms of this common type. Not only are all known living animals thus reducible to some five or six fundamental plans of structure, but amongst the vast series of fossil forms no one has yet been found—however unlike any existing animal—to possess peculiarities which would entitle it to be placed in a new sub-kingdom. All fossil animals, therefore, are capable of being referred to one or other of the primary divisions of the animal kingdom. Many fossil groups have no closely-related group

* In the Appendix a brief definition is given of the sub-kingdoms, and the chief divisions of each are enumerated.

now in existence; but in no case do we meet with any grand structural type which has not survived to the present day.

The old types of life differ in many respects from those now upon the earth; and the further back we pass in time, the more marked does this divergence become. Thus, if we were to compare the animals which lived in the Silurian seas with those inhabiting our present oceans, we should in most instances find differences so great as almost to place us in another world. This divergence is the most marked in the Palæozoic forms of life, less so in those of the Mesozoic period, and less still in the Tertiary period. Each successive formation has therefore presented us with animals becoming gradually more and more like those now in existence; and though there is an immense and striking difference between the Silurian animals and those of to-day, this difference is greatly reduced if we compare the Silurian fauna with the Devonian; *that* again with the Carboniferous; and so on till we reach the present.

It follows from the above that the animals of any given formation are more like those of the next formation below, and of the next formation above, than they are to any others; and this fact of itself is an almost inexplicable one, unless we believe that the animals of any given formation are, in part at any rate, the lineal descendants of the animals of the preceding formation, and the progenitors, also in part at least, of the animals of the succeeding formation. In fact, the palæontologist is so commonly confronted with the phenomenon of closely-allied forms of animal life succeeding one another in point of time, that he is compelled to believe that such forms have been developed from some common ancestral type by some process of "*evolution.*" On the other hand, there are many phenomena, such as the apparently sudden introduction of new forms throughout all past time, and the common occurrence of wholly isolated types, which cannot be explained in this way. Whilst it seems certain, therefore, that many of the phenomena of the succession of animal life in past periods can only be explained by some law of evolution, it seems at the same time certain that there has always been some other deeper and higher law at work, on the nature of which it would be futile to speculate at present.

Not only do we find that the animals of each successive formation become gradually more and more like those now existing upon the globe, as we pass from the older rocks into the newer, but we also find that there has been a gradual progression and development in the *types* of animal life which characterize the geological ages. If we take the earliest-known and the oldest examples of any given group of animals, it can sometimes be shown that these primitive forms, though in themselves highly organized, possessed certain characters such as are now only seen in the *young* of their existing representatives. In technical language, the early forms of life in some instances possess "*embryonic*" characters, though this does not prevent them often attaining a size much more gigantic than their nearest living relatives. Moreover, the ancient forms of life are often what is called "comprehensive types"—that is to say, they possess characters in combination such as we nowadays only find separately developed in different groups of animals. Now, this permanent retention of embryonic characters and this "comprehensiveness" of structural type are signs of what a zoologist considers to be a comparatively low grade of organization; and the prevalence of these features in the earlier forms of animals is a very striking phenomenon, though they are none the less perfectly organized so far as their own type is concerned. As we pass upwards in the geological scale, we find that these features gradually disappear, higher and ever higher forms are introduced, and "specialization" of type takes the place of the former comprehensiveness. We shall have occasion to notice many of the facts on which these views are based at a later period, and in connection with actual examples. In the meanwhile, it is sufficient to state, as a widely-accepted generalization of palæontology, that there has been in the past a general progression of organic types, and that the appearance of the lower forms of life has in the main preceded that of the higher forms in point of time.

PART II.

HISTORICAL PALÆONTOLOGY.

PART II.

CHAPTER VII.

THE LAURENTIAN AND HURONIAN PERIODS.

The *Laurentian Rocks* constitute the base of the entire stratified series, and are, therefore, the oldest sediments of which we have as yet any knowledge. They are more largely and more typically developed in North America, and especially in Canada, than in any known part of the world, and they derive their title from the range of hills which the old French geographers named the "Laurentides." These hills are composed of Laurentian Rocks, and form the watershed between the valley of the St Lawrence river on the one hand, and the great plains which stretch northwards to Hudson Bay on the other hand. The main area of these ancient deposits forms a great belt of rugged and undulating country, which extends from Labrador westwards to Lake Superior, and then bends northwards towards the Arctic Sea. Throughout this extensive area the Laurentian Rocks for the most part present themselves in the form of low, rounded, ice-worn hills, which, if generally wanting in actual sublimity, have a certain geological grandeur from the fact that they "have endured the battles and the storms of time longer than any other mountains" (Dawson). In some places, however, the Laurentian Rocks produce scenery of the most magnificent character, as in the great gorge cut through them by the river Saguenay, where they rise at times into vertical precipices 1500 feet in height. In the famous group of the Adirondack mountains, also, in the state of New York they form elevations no less than 6000 feet above the level of the sea. As a general rule, the character of the Laurentian region is that of a rugged, rocky, rolling country, often densely timbered, but rarely well fitted for agriculture, and chiefly attractive to the hunter and the miner.

As regards its mineral characters, the Laurentian series is

composed throughout of metamorphic and highly crystalline rocks, which are in a high degree crumpled, folded, and faulted. By the late Sir William Logan the entire series was divided into two great groups, the *Lower Laurentian* and the *Upper Laurentian,* of which the latter rests unconformably upon the truncated edges of the former, and is in turn unconformably overlaid by strata of Huronian and Cambrian age (fig. 20).

The *Lower Laurentian* series attains the enormous thickness of over 20,000 feet, and is composed mainly of great beds of gneiss, altered sandstones (quartzites), mica-schist, hornblende-schist, magnetic iron-ore, and hæmatite, together with masses of limestone. The limestones are especially interesting, and have an

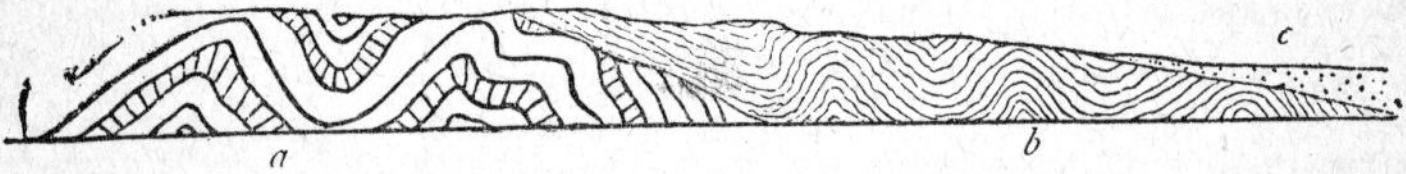

Fig. 20—Diagrammatic section of the Laurentian Rocks in Lower Canada. *a* Lower Laurentian; *b* Upper Laurentian, resting unconformably upon the lower series; *c* Cambrian strata (Potsdam Sandstone), resting unconformably on the Upper Laurentian.

extraordinary development—three principal beds being known, of which one is not less than 1500 feet thick; the collective thickness of the whole being about 3500 feet.

The *Upper Laurentian* series, as before said, reposes unconformably upon the Lower Laurentian, and attains a thickness of at least 10,000 feet. Like the preceding, it is wholly metamorphic, and is composed partly of masses of gneiss and quartzite; but it is especially distinguished by the possession of great beds of felspathic rock, consisting principally of "Labrador felspar."

Though typically developed in the great Canadian area already spoken of, the Laurentian Rocks occur in other localities, both in America and in the Old World. In Britain, the so-called "fundamental gneiss" of the Hebrides and of Sutherlandshire is probably of Lower Laurentian age, and the "hypersthene rocks" of the Isle of Syke may, with great probability, be regarded as referable to the Upper Laurentian. In other localities in Great Britain (as in St. David's, South Wales; the Malvern Hills; and the North of Ireland) occur ancient metamorphic deposits which also are probably referable to the Laurentian series. The so-called "primitive gneiss" of Norway appears to belong to the Laurentian, and the ancient metamorphic rocks of Bohemia and Bavaria may be regarded as being approximately of the same age.

By some geological writers the ancient and highly metamorphosed sediments of the Laurentian and the succeeding Huronian series have been spoken of as the "Azoic rocks" (Gr. *a*, without; *zoe*, life); but even if we were wholly destitute of any evidence of life during these periods, this name would be objectionable upon theoretical grounds. If a general name be needed, that of "Eozoic" (Gr. *eos*, dawn; *zoe*, life), proposed by Principal Dawson, is the most appropriate. Owing to their metamorphic condition, geologists long despaired of ever detecting any traces of life in the vast pile of strata which constitute the Laurentian System. Even before any direct traces were discovered, it was, however, pointed out that there were good reasons for believing that the Laurentian seas had been tenanted by an abundance of living beings. These reasons are briefly as follows:—(1) Firstly, the Laurentian series consists, beyond question, of marine sediments which originally differed in no essential respect from those which were subsequently laid down in the Cambrian or Silurian periods. (2) In all formations later than the Laurentian, any limestones which are present can be shown, with few exceptions, to be *organic* rocks, and to be more or less largely made up of the comminuted *débris* of marine or fresh-water animals. The Laurentian limestones, in consequence of the metamorphism to which they have been subjected, are so highly crystalline (fig. 21) that the microscope fails to detect any organic structure in the rock, and no fossils beyond those which will be spoken of immediately have as yet been discovered in them. We know, however, of numerous cases in which limestones, of later age, and undoubtedly organic to begin with, have been rendered so intensely crystalline by metamorphic action that all traces of organic structure have been obliterated. We have therefore, by analogy, the strongest possible ground for believing that the vast beds of Laurentian limestone have been originally organic in their origin, and primitively composed, in the main, of the calcareous skele-

Fig. 21.—Section of Lower Laurentian Limestone from Hull, Ottawa; enlarged five diameters. The rock is very highly crystalline, and contains mica and other minerals. The irregular black masses in it are graphite. (Original.)

tons of marine animals. It would, in fact, be a matter of great difficulty to account for the formation of these great calcareous masses on any other hypothesis. (3) The occurrences of phosphate of lime in the Laurentian Rocks in great abundance, and sometimes in the form of irregular beds, may very possibly be connected with the former existence in the strata of the remains of marine animals of whose skeletons this mineral is a constituent. (4) The Laurentian Rocks contain a vast amount of carbon in the form of black-lead or *graphite*. This mineral is especially abundant in the limestones, occurring in regular beds, in veins or strings, or disseminated through the body of the limestone in shape of crystals, scales, or irregular masses. The amount of graphite in some parts of the Lower Laurentian is so great that it has been calculated as equal to the quantity of carbon present in an equal thickness of the Coal-measures. The general source of solid carbon in the crust of the earth is, however, plant-life; and it seems impossible to account for the Laurentian graphite, except upon the supposition that it is metamorphosed vegetable matter. (5) Lastly, the great beds of iron-ore (peroxide and magnetic oxide) which occur in the Laurentian series interstratified with the other rocks, point with great probability to the action of vegetable life; since similar deposits in later formations can commonly be shown to have been formed by the deoxidizing power of vegetable matter in a state of decay.

In the words of Principal Dawson, "any one of these reasons might, in itself, be held insufficient to prove so great and, at first sight, unlikely a conclusion as that of the existence of abundant animal and vegetable life in the Laurentian; but the concurrence of the whole in a series of deposits unquestionably marine, forms a chain of evidence so powerful that it might command belief even if no fragment of any organic or living form or structure had ever been recognized in these ancient rocks." Of late years, however, there have been discovered in the Laurentian Rocks certain bodies which are believed to be truly the remains of animals, and of which by far the most important is the structure known under the now celebrated name of *Eozoön*. If truly organic, a very special and exceptional interest attaches itself to *Eozoön*, as being the most ancient fossil animal of which we have any knowledge; but there are some who regard it really a peculiar form of mineral structure, and a severe, protracted, and still unfinished controversy has been carried on as to its nature. Into this controversy it is wholly unnecessary to enter here; and it will be sufficient to briefly explain the structure of *Eozoön,* as elucidated by the elaborate and masterly investigations of Carpenter and Dawson, from the standpoint that it is a genuine organism—the balance of evidence up to this moment inclining decisively to this view.

The structure known as *Eozoön* is found in various localities in the Lower Laurentian limestones of Canada, in the form of isolated masses or spreading layers, which are composed of thin alternating laminæ, arranged more or less concentrically (fig. 22). The laminæ of these masses are usually of different

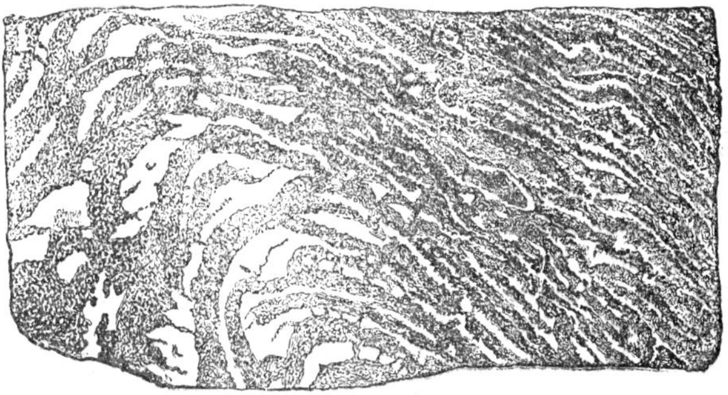

Fig. 22.—Fragment of *Eozoön*, of the natural size, showing alternate laminæ of loganite and dolomite. (After Dawson.)

colors and composition; one series being white, and composed of carbonate of lime—whilst the laminæ of the second series alternate with the preceding, are green in color, and are found by chemical analysis to consist of some silicate, generally serpentine or the closely-related "loganite." In some instances, however, all the laminæ are calcareous, the concentric arrangement still remaining visible in consequence of the fact that the laminæ are composed alternately of lighter and darker colored limestone.

When first discovered, the masses of *Eozoön* were supposed to be of a mineral nature; but their striking general resemblance to the undoubted fossils which will be subsequently spoken of under the name of *Stromatopora* was recognized by Sir William Logan, and specimens were submitted for minute examination, first to Principal Dawson, and subsequently to Dr. W. B. Carpenter. After a careful microscopic examination, these two distinguished observers came to the conclusion that *Eozoön* was truly organic, and in this opinion they were afterwards corroborated by other high authorities (Mr. W. K. Parker, Professor Rupert Jones, Mr. H. B. Brady, Professor Gümbel, &c.) Stated briefly, the structure of *Eozoön*, as exhibited by the microscope, is as follows:—

The concentrically-laminated mass of *Eozoön* is composed of numerous calcareous layers, representing the original skeleton of the organism (fig. 23, *b*). These calcareous layers serve to separate and define a series of chambers arranged in successive tiers, one above the other (fig. 23, A, B, C); and they are perforated not only by passages (fig. 23, *c*), which serve to place successive tiers of chambers in communication, but also by a system of delicate branching canals (fig. 23, *d*). Moreover, the central and principal portion of each calcareous layer, with the ramified canal-system just spoken of, is bound both above and below by a thin lamina which has a structure of its own, and which may be regarded as the proper shell-wall (fig. 23, *a a*). This proper wall forms the actual lining of the chambers, as well as the outer surface of the whole mass; and it is perforated with numerous fine vertical tubes (fig. 24, *a a*), opening into the chambers and on to the surface by corresponding fine pores. From the resemblance of this tubulated layer to similar structures in the shell of the Nummulite, it is often spoken of as the "Nummuline layer." The chambers are sometimes piled up one above the other in an irregular manner; but they are more commonly arranged in regular tiers, the separate chambers being marked off from one another by projections of the wall in the form of partitions, which are so far imperfect as to allow of a free communication between contiguous chambers. In the original condition

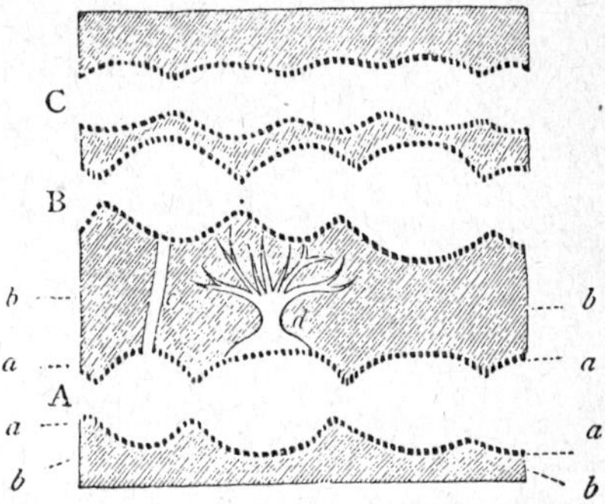

Fig. 23.—Diagram of a portion of *Eozoön* cut vertically. A, B, C, Three tiers of chambers communicating with one another by slightly constructed apertures: *a a*, The true shell-wall, perforated by numerous delicate tubes; *b b*, The main calcareous skeleton ("intermediate skeleton"); *c*, Passage of communication ("stolon-passage") from one tier of chambers to another; *d*, Ramifying tubes in the calcareous skeleton. (After Carpenter.)

of the organism, all these chambers, of course, must have been filled with living matter; but they are found in the present state of the fossil to be generally filled with some silicate, such as serpentine, which not only fills the actual chambers, but has

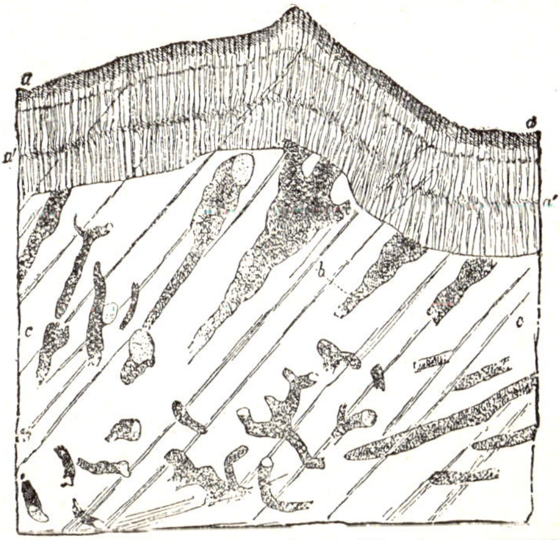

Fig. 24.—Portion of one of the calcareous layers of *Eozoön*, magnified 100 diameters. *a a*, The proper wall ("Nummuline layer") of one of the chambers, showing the fine vertical tubuli with which it is penetrated, and which are slightly bent along the line *a' a'*, *c c*. The intermediate skeleton, with numerous branched canals. The oblique lines are the cleavage planes of the carbonate of lime, extending across both the intermediate skeleton and the proper wall. (After Carpenter.)

also penetrated the minute tubes of the proper wall and the branching canals of the intermediate skeleton. In some cases the chambers are simply filled with crystalline carbonate of lime. When the originally porous fossil has been permeated by a silicate, it is possible to dissolve away the whole of the calcareous skeleton by means of acids, leaving an accurate and beautiful cast of the chambers and the tubes connected with them in the insoluble silicate.

The above are the actual appearances presented by *Eozoön* when examined microscopically, and it remains to see how far they enable us to decide upon its true position in the animal kingdom. Those who wish to study this interesting subject in detail must consult the admirable memoirs by Dr. W. B. Carpenter and Principal Dawson: it will be enough here to indicate the results which have been arrived at. The only animals at the present day which possess a continuous calcareous skeleton, perforated by pores and penetrated by canals, are certain organisms belonging to the group of the

Foraminifera. We have had occasion before to speak of these animals, and as they are not conspicuous or commonly-known forms of life, it may be well to say a few words as to the structure of the living representatives of the group. The *Foraminifera* are all inhabitants of the sea, and are mostly of small or even microscopic dimensions. Their bodies are composed of an apparently structureless animal substance of an albuminous nature ("sarcode"), of a gelatinous consistence, transparent, and exhibiting numerous minute granules or rounded particles. The body-substance cannot be said in

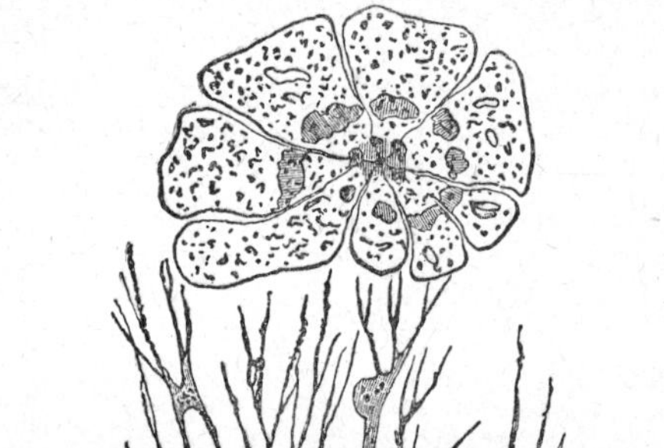

Fig. 25.—The animal of *Nonionina*, one of the *Foraminifera*, after the shell has been removed by a weak acid; *b*. *Gromia*, a single-chambered Foraminifer (after Schultze), showing the shell surrounded by a network of filaments derived from the body substance.

itself to possess any definite form, except in so far as it may be bounded by a shell; but it has the power, wherever it may be exposed, of emitting long thread-like filaments ("pseudopodia"), which interlace with one another to form a network (fig. 25, *b*). These filaments can be thrown out at will, and to considerable distances, and can be again retracted into the soft mass of the general body-substance, and they are the agents by which the animal obtains its food. The soft bodies of the *Foraminifera* are protected by a shell, which is usually

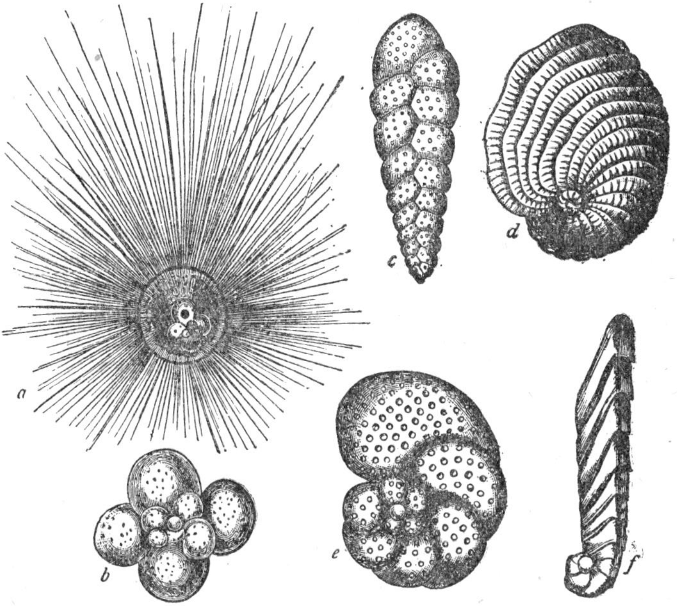

Fig. 26—Shells of living *Foraminifera*. *a*, *Orbulina universa*, in its perfect condition, showing the tubular spines which radiate from the surface of the shell; *b*, *Globigerina bulloides*, in its ordinary condition, the thin hollow spines which are attached to the shell when perfect having been broken off; *c*, *Textularia variabilis*; *d*, *Peneroplis planatus*; *e*, *Rotalia concamerata*; *f*, *Cristellaria subarcuatula*. [Fig. *a* is after Wyville Thomson; the others are after Williamson. All the figures are greatly enlarged.]

calcareous, but may be composed of sand-grains cemented together; and it may consist of a single chamber (fig. 26, *a*), or of many chambers arranged in different ways (fig. 26, *b*-*f*). Sometimes the shell has but one large opening into it—the mouth; and then it is from this aperture that the animal protrudes the delicate net of filaments with which it seeks its food. In other cases the entire shell is perforated with minute pores (fig. 26, *e*), through which the soft body-substance gains the exterior, covering the whole shell with a gelatinous film of animal matter, from which filaments can be emitted at any point. When the shell consists of many chambers, all of these are placed in direct communication with one another, and the actual substance of the shell is often traversed by minute canals filled with living matter (*e.g.*, in *Calcarina* and *Nummulina*). The shell, therefore, may be regarded, in such cases, as a more or less completely porous calcareous structure, filled to its minutest internal recesses with the substance of the living animals, and covered externally with a layer of the same substance, giving off a network of interlacing filaments.

Such, in brief, is the structure of the living *Foraminifera;* and it is believed that in *Eozoön* we have an extinct example of the same group, not only of special interest from its immemorial antiquity, but hardly less striking from its gigantic dimensions. In its original condition, the entire chamber-system of *Eozoön* is believed to have been filled with soft structureless living matter, which passed from chamber to chamber through the wide apertures connecting these cavities, and from tier to tier by means of the tubuli in the shell-wall and the branching canals in the intermediate skeleton. Through the perforated shell-wall covering the outer surface the soft body-substance flowed out, forming a gelatinous investment, from every point of which radiated an interlacing net of delicate filaments, providing nourishment for the entire colony. In its present state, as before said, all the cavities originally occupied by the body-substance have been filled with some mineral substance, generally with one of the silicates of magnesia; and it has been asserted that this fact militates strongly against the organic nature of *Eozoön,* if not absolutely disproving it. As a matter of fact, however—as previously noticed—it is by no means very uncommon at the present day to find the shells of living species of *Foraminifera* in which all the cavities primitively occupied by the body-substance, down to the minutest pores and canals, have been similarly injected by some analogous silicate, such as glauconite.

Those, then, whose opinions on such a subject deservedly carry the greatest weight, are decisively of opinion that we are presented in the *Eozoön* of the Laurentian Rocks of Canada with an ancient, colossal, and in some respects abnormal type of the *Foraminifera.* In the words of Dr. Carpenter, it is not pretended that "the doctrine of the Foraminiferal nature of *Eozoön* can be *proved* in the demonstrative sense;" but it may be affirmed "that the *convergence of a number of separate and independent probabilities,* all accordant with that hypothesis, while a separate explanation must be invented for each of them on any other hypothesis, gives it that *high probability* on which we rest in the ordinary affairs of life, in the verdicts of juries, and in the interpretation of geological phenomena generally."

It only remains to be added, that whilst *Eozoön* is by far the most important organic body hitherto found in the Laurentian, and has been here treated at proportionate length, other traces of life have been detected, which may subsequently prove of great interest and importance. Thus, Principal Dawson has recently described under the name of *Archæosphærinæ* certain singular rounded bodies which he has discovered in the Laurentian limestones, and which he believes to be casts of the shells of *Foraminifera* possibly somewhat allied to the existing *Globigerinæ*. The same eminent palæontologist has also described undoubted worm-burrows from rocks probably of Laurentian age. Further and more extended researches, we may reasonably hope, will probably bring to light other actual remains of organisms in these ancient deposits.

The Huronian Period.

The so-called *Huronian Rocks*, like the Laurentian, have their typical development in Canada, and derive their name from the fact that they occupy an extensive area on the borders of Lake Huron. They are wholly metamorphic, and consist principally of altered sandstones or quartzites, siliceous, felspathic, or talcose slates, conglomerates, and limestones. They are largely developed on the north shore of Lake Superior, and give rise to a broken and hilly country, very like that occupied by the Laurentians, with an abundance of timber, but rarely with sufficient soil of good quality for agricultural purposes. They are, however, largely intersected by mineral veins, containing silver, gold, and other metals, and they will ultimately doubtless yield a rich harvest to the miner. The Huronian Rocks have been identified, with greater or less certainty, in other parts of North America, and also in the Old World.

The total thickness of the Huronian Rocks in Canada is estimated as being not less than 18,000 feet, but there is considerable doubt as to their precise geological position. In their typical area they rest unconformably on the edge of strata of *Lower* Laurentian age; but they have never been seen in direct contact with the Upper Laurentian, and their exact relations to this series are therefore doubtful. It is thus open to question whether the Huronian Rocks constitute a distinct formation, to be intercalated in point of time between the Laurentian and the Cambrian groups; or whether, rather, they should not be considered as the metamorphosed representatives of the Lower Cambrian Rocks of other regions.

As regards the fossils of the Huronian Rocks, little can be said. some of the specimens of *Eozoön Canadense* which have been discovered in Canada are thought to come from rocks which are probably of Huronian age. In Bavaria, Dr. Gümbel has described a species of *Eozoön* under the name of *Eozoön Bavaricum,* from certain metamorphic limestones which he refers to the Huronian formation. Lastly, the late Mr. Billings described, from rocks in Newfoundland apparently referable to the Huronian, certain problematical limpet-shaped fossils, to which he gave the name of *Aspidella.*

LITERATURE.

Amongst the works and memoirs which the student may consult with regard to the Laurentian and Huronian deposits may be mentioned the following:*—

(1) 'Report of Progress of the Geological Survey of Canada from its Commencement to 1863,' pp. 38-49, and pp. 50-66.
(2) 'Manual of Geology.' Dana. 2nd Ed. 1875.
(3) 'The Dawn of Life.' J. W. Dawson. 1876.
(4) "On the Occurrence of Organic Remains in the Laurentian Rocks of Canada." Sir W. E. Logan. 'Quart. Journ. Geol. Soc.,' xxii. 45-50.
(5) "On the Structure of Certain Organic Remains in the Laurentian Limestones of Canada." J. W. Dawson. 'Quart. Journ. Geol Soc.,' xxi. 51-59.
(6) "Additional Note on the Structure and Affinities of Eozoön Canadense." W. B. Carpenter. 'Quart. Journ. Geol. Soc.,' xxi. 59-66.
(7) "Supplemental Notes on the Structure and Affinities of Eozoön Canadense." W. B. Carpenter. 'Quart. Journ. Geol. Soc., xxi. 219-228.
(8) "On the So-Called Eozoönal Rocks." King & Rowney. 'Quart. Journ. Geol. Soc.,' xxii. 185-218.
(9) 'Chemical and Geological Essays.' Sterry Hunt.

The above list only includes some of the more important memoirs which may be consulted as to the geological and chemical features of the Laurentian and Huronian Rocks, and as

* In this and in all subsequently following bibliographical lists, not only is the selection of works and memoirs quoted necessarily extremely limited; but only such have, as a general rule, been chosen for mention as are easily accessible to students who are in the position of being able to refer to a good library. Exceptions, however, are occasionally made to this rule, in favor of memoirs or works of special historical interest. It is also unnecessary to add that it has not been thought requisite to insert in these lists the well-known handbooks of geological and palæontological science, except in such instances as where they contain special information on special points.

to the true nature of *Eozoön*. Those who are desirous of studying the later phases of the controversy with regard to *Eozoön* must consult the papers of Carpenter, Carter, Dawson, King & Rowney, Hahn, and others, in the 'Quart. Journ. of the Geological Society,' the 'Proceedings of the Royal Irish Academy,' the 'Annals of Natural History,' the 'Geological Magazine.' &c. Dr. Carpenter's 'Introduction to the Study of the Foraminifera' should also be consulted.

CHAPTER VIII.

THE CAMBRIAN PERIOD.

The traces of life in the Laurentian period, as we have seen, are but scanty; but the *Cambrian Rocks*—so called from their occurrence in North Wales and its borders ("Cambria")—have yielded numerous remains of animals and some dubious plants. The Cambrian deposits have thus a special interest as being the oldest rocks in which occur any number of well-preserved and unquestionable organisms. We have here the remains of the first *fauna,* or assemblage of animals, of which we have at present knowledge. As regards their geographical distribution, the Cambrian Rocks have been recognized in many parts of the world, but there is some question as to the precise limits of the formation, and we may consider that their most typical area is in South Wales, where they have been carefully worked out chiefly by Dr. Henry Hicks. In this region, in the neighborhood of the promontory of St. David's, the Cambrian Rocks are largely developed, resting upon an ancient ridge of Pre-Cambrian (Laurentian?) strata, and overlaid by the lowest beds of the Lower Silurian. The subjoined sketch section (fig. 27) exhibits in a general manner the succession of strata in this locality.

From this section it will be seen that the Cambrian Rocks in Wales are divided in the first place into a lower and an upper group. The *Lower Cambrian* is constituted at the base by a great series of grits, sandstones, conglomerates, and slates, which are known as the "Longmynd group," from their vast development in the Longmynd Hills in Shropshire, and which attain in North Wales a thickness of 8000 feet or more. The Longmynd beds are succeeded by the so-called "Menevian group," a series of sandstones, flags, and grits, about 600 feet in thickness, and containing a considerable number of fossils. The *Upper Cambrian* series consists in its lower portion of nearly 5000 feet of strata, principally shaly and slaty, which are known as the "Lingula Flags," from the great abundance in them of a shell referable to the genus *Lingula*. These are followed by 1000 feet of dark shales and flaggy sandstones, which are known as the "Tremadoc slates," from their occurrence near Tremadoc in North Wales; and these in turn are surmounted, apparently quite conformably, by the basement beds of the Lower Silurian.

GENERALIZED SECTION OF THE CAMBRIAN ROCKS IN WALES.

Fig. 27.

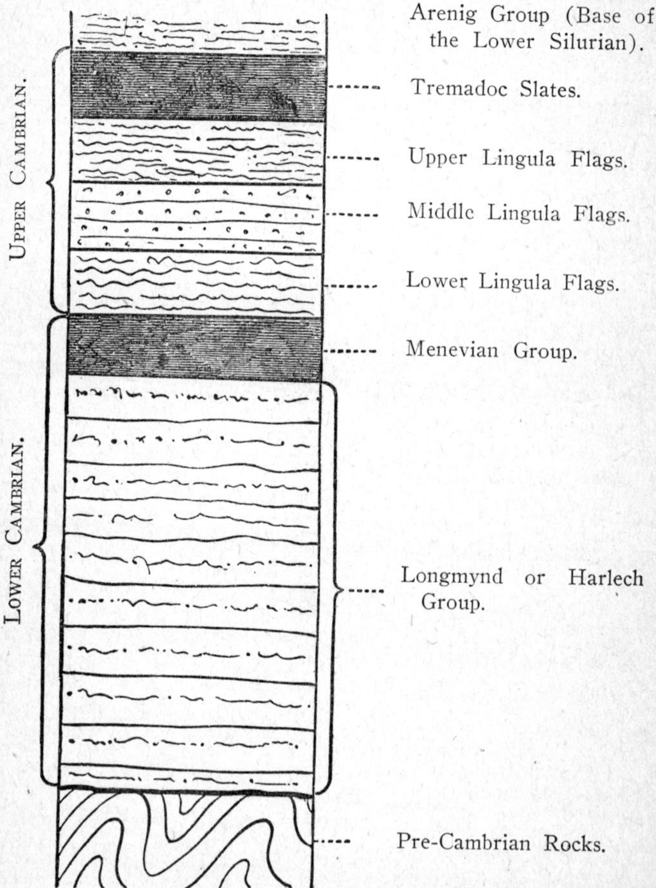

The opposite may be regarded as giving a typical series of the Cambrian Rocks in a typical locality; but strata of Cambrian age are known in many other regions, of which it is only possible here to allude to a few of the most important. In Scandinavia occurs a well-developed series of Cambrian deposits, representing both the lower and upper parts of the formation. In Bohemia, the Upper Cambrian, in particular, is largely developed, and constitutes the so-called "Primordial zone" of Barrande. Lastly, in North America, whilst the Lower Cambrian is only imperfectly developed, or is represented by the Huronian, the Upper Cambrian formation has a wide extension, containing fossils similar in character to the analogous strata in Europe, and known as the "Potsdam Sandstone." The subjoined table shows the chief areas where Cambrian Rocks are developed, and their general equivalency:

TABULAR VIEW OF THE CAMBRIAN FORMATION.

	Britain.	*Europe.*	*America.*
Upper Cambrian.	*a.* Tremadoc Slates. *b.* Lingula Flags.	*a.* Primordial zone of Bohemia. *b.* Paradoxides Schists, Olenus Schists, and Dictyonema schists of Sweden.	*a.* Potsdam Sandstone. *b.* Acadian group of New Brunswick.
Lower Cambrian.	*a.* Longmynd Beds. *b.* Llanberis Slates. *c.* Harlech Grits. *d.* *Oldhamia* Slates of Ireland. *e.* Conglomerates and Sandstones of Sutherlandshire? *f.* Menevian Beds.	*a.* Fucoidal Sandstone of Sweden. *b.* *Eophyton* Sandstone of Sweden.	Huronian Formation?

Like all the older Palæozoic deposits, the Cambrian Rocks, though by no means necessarily what would be called actually "metamorphic," have been highly cleaved, and otherwise altered from their original condition. Owing partly to their indurated state, and partly to their great antiquity, they are usually found in the heart of mountainous districts, which have undergone great disturbance, and have been subjected to an enormous amount of denudation. In some cases, as in the Longmynd Hills in Shropshire, they form low rounded elevations, largely covered by pasture, and with few or no elements of sublimity. In other cases, however, they rise into bold and rugged mountains, girded by precipitous cliffs. Industrially, the Cambrian Rocks are of interest, if only for the reason that the celebrated Welsh slates of Llanberis are derived from highly-cleaved beds of this age. Taken as a whole, the Cambrian formation is essentially composed of arenaceous and muddy sediments, the latter being sometimes red, but more commonly nearly black in color. It has often been supposed that the Cambrians are a deep-sea deposit, and that we may thus account for the few fossils contained in them; but the paucity of fossils is to a large extent imaginary, and some of the Lower Cambrian beds of the Longmynd Hills would appear to have been laid down in shallow water, as they exhibit rain-prints, sun-cracks, and ripple-marks—incontrovertible evidence of their having been a shore-deposit. The occurrence of innumerable worm-tracks and burrows in many Cambrian strata is also a proof of shallow-water conditions; and the general absence of limestones, coupled with the coarse mechanical nature of many of the sediments of the Lower Cambrian, may be taken as pointing in the same direction.

The *life* of the Cambrian, though not so rich as in the succeeding Silurian period, nevertheless consists of representatives of most of the great classes of invertebrate animals. The coarse sandy deposits of the formation, which abound more particularly toward its lower part, naturally are to a large extent barren of fossils; but the muddy sediments, when not too highly cleaved, and especially towards the summit of the group, are replete with organic remains. This is also the case, in many localities at any rate, with finer beds of the Potsdam Sandstone in America. Limestones are known to occur in only a few areas (chiefly in America), and this may account for the apparent total absence of corals. It is, however, interesting to note that, with this exception, almost all the other leading groups of Invertebrates are known to have come into existence during the Cambrian period.

Of the land-surfaces of the Cambrian period we know nothing; and there is, therefore, nothing surprising in the fact that our acquaintance with the Cambrian vegetation is confined to some marine plants or sea-weeds, often of a very obscure and problematical nature. The "Fucoidal Sandstone" of Sweden, and the "Potsdam Sandstone" of North America, have both yielded numerous remains which have been regarded as markings left by sea-weeds or "Fucoids;" but these are highly enigmatical in their characters, and would, in many instances, seem to be rather referable to the tracks and burrows of marine worms. The first-mentioned of these formations has also yielded the curious, furrowed and striated stems which have been described as a kind of land-plant under the name of *Eophyton* (fig. 28). It cannot be said, however, that the vege-

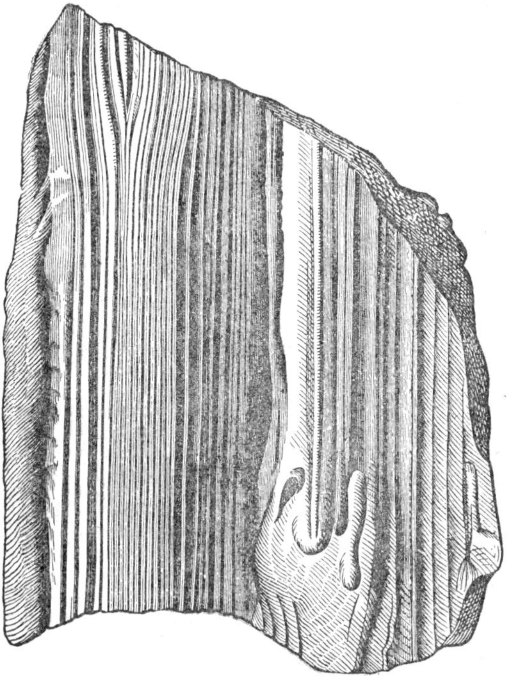

Fig. 28.—Fragment of *Eophyton Linneanum*, a supposed land-plant, Lower Cambrian, Sweden, of the natural size.

table origin of these singular bodies has been satisfactorily proved. Lastly, there are found in certain green and purple beds of Lower Cambrian age at Bray Head, Wicklow, Ireland, some very remarkable fossils, which are well known under the name of *Oldhamia,* but the true nature of which is very doubtful. The commonest form of *Oldhamia* (fig. 29) consists of a thread-like stem or axis, from which spring at regular intervals bundles of short filamentous branches in a fan-like manner.

In the locality where it occurs, the fronds of *Oldhamia* are very abundant, and are spread over the surfaces of the strata in tangled layers. That it is organic is certain, and that it is a calcareous sea-weed is probable; but it may possibly belong to the sea-mosses (*Polyzoa*), or to the sea-firs (*Sertularians*).

Amongst the lower forms of animal life (*Protozoa*), we find the Sponges represented by the curious bodies, composed of netted fibres, to which the name of *Protospongia* has been given (fig. 32, *a*); and the comparatively gigantic, conical, or cylindrical fossils termed *Archæocyathus* by Mr. Billings are certainly referable either to the *Foraminifera* or to the Sponges. The almost total absence of limestones in the formation may be regarded as a sufficient explanation of the fact that the *Foraminifera* are not more largely and unequivocally represented; though the existence of greensands in the Cambrian beds of Wisconsin and Tennessee may be taken as an indication that this class of animals was by no means wholly wanting. The same fact may explain the total absence of corals, so far as at present known.

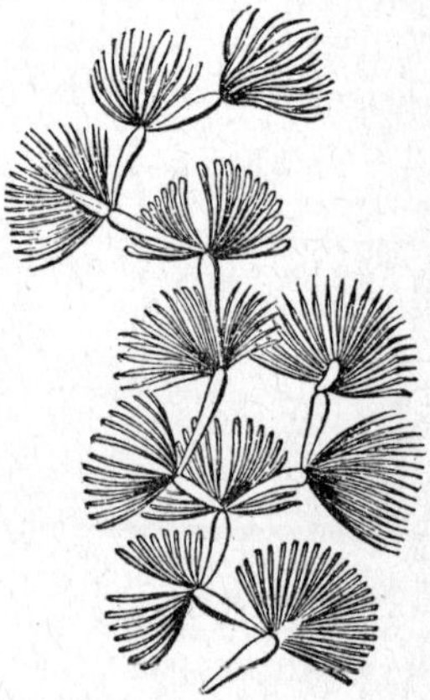

Fig. 29.—A portion of *Oldhamia antiqua*, Lower Cambrian, Wicklow, Ireland, of the natural size. (After Salter.)

The group of the *Echinodermata* (Sea-lilies, Sea-urchins, and their allies) is represented by a few forms, which are principally of interest as being the earliest-known examples of the class. It is also worthy of note that these precursors of a group which subsequently attains such geological importance, are referable to no less than three distinct *orders*—the Crinoids or Sea-lilies, represented by a species of *Dendrocrinus;* the Cystideans by *Protocystites;* and the Starfishes by *Palasterina* and some other forms. Only the last of these groups, however, appears to occur in the Lower Cambrian.

The Ringed-worms (*Annelida*), if rightly credited with all the remains usually referred to them, appear to have swarmed in the Cambrian seas. Being soft-bodied, we do not find the actual worms themselves in the fossil condition, but we have, nevertheless, abundant traces of their existence. In some cases we find vertical burrows of greater or less depth, often expanded towards their apertures, in which the worm must have actually lived (fig. 30), as various species do at the pres-

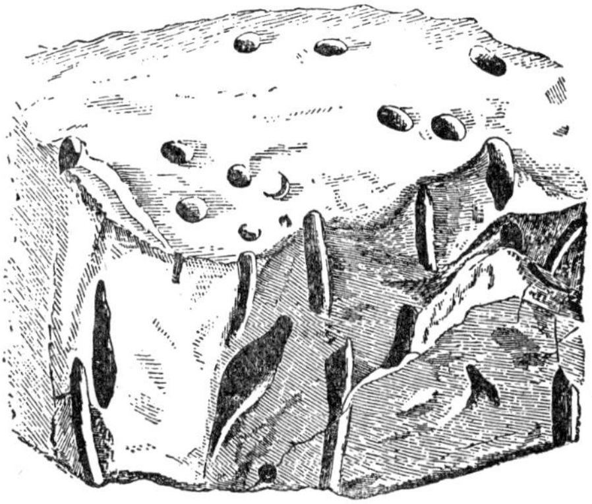

Fig. 30.—Annelide-burrows (*Scolithus linearis*), from the Potsdam Sandstone of Canada, of the natural size. (After Billings.)

ent day. In these cases, the tube must have been rendered more or less permanent by receiving a coating of mucus, or perhaps a genuine membraneous secretion, from the body of the animal, and it may be found quite empty, or occupied by a cast of sand or mud. Of this nature are the burrows which have been described under the names of *Scolithus* and *Scoleco-derma,* and probably the *Histioderma* of the Lower Cambrian of Ireland. In other cases, as in *Arenicolites* (fig. 32, *b*), the worm seems to have inhabited a double burrow, shaped like the letter U, and having two openings placed close together on the surface of the stratum. Thousands of these twin-burrows occur in some of the strata of the Longmynd, and it is supposed that the worm used one opening to the burrow as an aperture of entrance, and the other as one of exit. In other cases, again, we find simply the meandering trails caused by the worm dragging its body over the surface of the mud. Markings of this kind are commoner in the Silurian Rocks, and it is generally more or less doubtful whether they may not have been caused by other marine animals, such as shell-fish, whilst some of them have certainly nothing whatever to do with the worms. Lastly, the Cambrian beds often show twining cylindrical bodies, commonly more or less matted together, and not confined to the surfaces of the strata, but passing through them. These have often been regarded as the remains of sea-weeds, but it is more probable that they represent casts of the underground burrows of worms of similar habits to the common lob-worm (*Arenicola*) of the present day.

The *Articulate* animals are numerously represented in the Cambrian deposits, but exclusively by the class of *Crustaceans*. Some of these are little double-shelled creatures, resembling our living water-fleas (*Ostracoda*). A few are larger forms, and belong to the same group as the existing brine-shrimps and fairy-shrimps (*Phyllopoda*). One of the most characteristic of these is the *Hymenocaris vermicauda* of the Lingula Flags (fig. 32, *d*). By far the larger number of the Cambrian *Crustacea* belong, however, to the remarkable and wholly extinct group of the *Trilobites*. These extraordinary animals must have literally swarmed in the seas of the later portion of this and the whole of the succeeding period; and they survived in greatly diminished numbers till the earlier portion of the Carboniferous period. They died out, however, wholly before the close of the Palæozoic epoch, and we have no Crustaceans at the present day which can be considered as their direct representatives. They have, however, relationships of a more or less intimate character with the existing groups of the Phyllopods, the King-crabs (*Limulus*), and the Isopods ("Slaters," Wood-lice, &c.) Indeed, one member of the last-mentioned order, namely, the *Serolis* of the coasts of Patagonia, has been regarded as the nearest living ally of the Trilobites. Be this as it may, the Trilobites possessed a skeleton which, though capable of undergoing almost endless variations, was wonderfully constant in its pattern of structure, and we may briefly describe here the chief features of this.

The upper surfaces of the body of a Trilobite was defended by a strong shell or "crust," partly horny and partly calcareous in its composition. This shell (fig. 31) generally exhibits a very distinct "trilobation" or division into three longitudinal lobes, one central and two lateral. It also exhibits a more important and more fundamental division into three transverse portions, which are so loosely connected with one another as very commonly to be found separate. The first and most anterior of these divisions is a shield or buckler which covers the head; the second or middle portion is composed of movable rings covering the trunk ("thorax"); and the third is a shield which covers the tail or "abdomen." The head-shield (fig. 31, *e*) is generally more or less semicircular in shape; and its central portion, covering the stomach of the animal, is usually strongly elevated, and generally marked by lateral furrows. A little on each side of the head are placed the eyes, which are generally crescentic in shape, and resemble the eyes of insects and many existing Crustaceans in being "compound," or made up of numerous simple eyes aggregated together. So excellent is the state of preservation of many specimens of Trilobites, that the numerous individual lenses of the eyes have been uninjured, and as many as four hundred have been counted in each eye of some forms. The eyes may be supported upon prominences, but they are never carried on mov-

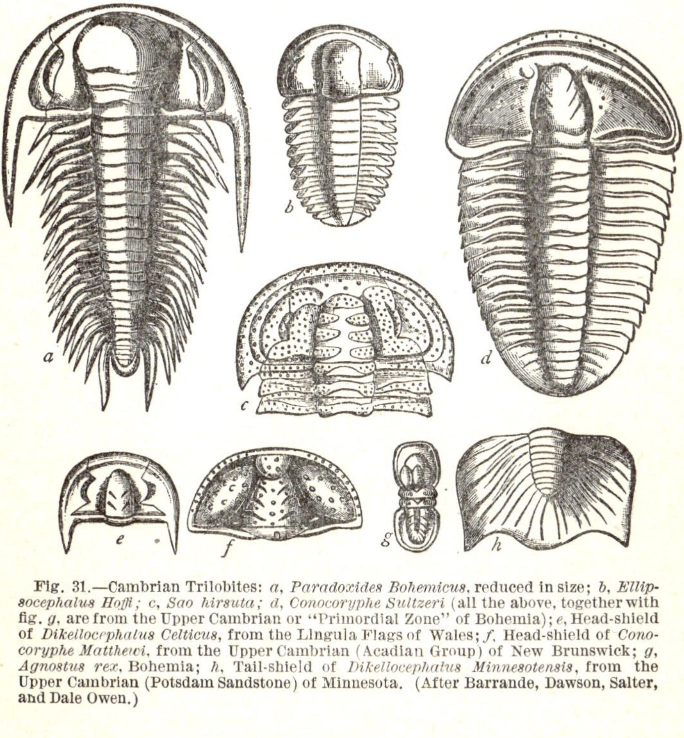

Fig. 31.—Cambrian Trilobites: *a*, *Paradoxides Bohemicus*, reduced in size; *b*, *Ellipsocephalus Hoffi*; *c*, *Sao hirsuta*; *d*, *Conocoryphe Sultzeri* (all the above, together with fig. *g*, are from the Upper Cambrian or "Primordial Zone" of Bohemia); *e*, Head-shield of *Dikellocephalus Celticus*, from the Lingula Flags of Wales; *f*, Head-shield of *Conocoryphe Matthewi*, from the Upper Cambrian (Acadian Group) of New Brunswick; *g*, *Agnostus rex*, Bohemia; *h*, Tail-shield of *Dikellocephalus Minnesotensis*, from the Upper Cambrian (Potsdam Sandstone) of Minnesota. (After Barrande, Dawson, Salter, and Dale Owen.)

able stalks (as they are in the existing lobsters and crabs); and in some of the Cambrian Trilobites, such as the little *Agnosti* (fig. 31, *g*), the animal was blind. The lateral portion of the head-shield are usually separated from the central portion by a peculiar line of division (the so-called "facial suture") on each side; but this is also wanting in some of the Cambrian species. The backward angles of the head-shield, also, are often prolonged into spines, which sometimes reach a great length. Following the head-shield behind, we have a portion of the body which is composed of movable segments or "body-rings," and which is technically called the "thorax." Ordinarily, this region is strongly trilobed, and each ring consists of a central convex portion, and of two flatter side-lobes. The number of body-rings in the thorax is very variable (from two to twenty-six), but is fixed for the adult forms of each group of the Trilobites. The young forms have much fewer rings than the full-grown ones; and it is curious to find that the Cambrian Trilobites very commonly have either a great many rings (as in *Paradoxides,* fig. 31, *a*), or else very few (as in *Agnostus,* fig. 31, *g*). In some instances, the body-rings do not seem to have been so constructed as to allow of much movement, but in other cases this region of the body is so flexible that the animal possessed the power of rolling itself up completely, like a hedgehog; and many individuals have been permanently preserved as fossils in this defensive condition. Finally, the body of the Trilobite was completed by a tail-shield (technically termed the "pygidium"), which varies much in size and form, and is composed of a greater or less number of rings, similar to those which form the thorax, but immovably amalgamated with one another (fig. 31, *h*).

The under surface of the body in the Trilobites appears to have been more or less entirely destitute of hard structures, with the exception of a well-developed upper lip, in the form of a plate attached to the inferior side of the head-shield in front. There is no reason to doubt that the animal possessed legs; but these structures seem to have resembled those of many living Crustaceans in being quite soft and membranous. This, at any rate, seems to have been generally the case; though structures which have been regarded as legs have been detected on the under surface of one of the larger species of Trilobites. There is also, at present, no direct evidence that the Trilobites possessed two pairs of jointed feelers ("antennæ") which are so characteristic of recent Crustaceans.

The Trilobites vary much in size, and the Cambrian formation presents examples of both the largest and the smallest members of the order. Some of the young forms may be little bigger than a millet-seed, and some adult examples of the smaller species (such as *Agnostus*) may be only a few lines in length; whilst such giants of the order as *Paradoxides* and *Asaphus* may reach a length of from one to two feet. Judging from what we actually know as to the structure of the Trilobites, and also from analogous recent forms, it would seem that these ancient Crustaceans were mud-haunting creatures, denizens of shallow seas, and affecting the soft silt of the bottom rather than the clear water above. Whenever muddy sediments are found in the Cambrian and Silurian formations, there we are tolerably sure to find Trilobites, though they are by no means absolutely wanting in limestones. They appear to have crawled about upon the sea-bottom, or burrowed in the yielding mud, with the soft under surface directed downwards; and it is probable that they really derived their nutriment from the organic matter contained in the ooze amongst which they lived. The vital organs seem to have occupied the central lobe of the skeleton, by which they were protected; and a series of delicate leaf-like paddles, which probably served as respiratory organs, would appear to have been carried on the under surface of the thorax. That they had their enemies may be regarded as certain; but we have no evidence that they were furnished with any offensive weapons, or, indeed, with any means of defence beyond their hard crust, and the power, possessed by so many of them, of rolling themselves into a ball. An additional proof of the fact that they for the most part crawled along the sea-bottom is found in the occurrence of tracks and markings of various kinds, which can hardly be ascribed to any other creatures with any show of probability. That this is the true nature of some of the markings in question cannot be doubted at all; and in other cases no explanation so probable has yet suggested. If, however, the tracks which have been described from the Potsdam Sandstone of North America under the name of *Protichnites* are really due to the peregrinations of some Trilobite, they must have been produced by one of the largest examples of the order.

As already said, the Cambrian Rocks are very rich in the remains of Trilobites. In the lowest beds of the series (Longmynd Rocks), representatives of some half-dozen genera have now been detected, including the dwarf *Agnostus* and the giant

Paradoxides. In the higher beds, the number both of genera and species is largely increased; and from the great comparative abundance of individuals, the Trilobites have every right to be considered as the most characteristic fossils of the Cambrian period,—the more so as the Cambrian species belong to peculiar types, which, for the most part, died out before the commencement of the Silurian epoch.

All the remaining Cambrian fossils which demand any notice here are members of one or other division of the great class of the *Mollusca*, or "Shell-fish" properly so called. In the Lower Cambrian Rocks the Lamp-shells (*Brachiopoda*) are the principal or sole representatives of the class, and appear chiefly in three interesting and important types—namely, *Lingulella*,

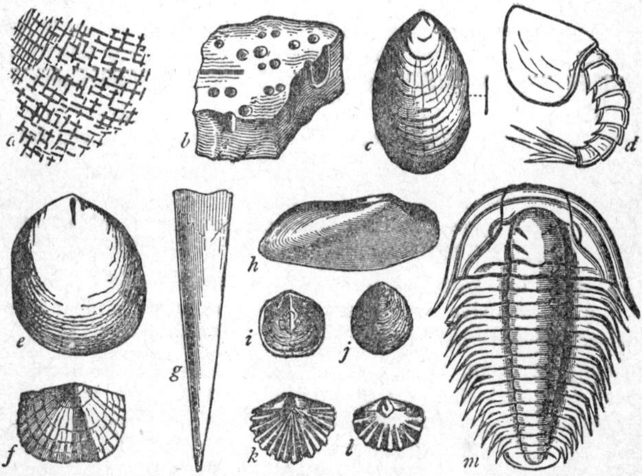

Fig. 32.—Cambrian Fossils: *a*, *Protospongia fenestrata*, Menevian Group; *b*, *Arenicolites didymus*, Longmynd Group; *c*, *Lingulella ferruginea*, Longmynd and Menevian, enlarged; *d*, *Hymenocaris vermicauda*, Lingula Flags; *e*. *Lingulella Davisii*, Lingula Flags; *f*, *Orthis lenticularis*, Lingula Flags; *g*, *Theca Davidii*, Tremadoc Slates; *h*, *Modiolopsis Solvensis*, Tremadoc Slates; *i*, *Obolella sagittalis*, interior of valve, Menevian; *j*, Exterior of the same; *k*, *Orthis Hicksii*, Menevian; *l*, Cast of the same; *m*, *Olenus micrurus*, Lingula Flags. (After Salter, Hicks and Davidson.)

Discina, and *Obolella*. Of these the last (fig. 32, *i*) is highly characteristic of these ancient deposits; whilst *Discina* is one of those remarkable persistent types which, commencing at this early period, has continued to be represented by varying forms through all the intervening geological formations up to the present day. *Lingulella* (fig. 32, *c*), again, is closely allied to the existing "Goose-bill" Lamp-shell (*Lingula anatina*), and thus presents us with another example of an extremely long-lived type. The *Lingulellæ* and their successors, the *Lingulæ,* are singular in possessing a shell which is of a horny texture, and contains but a small proportion of calcareous matter. In the Upper Cambrian Rocks, the *Lingulellæ* become much more abundant, the broad satchel-shaped species known as *L. Davisii* (fig. 32, *e*) being so abundant that one of the great divisions of the Cambrian is termed the "Lingula Flags." Here, also, we meet for the first time with examples of the genus *Orthis* (fig. 32, *f, k, l*) a characteristic Palæozoic type of the Brachiopods, which is destined to undergo a vast extension in later ages.

Of the higher groups of the *Mollusca* the record is as yet but scanty. In the Lower Cambrian, we have but the thin, fragile, dagger-shaped shells of the free-swimming oceanic Molluscs or "Winged-snails" (*Pteropoda*), of which the most characteristic is the genus *Theca* (fig. 32, *g*). In the upper Cambrian, in addition to these, we have a few Univalves (*Gasteropoda*), and, thanks to the researches of Dr. Hicks, quite a small assemblage of Bivalves (*Lamellibranchiata*), though these are mostly of no great dimensions (fig. 32, *h*). Of the chambered *Cephalopoda* (Cuttle-fishes and their allies), we have but few traces, and these wholly confined to the higher beds of the formation. We meet, however, with examples of the wonderful genus *Orthoceras,* with its straight, partitioned shell, which we shall find in an immense variety of forms in the Silurian rocks. Lastly, it is worthy of note that the lowest of all the groups of the *Mollusca*—namely, that of the Sea-mats, Sea-mosses, and Lace-corals (*Polyzoa*)—is only doubtfully known to have any representatives in the Cambrian, though undergoing a large and varied development in the Silurian deposits.

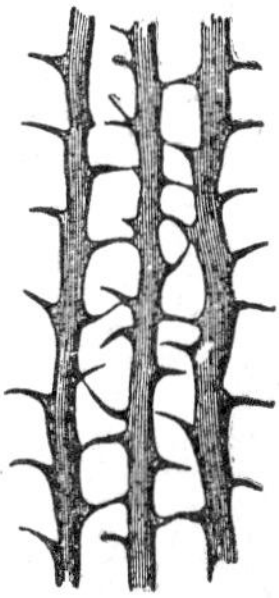

Fig. 33.—Fragment of *Dictyonema sociale*, considerably enlarged, showing the horny branches, with their connecting cross-bars, and with a row of cells on each side. (Original.)

An exception, however, may with much probability be made to this statement in favor of the singular genus *Dictyonema* (fig. 33), which is highly characteristic of the highest Cambrian beds (Tremadoc Slates). This curious fossil occurs in the form of fan-like or funnel-shaped expansions, composed of slightly-diverging horny branches, which are united in a net-like manner by numerous delicate cross-bars, and exhibit a row of little cups or cells, in which the animals were contained, on each side. *Dictyonema* has generally been referred to the *Graptolites;* but it has a much greater affinity with the plant-like Sea-firs (*Sertularians*) or the Sea-mosses (*Polyzoa*), and the balance of evidence is perhaps in favor of placing it with the latter.

Literature.

The following are the more important and accessible works and memoirs which may be consulted in studying the stratigraphical and palæontological relations of the Cambrian Rocks:—

(1) 'Siluria.' Sir Roderick Murchison. 5th ed., pp. 21-46.

(2) 'Synopsis of the Classification of the British Palæozoic Rocks.' Sedgwick. Introduction to the 3d Fasciculus of the 'Descriptions of British Palæozoic Fossils in the Woodwardian Museum,' By F. M'Coy, pp. i-xcviii, 1855.

(3) 'Catalogue of the Cambrian and Silurian Fossils in the Geological Museum of the University of Cambridge.' Salter. With a Preface by Prof. Sedgwick. 1873.

(4) 'Thesaurus Siluricus.' Bigsby. 1868.

(5) "History of the Names Cambrian and Silurian." Sterry Hunt.—'Geological Magazine.' 1873.

(6) 'Système Silurien du Centre de la Bohême.' Barrande. Vol. I.

(7) 'Report of Progress of the Geological Survey of Canada, from its Commencement to 1863,' pp. 87-109.

(8) 'Acadian Geology.' Dawson. Pp. 641-657.

(9) "Guide to the Geology of New York," Lincklaen; and "Contributions to the Palæontology of New York," James Hall.—'Fourteenth Report on the State Cabinet.' 1861.

(10) 'Palæozoic Fossils of Canada.' Billings. 1865.

(11) 'Manual of Geology.' Dana. Pp. 166-182. 2nd ed. 1875.

(12) "Geology of North Wales," Ramsay; with Appendix on the Fossils, Salter.—'Memoirs of the Geological Survey of Great Britain,' vol. iii, 1866.

(13) "On the Ancient Rocks of the St. David's Promontory, South Wales, and their Fossil Contents." Harkness and Hicks.—'Quart. Journ. Geol. Soc.,' xxvii. 384-402. 1871.

(14) "On the Tremadoc Rocks in the Neighborhood of St. David's, South Wales, and their Fossil Contents." Hicks.—'Quart Journ. Geol. Soc.,' xxix. 39-52 1873.

In the above list, allusion has necessarily been omitted to numerous works and memoirs on the Cambrian deposits of Sweden and Norway, Central Europe, Russia, Spain, and various parts of North America, as well as to a number of important papers on the British Cambrian strata by various well-known observers. Among these latter may be mentioned memoirs by Prof. Phillips, and Messrs Salter, Hicks, Belt, Plant, Homfray, Ash, Holl, &c.

CHAPTER IX.

THE LOWER SILURIAN PERIOD.

The great system of deposits to which Sir Roderick Murchison applied the name of "Silurian Rocks" reposes directly upon the highest Cambrian beds, apparently without any marked unconformity, though with a considerable change in the nature of the fossils. The name "Silurian" was originally proposed by the eminent geologist just alluded to for a great series of strata lying below the Old Red Sandstone, and occupying districts in Wales and its borders which were at one time inhabited by the "Silures," a tribe of ancient Britons. Deposits of a corresponding age are now known to be largely developed in other parts of England, in Scotland, and in Ireland, in North America, in Australia, in India, in Bohemia, Saxony, Bavaria, Russia, Sweden and Norway, Spain, and in various other regions of less note. In some regions, as in the neighborhood of St. Petersburg, the Silurian strata are found not only to have preserved their original horizontality, but also to have retained almost unaltered their primitive soft and incoherent nature. In other regions, as in Scandinavia and many parts of North America, similar strata, now consolidated into shales, sandstones, and limestones, may be found resting with a very slight inclination on still older sediment. In a great many regions, however, the Silurian deposits are found to have undergone more or less folding, crumpling, and dislocation, accompanied by induration and "cleavage" of the finer and softer sediments; whilst in some regions, as in the Highlands of Scotland, actual "metamorphism" has taken place. In consequence of the above, Silurian districts usually present the bold, rugged, and picturesque outlines which are characteristic of the older "Primitive" rocks of the earth's crust in general. In many instances, we find Silurian strata rising into mountain-chains of great grandeur and sublimity, exhibiting the utmost diversity of which rock-scenery is capable, and delighting the artist with endless changes of valley, lake, and cliff. Such districts are little suitable for agriculture, though this is often compensated for by the valuable mineral products contained in the rocks. On the other hand, when the rocks are tolerably soft and uniform in their nature, or when few disturbances of the crust of the earth have taken place, we may find Silurian areas to be covered with an abundant pasturage or to be heavily timbered.

Under the head of "Silurian Rocks," Sir Roderick Murchison included all the strata between the summit of the "Longmynd" beds and the Old Red Sandstone, and he divided these into the two great groups of the *Lower* Silurian and *Upper* Silurian. It is, however, now generally admitted that a considerable portion of the basement beds of Murchison's Silurian series must be transferred—if only upon palæontological grounds—to the Upper Cambrian, as has here been done; and much controversy has been carried on as to the proper nomenclature of the Upper Silurian and of the remaining portion of Murchison's Lower Silurian. Thus, some would confine the name "Silurian" exclusively to the Upper Silurian, and would apply the name of "Cambro-Silurian" to the Lower Silurian, or would include all beds of the latter age in the "Cambrian" series of Sedgwick. It is not necessary to enter into the merits of these conflicting views. For our present purpose, it is sufficient to recognize that there exist two great groups of rocks between the highest Cambrian beds, as here defined, and the base of the Devonian or Old Red Sandstone. These two great groups are so closely allied to one another, both physically and palæontologically, that many authorities have established a third or intermediate group (the "Middle Silurian"), by which a passage is made from one into the other. This method of procedure involves disadvantages which appear to outweigh its advantages; and the two groups in question are not only generally capable of very distinct stratigraphical separation, but at the same time exhibit, together with the alliances above spoken of, so many and such important palæontological differences, that it is best to consider them separately. We shall therefore follow this course in the present instance; and pending the final solution of the controversy as to Cambrian and Silurian nomenclature, we shall distinguish these two groups of strata as the "Lower Silurian" and the "Upper Silurian."

The *Lower Silurian Rocks* are known already to be developed in various regions; and though their *general* succession in these areas is approximately the same, each area exhibits peculiarities of its own, whilst the subdivisions of each are known by special names. All, therefore, that can be attempted here, is to select two typical areas—such as Wales and North America—and to briefly consider the grouping and divisions of the Lower Silurian in each.

In Wales, the line between the Cambrian and Lower Silurian is somewhat ill-defined, and is certainly not marked by any strong unconformity. There are, however, grounds for accepting the line proposed, for palæontological reasons, by Dr. Hicks, in accordance with which the Tremadoc Slates ("Lower Tremadoc" of Salter) become the highest of the Cambrian deposits of Britain. If we take this view, the Lower Silurian rocks of Wales and adjoining districts are found to have the following *general* succession from below upwards (fig. 34):—

1. The *Arenig Group*.—This group derives its name from the Arenig mountains, where it is extensively developed. It consists of about 4000 feet of slates, shales, and flags, and is divisible into a lower, middle, and upper division, of which the former is often regarded as Cambrian under the name of "Upper Tremadoc Slates."

2. The *Llandeilo Group*.—The thickness of this group varies from about 4000 to as much as 10,000 feet; but in this latter case a great amount of the thickness is made up of volcanic ashes and interbedded traps. The sedimentary beds of this group are principally slates and flags, the latter occasionally with calcareous bands; and the whole series can be divided into a lower, middle, and upper Llandeilo division, of which the last is the most important. The name of "Llandeilo" is derived from the town of the same name in Wales, where strata of this age were described by Murchison.

3. The *Caradoc* or *Bala Group*.—The alternative names of this group are also of local origin, and are derived, the one

from Caer Caradoc in Shropshire, the other from Bala in Wales, strata of this age occurring in both localities. The series is divided into a lower and upper group, the latter chiefly composed of shales and flags, and the former of sandstones and shales, together with the important and interesting calcareous band known as the "Bala Limestone." The thickness of the entire series varies from 4000 to as much as 12,000 feet, according as it contains more or less of interstratified igneous rocks.

4. The *Llandovery Group* (Lower Llandovery of Murchison).—This series, as developed near the town of Llandovery, in Caermarthenshire, consists of less than 1000 feet of conglomerates, sandstones, and shales. It is probable, however, that the little calcareous band known as the "Hirnant Limestone," together with certain pale-colored slates which lie above the Bala Limestone, though usually referred to the Caradoc series, should in reality be regarded as belonging to the Llandovery group.

The general succession of the Lower Silurian strata of Wales and its borders, attaining a maximum thickness (along with contemporaneous igneous matter) of nearly 30,000 feet, is diagramatically represented in the annexed sketch-section (fig. 34):—

[GENERALIZED SECTION

Generalized Section of the Lower Silurian Rocks of Wales.

Fig. 34.

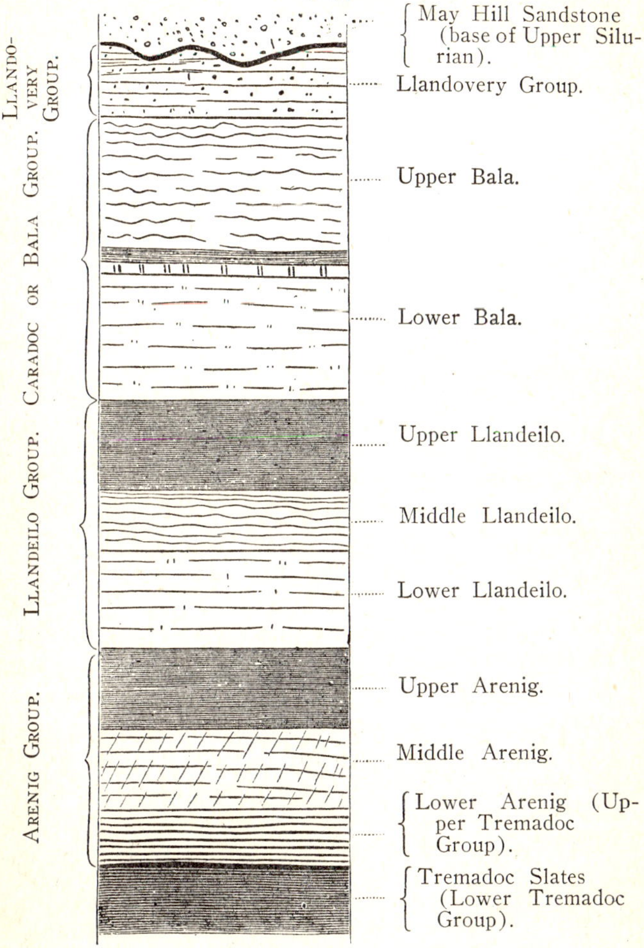

In North America, both in the United States and in Canada, the Silurian rocks are very largely developed, and may be regarded as constituting an exceedingly full and typical series of the deposits of this period. The chief groups of the Silurian rocks of North America are as follows, beginning, as before, with the lowest strata, and proceeding upwards (fig. 35) :—

1. *Quebec Group.*—This group is typically developed in the vicinity of Quebec, where it consists of about 5000 feet of strata, chiefly variously-colored shales, together with some sandstones and a few calcareous bands. It contains a number of peculiar Graptolites, by which it can be identified without question with the Arenig group of Wales and the corresponding Skiddaw Slates of the North of England. It is also to be noted that numerous Trilobites of a distinct Cambrian *facies* have been obtained in the limestones of the Quebec group, near Quebec. These fossils, however, have been exclusively obtained from the limestones of the group; and as these limestones are principally calcareous breccias or conglomerates, there is room for believing that these primordial fossils are really derived, in part at any rate, from fragments of an upper Cambrian limestone. In the State of New York, the Graptolitic shales of Quebec are wanting; and the base of the Silurian is constituted by the so-called "Calciferous Sand-rock" and "Chazy Limestone."* The first of these is essentially and typically calcareous, and the second is a genuine limestone.

2. The *Trenton Group.*—This is an essentially calcareous group, the various limestones of which it is composed being known as the "Bird's-eye," "Black River," and "Trenton" Limestones, of which the last is the thickest and most important. The thickness of this group is variable, and the bands of limestone in it are often separated by beds of shale.

3. The *Cincinnati Group* (Hudson River Formation†).—This group consists essentially of a lower series of shales, often black in color and highly charged with bituminous matter (the "Utica Slates"), and of an upper series of shales, sandstones, and limestones (the "Cincinnati" rocks proper). The exact parallelism of the Trenton and Cincinnati groups with

* The precise relations of the Quebec shales with Graptolites (Levis Formation) to the Calciferous and Chazy beds are still obscure, though there seems little doubt but that the Quebec Shales are superior to the Calciferous Sand-rock.

† There is some difficulty about the precise nomenclature of this group. It was originally called the "Hudson River Formation;" but this name is inappropriate, as rocks of this age hardly touch anywhere the actual Hudson River itself, the rocks so called formerly being now known to be of more ancient date. There is also some want of propriety in the name of "Cincinnati Group," since the rocks which are known under this name in the vicinity of Cincinnati itself are the representatives of the Trenton Limestone, Utica Slates, and the Old Hudson River group, inseparably united in what used to be called the "Blue Limestone Series."

the subdivisions of the Welsh Silurian series can hardly be stated positively. Probably no precise equivalency exists; but there can be no doubt but that the Trenton and Cincinnati groups correspond, as a whole, with the Llandeilo and Caradoc groups of Britain. The subjoined diagrammatic section (fig. 35) gives a general idea of the succession of the Lower Silurian rocks of North America:—

GENERALIZED SECTION OF THE LOWER SILURIAN ROCKS OF NORTH AMERICA.

Fig. 35.

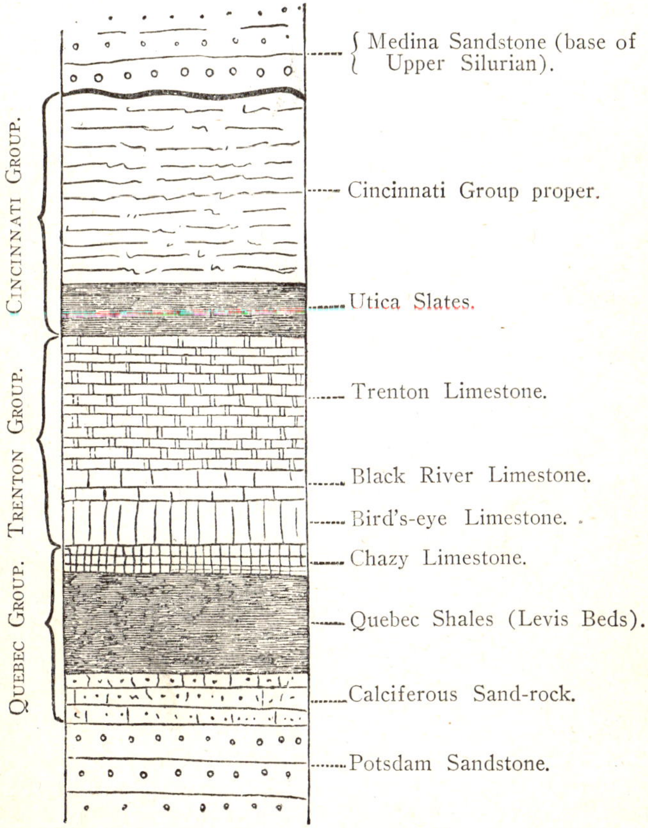

Of the *life* of the Lower Silurian period we have record in a vast number of fossils, showing that the seas of this period were abundantly furnished with living denizens. We have, however, in the meanwhile, no knowledge of the land-surfaces of the period. We have therefore no means of speculating as to the nature of the terrestial animals of this ancient age, nor is anything known with certainty of any land-plants which may have existed. The only relics of vegetation upon which a positive opinion can be expressed belong to the obscure group of the "Fucoids," and are supposed to be the remains of sea-weeds. Some of the fossils usually placed under this head are probably not of a vegetable nature at all, but others

Fig. 36.—*Licrophycus Ottawaensis*, a "Fucoid," from the Trenton Limestone (Lower Silurian) of Canada. (After Billings.)

(fig. 36) appear to be unquestionable plants. The true affinities of these, however, are extremely dubious. All that can be said is, that remains which appear to be certainly vegetable, and which are most probably due to marine plants, have been recognized nearly at the base of the Lower Silurian (Arenig), and that they are found throughout the series whenever suitable conditions recur.

The Protozoans appear to have flourished extensively in the Lower Silurian seas, though to a large extent under forms which are still little understood. We have here for the first time the appearance of *Foraminifera* of the ordinary type—one of the most interesting observations in this connection being that made by Ehrenberg, who showed that the lower Silurian sandstones of the neighborhood of St. Petersburg contained casts in glauconite of Foraminiferous shells, some of which are referable to the existing genera *Rotalia* and *Textularia*. True *Sponges*, belonging to that section of the group in which the skeleton is calcareous, are also not unknown, one of the most characteristic genera being *Astylospongia* (fig. 37). In this genus are included more or less globular, often lobed sponges, which are believed not to have been attached to foreign bodies. In the form here figured there is a funnel-shaped cavity at the summit; and the entire mass of the sponge is perforated, as in living examples, by a system of canals which convey the sea-water to all parts of the organism. The canals by which the sea-water gains entrance open on the exterior of the sphere, and those by which it again escapes from the sponge open into the cup-shaped depression at the summit.

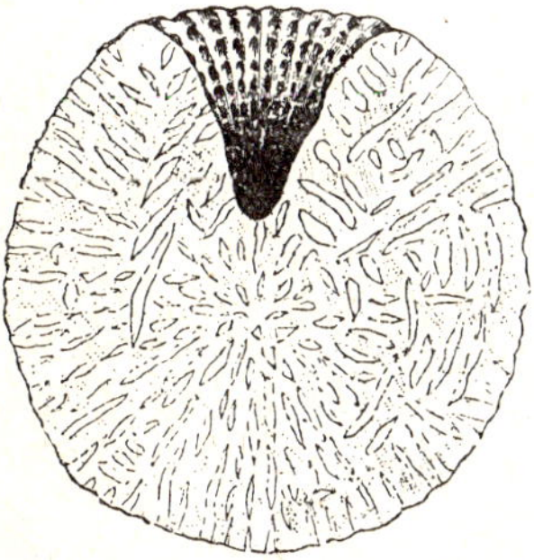

Fig. 37.—*Astylospongia præmorsa*, cut vertically so as to exhibit the canal-system in the interior. Lower Silurian, Tennessee. (After Ferdinand Rœmer.)

The most abundant, and at the same time the least understood, of Lower Silurian Protozoans belonging, however, to the genera *Stromatopora* and *Receptaculites*, the structure of which can merely be alluded to here. The specimens of *Stromatopora* (fig. 38) occur as hemispherical, pear-shaped, globular, or irregular masses, often of very considerable size, and sometimes demonstrably attached to foreign bodies. In their structure these masses consist of numerous thin calcareous laminæ, usually arranged concentrically, and separated by narrow interspaces. These interspaces are generally crossed by numerous vertical calcareous pillars, giving the vertical section

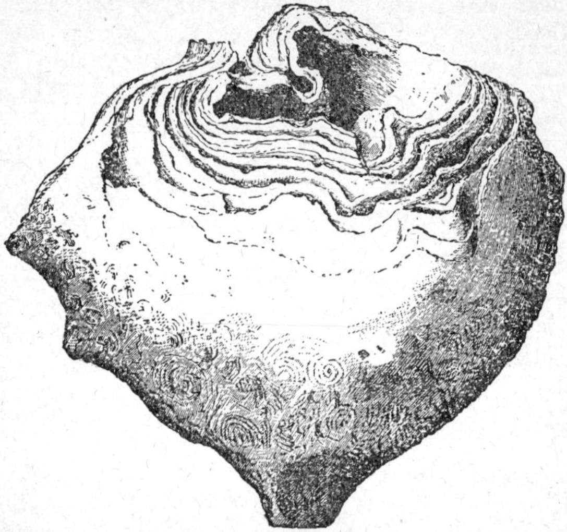

Fig. 38—A small and perfect specimen of *Stromatopora rugosa*, of the natural size, from the Trenton Limestone of Canada. (After Billings.)

of the fossil a lattice-like appearance. There are also usually minute pores in the concentric laminæ, by which the successive interspaces are placed in communication; and sometimes the surface presents large rounded openings, which appear to correspond with the water-canals of the Sponges. Upon the whole, though presenting some curious affinities to the calcareous Sponges, *Stromatopora* is perhaps more properly regarded as a gigantic *Foraminifer.* If this view be correct, it is of special interest as being probably the nearest ally of *Eozoön,* the general appearance of the two being strikingly similar, though their minute structure is not at all the same. Lastly, in the fossils known as *Receptaculites* and *Ischadites* we are also presented with certain singular Lower Silurian Protozoans, which may with great probability be regarded as gigantic *Foraminifera.* Their structure is very complex; but fragments are easily recognized by the fact that the exterior is covered with numerous rhomboidal calcareous plates, closely fitting together, and arranged in peculiar intersecting curves, presenting very much the appearance of the engine-turned case of a watch.

Passing next to the sub-kingdom of *Cœlenterate* animals (Zoophytes, Corals, &c.), we find that this great group, almost or wholly absent in the Cambrian, is represented in Lower Silurian deposits by a great number of forms belonging on the one hand to the true Corals, and on the other hand to the singular family of the *Graptolites*. If we except certain plant-like fossils which probably belong rather to the Sertularians or the Polyzoans (*e. g., Dictyonema, Dendrograptus,* &c.), the family of the *Graptolites* may be regarded as exclusively Silurian in its distribution. Not only is this the case, but it attained its maximum development almost upon its first appearance, in the Arenig Rocks; and whilst represented by a great variety of types in the Lower Silurian, it only exists in the Upper Silurian in a much diminished form. The *Graptolites* (Gr. *grapho,* I write; *lithos,* stone) were so named by Linnæus, from the resemblance of some of them to written or pencilled marks upon the stone, though the great naturalist himself did not believe them to be true fossils at all. They occur as linear or leaf-like bodies, sometimes simple, sometimes compound and branched; and no doubt whatever can be entertained as to their being the skeletons of composite organisms, or colonies of semi-independent animals united together by a common fleshy trunk, similar to what is observed in the colonies of the existing Sea-firs (Sertularians). This fleshy trunk or common stem of the colony was protected by a delicate horny sheath, and it gave origin to the little flower-like "polypites," which constituted the active element of the whole assemblage. These semi-independent beings were, in turn, protected each by a little horny cup or cell, directly connected with the common sheath below, and terminating above in an opening through which the polypite could protrude its tentacled head or could again withdraw itself for safety. The entire skeleton, again, was usually, if not universally, supported by a delicate horny rod or "axis," which appears to have been hollow, and which often protrudes to a greater or less extent beyond one or both of the extremities of the actual colony.

The above gives the elementary constitution of any *Graptolite,* but there are considerable differences as to the manner in which these elements are arranged and combined. In some forms the common stem of the colony gives origin to but a single row of cells on one side. If the common stem is a simple, straight, or slightly-curved linear body, then we have the simplest form of Graptolite known (the genus *Monograptus*);

and it is worthy of note that these simple types do not come into existence till comparatively late (Llandeilo), and last nearly to the very close of the Upper Silurian. In other cases, whilst there is still but a single row of cells, the colony may consist of two of these simple stems springing from a common point, as in the so-called "twin Graptolites" (*Didymograptus,* fig. 40). This type is entirely confined to the earlier portion of the Lower Silurian period (Arenig and Llandeilo). In other cases, again, there may be four of such stems springing from a central point. (*Tetragraptus*). Lastly, there are numerous complex forms (such as *Dichograptus, Loganograptus,* &c.) in which there are eight or more of these simple branches, all arising from a common center (fig. 39), which is sometimes furnished with a singular horny disc. These complicated branching forms, as well as the *Tetragrapti,* are characteristic of the horizon of the Arenig group. Similar forms, often specifically identical, are found at this horizon in Wales, in the great series of the Skiddaw Slates of the north of England, in the Quebec group in Canada, in equivalent beds in Sweden, and in certain gold-bearing slates of the same age in Victoria in Australia.

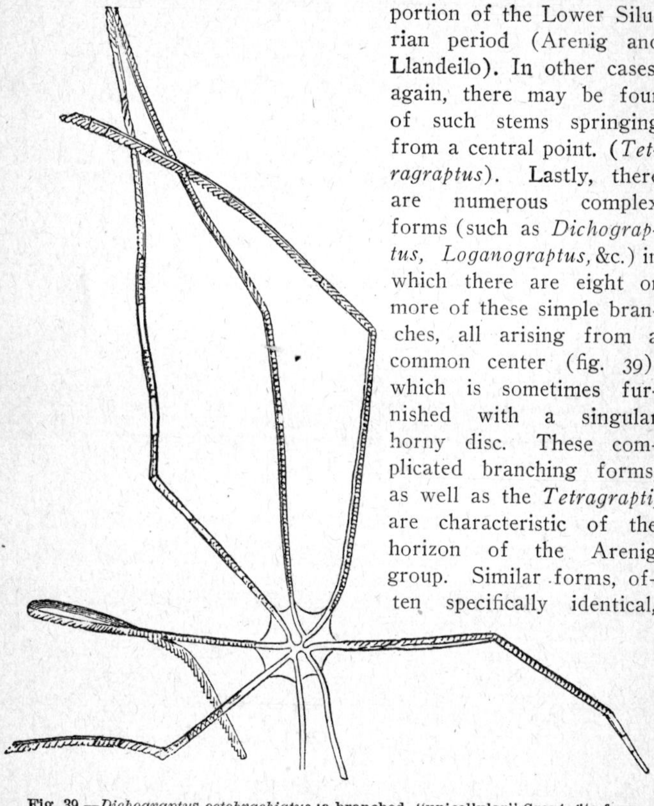

Fig. 39.—*Dichograptus octobrachiatus*, **a branched, "unicellular" Graptolite from the Skiddaw and Quebec Groups (Arenig). (After Hall.)**

In another great group of Graptolites (including the genera *Diplograptus, Dicranograptus, Climacograptus*, &c.) the common stem of the colony gives origin, over part or the whole of its

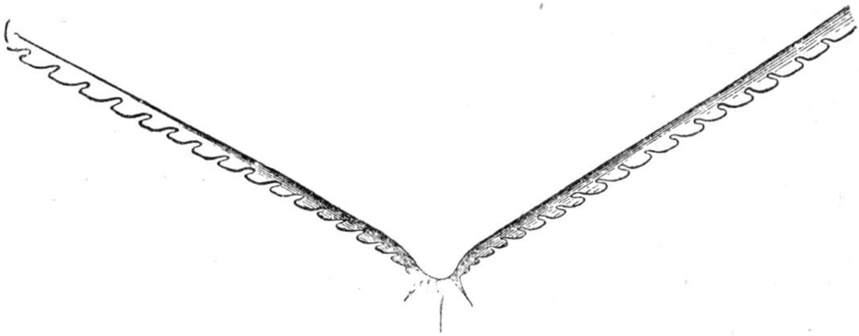

Fig. 40.—Central portion of the colony of *Didymograptus divaricatus*, Upper Llandeilo, Dumfriesshire. (Original.)

length, to *two* rows of cells, one on each side (fig. 41). These "double-celled" Graptolites are highly characteristic of the

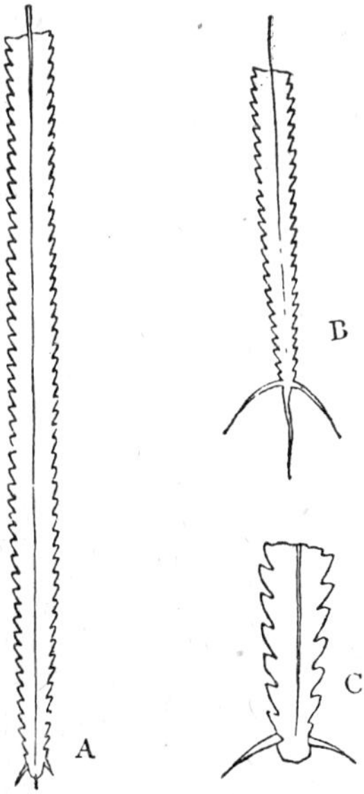

Fig. 41.—Examples of *Diplograptus pristis*, showing variations in the appendages at the base. Upper Llandeilo, Dnmfreisshire. (Original.)

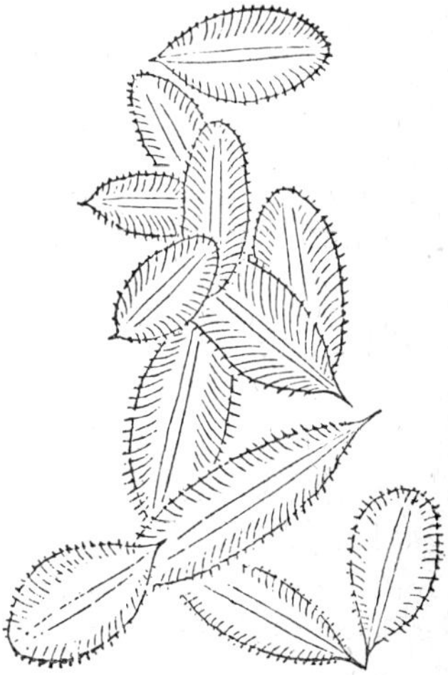

Fig. 42.—Group of individuals of *Phyllograptus typus*, from the Quebec group of Canada. (After Hall.) One of the four rows of cells is hidden on the under surface.

Lower Silurian deposits; and, with an exception more apparent than real in Bohemia, they are exclusively confined to strata of Lower Silurian age, and are not known to occur in the Upper Silurian. Lastly, there is a group of Graptolites (*Phyllograptus*, fig. 42) in which the colony is leaf-like in form, and is composed of *four* rows of cells springing in a cross-like manner from the common stem. These forms are highly characteristic of the Arenig group.

The Graptolites are usually found in dark-colored, often black shales, which sometimes contain so much carbon as to become "anthracitic." They may be simply carbonaceous; but they are more commonly converted into iron-pyrites, when they glitter with the brilliant luster of silver as they lie scattered on the surface of the rock, fully deserving in their metallic tracery the name of "written stones." They constitute one of the most important groups of Silurian fossils, and are of the greatest value in determining the precise stratigraphical position of the beds in which they occur. They present, however, special difficulties in their study; and it is still a moot point as to the precise position in the zoological scale. The balance of evidence is in favor of regarding them as an ancient and peculiar group of the Sea-firs (Hydroid Zoophytes), but some regard them as belonging rather to the Sea-mosses (*Polyzoa*). Under any circumstances, they cannot be directly compared either with the ordinary Sea-firs or the ordinary Sea-mosses; for these two groups consist of fixed organisms, whereas the Graptolites were certainly free-floating creatures, living at large in the open sea. The only Hydroid Zoophytes or Polyzoans which have a similar free mode of existence, have either no skeleton at all, or have hard structures quite unlike the horny sheaths of the Graptolites.

The second great group of Cœlenterate animals (*Actinozoa*) is represented in the Lower Silurian rocks by the numerous Corals. These, for obvious reasons, are much more abundant in regions where the Lower Silurian series is largely calcareous (as in North America) than in districts like Wales, where limestones are very feebly developed. The Lower Silurian Corals, though the first of their class, and presenting certain peculiarities, may be regarded as essentially similar in nature to existing Corals. These, as is well known, are the calcareous skeletons of animals — the so-called "Coral-Zoophytes" — closely allied to the common Sea-anemones in structure and habit. A *simple* coral (fig. 43) consists of a calcareous cup embedded in the soft tissues of the flower-like polype, and having at its summit a more or less deep depression (the "calice") in which the digestive organs are contained. The space within the corals is divided into compartments by numerous vertical calcareous plates (the "septa"), which spring from the inside of the wall of the cup, and of which some generally reach the

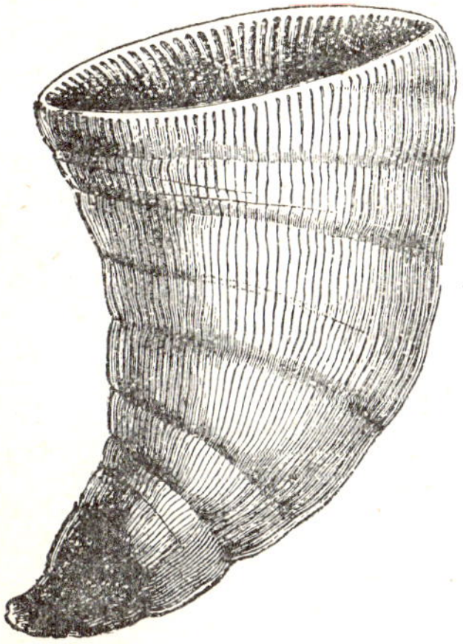

Fig. 43.—*Zaphrentis Stokesi*, a simple "cup-coral," Upper Silurian, Canada. (After Billings.)

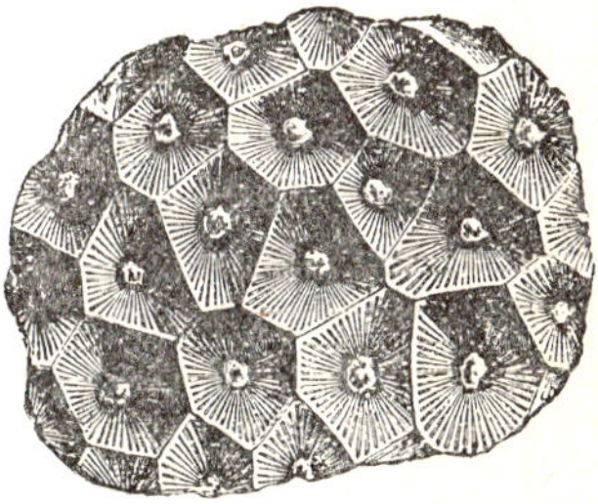

Fig. 44.—Upper surface of a mass of *Strombodes pentagonus*, Upper Silurian, Canada. (After Billings.)

center. *Compound* corals, again (fig. 44), consist of a greater or less number of structures similar in structure to the above, but united together in different ways into a common mass. *Simple* corals, therefore, are the skeletons of *single* and independent polypes; whilst *compound* corals are the skeletons of assemblages or *colonies* of similar polypes, living united with one another as an organic community.

In the general details of their structure, the Lower Silurian Corals do not differ from the ordinary Corals of the present day. The latter, however, have the vertical calcareous plates of the coral ("septa") arranged in multiples of six or five; whereas the former have these structures arranged in multiples of four, and often showing a cross-like disposition. For this reason, the common Lower Silurian Corals are separated to form a distinct group under the name of *Rugose* Corals or *Rugosa.* They are further distinguished by the fact that the cavity of the coral ("visceral chamber") is usually subdivided by more or less numerous *horizontal* calcareous plates or partitions, which divide the coral into so many tiers or storeys, and which are known as the "tabulæ" (fig. 45).

In addition to the Rugose Corals, the Lower Silurian rocks contain a number of curious compound corals, the tubes of which have either no septa at all or merely rudimentary ones, but which have the transverse partitions or "tabulæ" very highly developed. These are known as the *Tabulate*

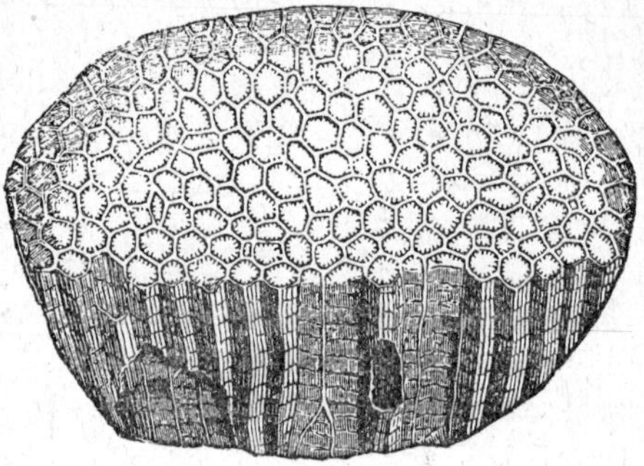

Fig. 45.—*Columnaria alveolata*, a Rugose compound coral, with imperfect septa, but having the corallites partitioned off into storeys by "tabulæ." Lower Silurian, Canada. (After Billings.)

Corals; and recent researches on some of their existing allies (such as *Heliopora*) have shown that they are really allied to the modern Sea-pens, Organ-pipe Corals, and Red Coral, rather than to the typical stony Corals. Amongst the characteristic Rugose Corals of the Lower Silurian may be mentioned species belonging to the genera *Columnaria, Favistella, Streptelasma,* and *Zaphrentis;* whilst amongst the "Tabulate" Corals, the principal forms belong to the genera *Chætetes, Halysites* (the Chain-coral), *Constellaria,* and *Heliolites.* These groups of the Corals, however, attain a greater development at a later period, and they will be noticed more particularly hereafter.

Passing on to higher animals, we find that the class of the *Echinodermata* is represented by examples of the Star-fishes (*Asteroidea*), the Sea-lilies (*Crinoidea*), and the peculiar extinct group of the Cystideans (*Cystoidea*), with one or two of the Brittle-stars (*Ophiuroidea*)—the Sea-urchins (*Echinoidea*) being still wanting. The Crinoids, though in some places extremely numerous, have not the varied development that they possess in the Upper Silurian, in connection with which their structure will be more fully spoken of. In the meanwhile, it is sufficient to note that many of the calcareous deposits of the Lower Silurian are strictly entitled to the name of "Crinoidal limestones," being composed in great part of the detached joints,

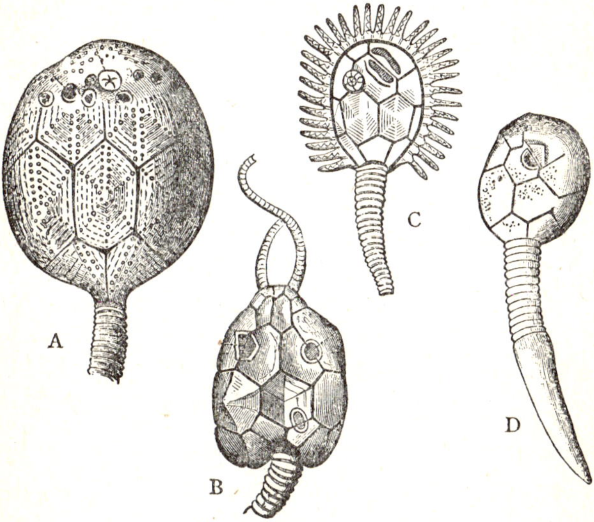

Fig. 46.—Group of Cystideans. A, *Caryocrinus ornatus*,* Upper Silurian, America; B, *Pleurocystites squamosus*, showing two short "arms," Lower Silurian, Canada; C, *Pseudocrinus bifasciatus*, Upper Silurian, England; D, *Lepadocrinus Gebhardi*, Upper Silurian, America. (After Hall, Billings and Salter.)

and plates, and broken stems, of these beautiful but fragile organisms (see fig. 12). Allied to the Crinoids are the singular creatures which are known as *Cystideans* (fig. 46). These are generally composed of a globular or ovate body (the "calyx"), supported upon a short stalk (the "column"), by which the organism was usually attached to some foreign body. The body was enclosed by closely-fitting calcareous plates, accurately jointed together; and the stem was made up of numerous distinct pieces or joints, flexibly united to each other by membrane. The chief distinction which strikes one in comparing the Cystideans with the Crinoids is, that the latter are always furnished, as will be subsequently seen, with a beautiful crown of branched and feathery appendages, springing from the sum-

* The genus *Caryocrinus* is sometimes regarded as properly belonging to the *Crinoids*, but there seem to be good reasons for rather considering it as an abnormal form of *Cystidean*.

mit of the calyx, and which are composed of innumerable calcareous plates or joints, and are known as the "arms." In the Cystideans, on the other hand, there are either no "arms" at all, or merely short, unbranched, rudimentary arms. The Cystideans are principally, and indeed nearly exclusively,

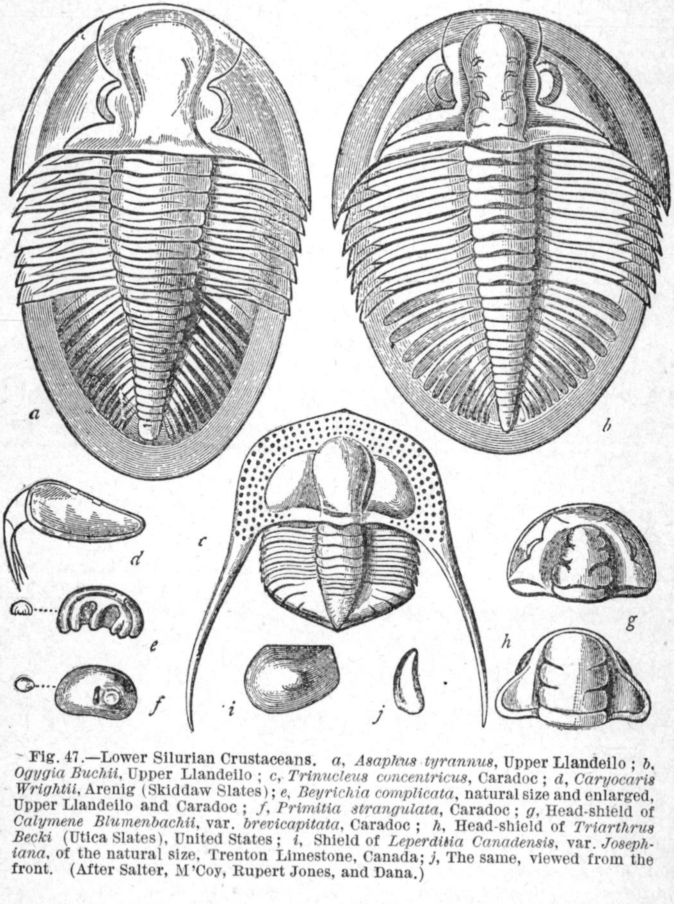

Fig. 47.—Lower Silurian Crustaceans. *a*, *Asaphus tyrannus*, Upper Llandeilo; *b*, *Ogygia Buchii*, Upper Llandeilo; *c*, *Trinucleus concentricus*, Caradoc; *d*, *Caryocaris Wrightii*, Arenig (Skiddaw Slates); *e*, *Beyrichia complicata*, natural size and enlarged, Upper Llandeilo and Caradoc; *f*, *Primitia strangulata*, Caradoc; *g*, Head-shield of *Calymene Blumenbachii*, var. *brevicapitata*, Caradoc; *h*, Head-shield of *Triarthrus Becki* (Utica Slates), United States; *i*, Shield of *Leperditia Canadensis*, var. *Josephiana*, of the natural size, Trenton Limestone, Canada; *j*, The same, viewed from the front. (After Salter, M'Coy, Rupert Jones, and Dana.)

Silurian fossils; and though occurring in the Upper Silurian in no small numbers, they are pre-eminently characteristic of the Llandeilo-Caradoc period of Lower Silurian time. They commenced their existence, so far as known, in the Upper Cambrian; and though examples are not absolutely unknown in later periods, they are pre-eminently characteristic of the earlier portion of the Palæozoic epoch.

The Ringed Worms (*Annelides*) are abundantly represented in the Lower Silurian, but principally by tracks and burrows similar in essential respects to those which occur so commonly in the Cambrian formation, and calling for no special comment. Much more important are the *Articulate* animals, represented, as heretofore, wholly by the remains of the aquatic group of the *Crustaceans*. Amongst these are numerous little bivalved forms—such as species of *Primitia* (fig. 47 *f*), *Beyrichia* (fig. 47, *e*), and *Leperditia* (fig. 47, *i* and *j*). Most of these are very small, varying from the size of a pin's head up to that of a hemp seed; but they are sometimes as large as a small bean (fig. 47, *i*), and they are commonly found in myriads together in the rock. As before said, they belong to the same great group as the living Water-fleas (*Ostracoda*). Besides these, we find the pod-shaped head-shields of the shrimp-like Phyllopods—such as *Caryocaris* (fig. 47, *d*) and *Ceratiocaris*. More important, however, than any of these are the *Trilobites*, which may be considered as attaining their maxi-

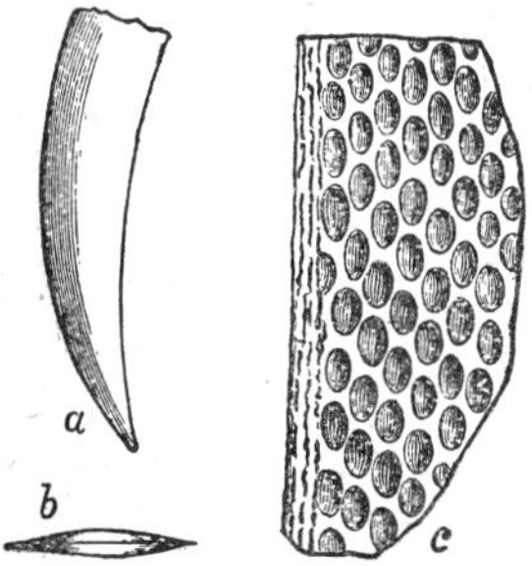

Fig. 48.—*Ptilodictya falciformis*. *a*, Small specimen of the natural size; *b*, Cross-section, showing the shape of the front; *c*, Portion of the surface, enlarged, Trenton Limestone and Cincinnati Group, America. (Original.)

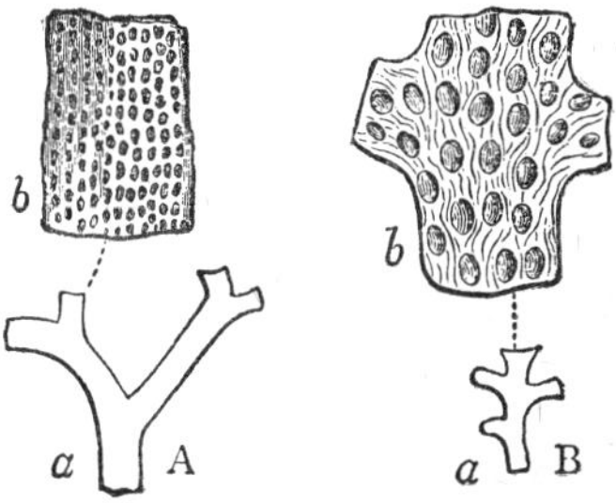

Fig. 49.—A, *Ptilodictya acuta*; B, *Ptilodictya Schafferi*. *a*, Fragment, of the natural size; *b*, portion, enlarged to show the cells. Cincinnati Group of Ohio and Canada. (Original.)

mum development in the Lower Silurian. The huge *Paradoxides* of the Cambrian have now disappeared, and with them almost all the principal and characteristic "primordial" genera, save *Olenus* and *Agnostus*. In their place we have a great number of new forms—some of them, like the great *Asaphus tyrannus* of the Upper Llandeilo (fig. 47, *a*), attaining a length of a foot or more, and thus hardly yielding in the matter of size to their ancient rivals. Almost every subdivision of the Lower Silurian series has its own special and characteristic species of Trilobites; and the study of these is therefore of great importance to the geologist. A few widly-dispersed and characteristic

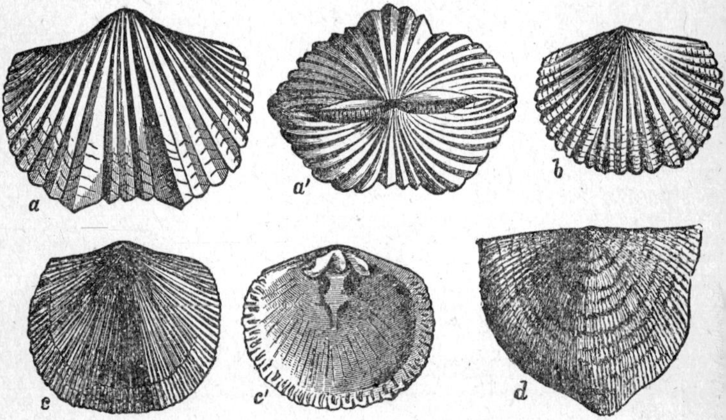

Fig. 50.—Lower Silurian Brachiopods *a*, and *a'* *Orthis biforata*, Llandeilo-Caradoc, Britain and America; *b*, *Orthis flabellulum*, Caradoc, Britain; *c*, *Orthis subquadrata*, Cincinnati Group, America; *c'* Interior of the dorsal valve of the same; *d*, *Strophomena deltoidea*, Llandeilo-Caradoc, Britain and America. (After Meek, Hall, and Salter.)

species have been here figured (fig. 47); and the following may be considered as the principal Lower Silurian genera—*Asaphus, Ogygia, Cheirurus, Ampyx, Calymene, Trinucleus, Lichas, Illænus, Æglina, Harpes, Remopleurides, Phacops, Acidaspis,* and *Homalonotus,* a few of them passing upwards under new forms into the Upper Silurian.

Coming next to the *Mollusca,* we find the group of the Sea-mosses and Sea-mats (*Polyzoa*) represented now by quite a number of forms. Amongst these are examples of the true Lace-corals (*Retepora* and *Fenestella*), with their netted fan-like or funnel-shaped fronds; and along with these are numerous delicate encrusting forms, which grew parasitically attached to shells and corals (*Hippothoa, Alecto,* &c.); but perhaps the most characteristic forms belong to the genus *Ptilodictya* (figs. 48 and 49). In this group the frond is flattened, with thin striated edges, sometimes sword-like or scimitar-shaped, but often more or less branched; and it consists of two layers of cells, separated by a delicate membrane, and opening upon opposite sides. Each of these little chambers or "cells" was originally tenanted by a minute animal, and the whole thus constituted a compound organism or colony.

The Lamp-shells or *Brachiopods* are so numerous, and present such varied types, both in this and the succeeding period of the Upper Silurian, that the name of "Age of Brachiopods" has with justice been applied to the Silurian period as a whole. It would be impossible here to enter into details as to the many different forms of Brachiopods which present themselves in the Lower Silurian deposits; but we may select the three genera *Orthis, Strophomena,* and *Leptæna* for illustration, as being specially characteristic of this period, though not exclusively confined to it. The numerous shells which belong to the extensive and cosmopolitan genus *Orthis* (fig. 50, *a, b, c,*

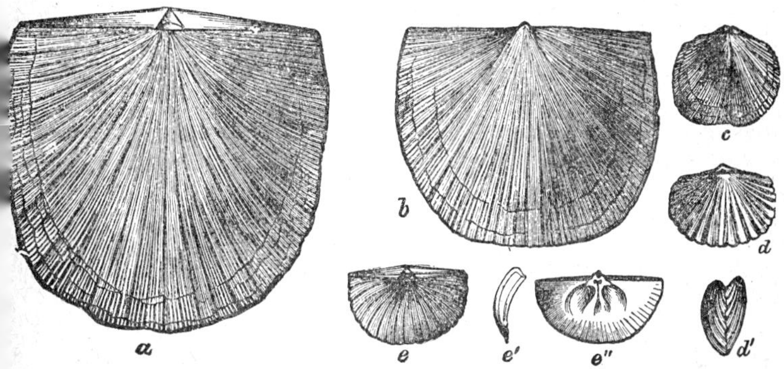

Fig. 51.—Lower Silurian Brachiopods. *a, Strophomena alternata,* **Cincinnati Group, America ;** *b, Strophomena filitexta,* **Trenton and Cincinnati Groups, America ;** *c, Orthis testudinaria,* **Caradoc, Europe, and America ;** *d, d′, Orthis plicatella,* **Cincinnati Group, America ;** *e, e′, e″, Leptæna sericea,* **Llandeilo and Caradoc, Europe and America. (After Meek, Hall, and the Author.)**

and fig. 51, *c* and *d*), are usually more or less transversely-oblong or subquadrate, the two valves (as more or less in all the Brachiopods) of unequal sizes, generally more or less convex, and marked with radiating ribs or lines. The valves of the shell are united to one another by teeth and sockets, and there is a straight hinge-line. The beaks are also separated by a distinct space ("hinge-area"), formed in part by each valve, which is perforated by a triangular opening, through which, in the living condition, passed a muscular cord attaching the shell to some foreign object. The genus *Strophomena* (fig. 50, *d*, and 51, *a* and *b*) is very like *Orthis* in general character; but the shell is usually much flatter, one or other valve often being concave, the hinge-line is longer, and the aperture for the emission of the stalk of attachment is partially closed by a calcareous plate. In *Leptæna*, again (fig. 51, *e*), the shell is like *Strophomena* in many respects, but generally comparatively longer, often completely semicircular, and having one valve convex and the other valve concave. Amongst other genera of Brachiopods which are largely represented in the Lower Silurian rocks may be mentioned *Lingula, Crania, Discina, Trematis, Siphonotreta, Acrotreta, Rhynchonella,* and *Athyris;* but none of these can claim the importance to which the three previously-mentioned groups are entitled.

The remaining Lower Silurian groups of *Mollusca* can be but briefly glanced at here. The Bivalves (*Lamellibranchiata*) find numerous representatives, belonging to such genera as *Modiolopsis, Ctenodonta, Orthonota, Palæarca, Lyrodesma, Ambonychia,* and *Cleidophorus.* The Univalves (*Gasteropoda*) are also very numerous, the two most important genera being *Murchisonia* (fig. 52) and *Pleurotomaria.* In both these groups the outer lip of the shell is notched; but the shell in the former is elongated and turreted, whilst in the latter it is depressed. The curious oceanic Univalves known as the *Heteropods* are also very abundant, the principal forms belonging to *Bellerophon* and *Maclurea.* In the former (fig. 53) there is a symmetrical convoluted shell, like that of the Pearly Nautilus in shape, but without any internal partitions, and having the aperture often expanded and notched behind. The species of *Maclurea* (fig. 54) are found both in North America and in Scotland, and are exclusively confined to the Lower Silurian period, so far as known. They have the shell coiled into a flat spiral, the mouth being furnished with a very curious, thick, and solid lid or "operculum." The Lower Silurian *Pteropods,* or "Winged Snails," are numerous, and belong principally to the genera *Theca, Conularia,* and *Tentaculites,* the last-mentioned of these often being extremely abundant in certain strata.

Fig. 52.—*Murchisonia gracilis,* Trenton Limestone, America. (After Billings.)

Lastly, the Lower Silurian Rocks have yielded a vast number of chambered shells, referable to animals which belong to the same great division as the Cuttle-fishes (the *Cephalopoda*), and of which the Pearly Nautilus is the only living representative at the present day. In this group of *Cephalopods* the animal possesses a well-developed external shell, which is divided into chambers by shelly partitions ("septa"). The animal lives in the last-formed and largest chamber of the

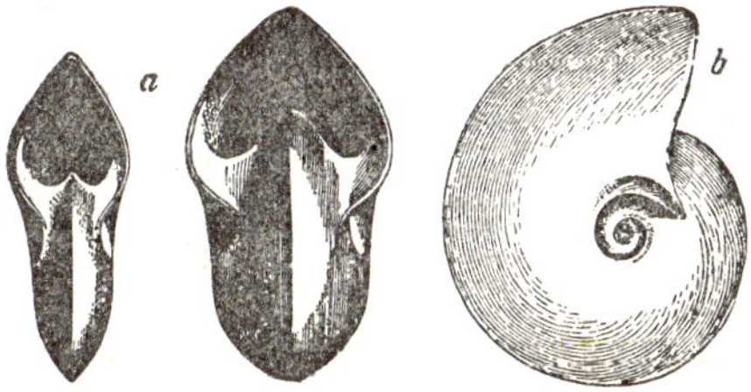

Fig. 53.—Different views of *Bellerophon Argo*, Trenton Limestone, Canada. (After Billings.)

shell, to which it is organically connected by muscular attachments. The head is furnished with long muscular processes or "arms," and can be protruded from the mouth of the shell at will, or again withdrawn within it. We learn, also, from the Pearly Nautilus, that these animals must have possessed two pairs of breathing organs or "gills;" hence all these forms are grouped together under the name of the "Tetrabranchiate"

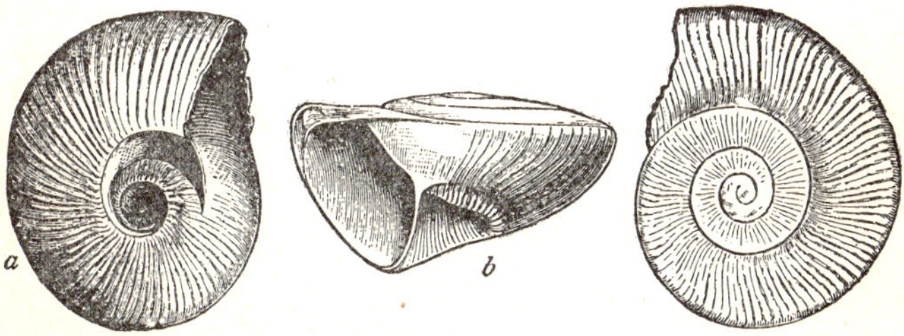

Fig. 54.—Different views of *Maclurea crenulata*, Quebec Group, Newfoundland. (After Billings.)

Cephalopods (Gr. *tetra*, four; *branghia*, gills). On the other hand, the ordinary Cuttle-fishes and Calamaries either possess an internal skeleton, or if they have an external shell, it is not chambered; their "arms" are furnished with powerful organs of adhesion in the form of suckers; and they possess only a single pair of gills. For this last reason they are termed the "Dibranchiate" Cephalopods (Gr. *dis,* twice; *branghia,* gills). No trace of the true Cuttle-fishes has yet been found in Lower Silurian deposits; but the Tetrabranchiate group is represented

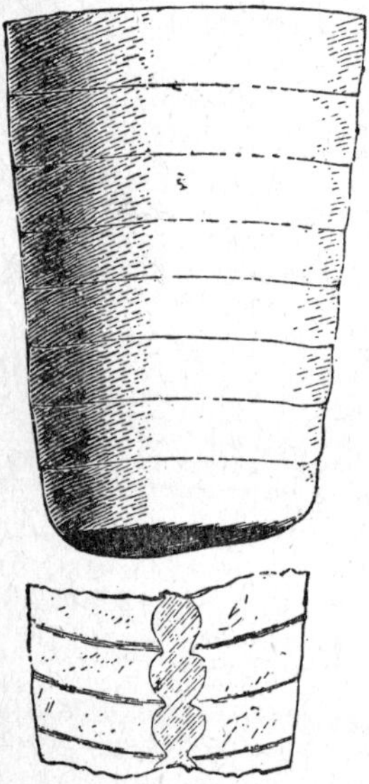

Fig. 55.—Fragment of *Orthoceras crebriseptum*, Cincinnati Group, North America, of the natural size. The lower figure is a section showing the air-chambers, and the form and position of the siphuncle. (After Billings.)

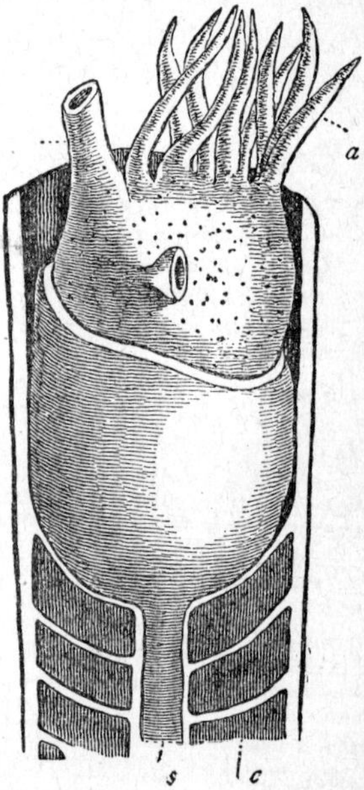

Fig. 56.—Restoration of *Orthoceras*, the shell being supposed to be divided vertically, and only its upper part being shown. *a*, Arms; *f*, Muscular tube ("funnel") by which water is expelled from the mantle-chamber; *c*, Air-chambers; *s*, Siphuncle.

by a great number of forms, sometimes of great size. The principal Lower Silurian genus is the well-known and widely-distributed *Orthoceras* (fig. 55). The shell in this genus agrees with that of the existing *Pearly Nautilus,* in consisting of num-

* This illustration is taken from a rough sketch made by the author many years ago, but he is unable to say from what original source it was copied.

erous chambers separated by shelly partitions (or septa), the latter being perforated by a tube which runs the whole length of the shell after the last chamber, and is known as the "siphuncle" (fig. 56, *s*). The last chamber formed is the largest, and in it the animal lives. The chambers behind this are apparently filled with some gas secreted by the animal itself; and these are supposed to act as a kind of float, enabling the creature to move with ease under the weight of its shell. The various air-chambers, though the siphuncle passes through them, have no direct connection with one another; and it is believed that the animal has the power of slightly altering its specific gravity, and thus of rising or sinking in the water by driving additional fluid into the siphuncle or partially emptying it. The *Orthoceras* further agrees with the Pearly Nautilus in the fact that the partitions or septa separating the different air-chambers are simple and smooth, concave in front and convex behind, and devoid of the elaborate lobation which they exhibit in the Ammonites; whilst the siphuncle pierces the septa either in the center or near it. In the Nautilus, however, the shell is coiled into a flat spiral; whereas in *Orthoceras* the shell is a straight, longer or shorter cone, tapering behind, and gradually expanding towards its mouth in front. The chief objections to the belief that the animal of the *Orthoceras* was essentially like that of the Pearly Nautilus are—the comparatively small size of the body-chamber, the often contracted aperture of the mouth, and the enormous size of some specimens of the shell. Thus, some *Orthocerata* have been discovered measuring ten or twelve feet in length, with a diameter of a foot at the larger extremity. These colossal dimensions certainly make it difficult to imagine that the comparatively small body-chamber could have held an animal large enough to move a load so ponderous as its own shell. To some, this difficulty has appeared so great that they prefer to believe that the *Orthoceras* did not live in its shell at all, but that its shell was an internal skeleton similar to what we shall find to exist in many of the true Cuttle-fishes. There is something to be said in favor of this view, but it would compel us to believe in the existence in Lower Silurian times of Cuttle-fishes fully equal in size to the giant "Kraken" of fable. It need only be added in this connection that the Lower Silurian rocks have yielded the remains of many other Tetrabranchiate Cephalopods besides *Orthoceras*. Some of these belong to *Cyrtoceras*, which only differs from *Orthoceras* in the bow-shaped form of the shell; others belong to *Phragmoceras, Lituites,* &c.; and, lastly, we have true *Nautili,* with their spiral shells, closely resembling the existing Pearly Nautilus.

Whilst all the sub-kingdoms of the Invertebrate animals are represented in the Lower Silurian rocks, no traces of Vertebrate animals have ever been discovered in these ancient deposits, unless the so-called "Conodonts" found by Pander in vast numbers in strata of this age * in Russia should prove to be really of this nature. These problematical bodies are of microscopic size, and have the form of minute, conical, tooth-shaped spines, with sharp edges, and hollow at the base. Their original discoverer regarded them as the horny teeth of fish allied to the Lampreys; but Owen came to the conclusion that they probably belonged to Invertebrates. The recent investigation of a vast number of similar but slightly larger bodies, of very various forms, in the Carboniferous rocks of Ohio, has led Professor Newberry to the conclusion that these singular fossils really are, as Pander thought, the teeth of Cyclostomatous fishes. The whole of this difficult question has thus been reopened, and we may yet have to record the first advent of Vertebrate animals in the Lower Silurian.

CHAPTER X.

THE UPPER SILURIAN PERIOD.

Having now treated of the Lower Silurian period at considerable length, it will not be necessary to discuss the succeeding group of the *Upper Silurian* in the same detail—the more so, as with a general change of *species* the Upper Silurian animals belong for the most part to the same great *types* as those which

* According to Pander, the "Conodonts" are found not only in the Lower Silurian beds, but also in the "Ungulite Grit" (Upper Cambrian), as well as in the Devonian and Carboniferous deposits of Russia. Should the Conodonts prove to be truly the remains of fishes, we should thus have to transfer the first appearance of Vertebrates to, at any rate, as early a period as the Upper Cambrian.

distinguish the Lower Silurian. As compared, also, as regards the total bulk of strata concerned, the thickness of the Upper Silurian is generally very much below that of the Lower Silurian, indicating that they represent a proportionately shorter period of time. In considering the general succession of the Upper Silurian beds, we shall, as before, select Wales and America as being two regions where these deposits are typically developed.

In Wales and its borders the general succession of the Upper Silurian rocks may be taken to be as follows, in ascending order (fig. 57):—

(1) The base of the Upper Silurian series is constituted by a series of arenaceous beds, to which the name of "May Hill Sandstone" was applied by Sedgwick. These are succeeded by a series of greenish-grey or pale-grey slates ("Tarannon Shales"), sometimes of great thickness; and these two groups of beds together form what may be termed the "*May Hill Group*" (Upper Llandovery of Murchison). Though not very extensively developed in Britain, this zone is one very well marked by its fossils; and it corresponds with the "Clinton Group" of North America, in which similar fossils occur. In South Wales this group is clearly unconformable to the highest member of the subjacent Lower Silurian (the Llandovery group); and there is a reason to believe that a similar, though less conspicuous, physical break occurs very generally between the base of the Upper and the summit of the Lower Silurian.

(2) The *Wenlock Group* succeeds the May Hill group, and constitutes the middle member of the Upper Silurian. At its base it may have an irregular limestone ("Woolhope Limestone"), and its summit may be formed by a similar but thicker calcareous deposit ("Wenlock Limestone"); but the bulk of the group is made up of the argillaceous and shaly strata known as the "Wenlock Shale." In North Wales the Wenlock group is represented by a great accumulation of flaggy and gritty strata (the "Denbighshire Flags and Grits"), and similar beds (the "Coniston Flags" and "Coniston Grits") take the same place in the north of England.

(3) The *Ludlow Group* is the highest member of the Upper Silurian, and consists typically of a lower arenaceous and shaly series (the "Lower Ludlow Rock"), a middle calcareous member (the "Aymestry Limestone"), and an upper shaly and sandy series (the "Upper Ludlow Rock"), and "Downton Sandstone"). At the summit, or close to the summit, of the Upper

Ludlow, is a singular stratum only a few inches thick (varying from an inch to a foot), which contains numerous remains of crustaceans and fishes, and is well known under the name of the "bone-bed." Finally, the Upper Ludlow rock graduates invariably into a series of red sandy deposits, which, when of a flaggy character, are known locally as the "Tile-stones." These beds are probably to be regarded as the highest member of the Upper Silurian; but they are sometimes looked upon as passage-beds into the Old Red Sandstone, or as the base of this formation. It is, in fact, apparently impossible to draw any actual line of demarcation between the Upper Silurian and the overlying deposits of the Devonian or Old Red Sandstone series. Both in Britain and in America the Lower Devonian beds repose with perfect conformity upon the highest Silurian beds, and the two formations appear to pass into one another by gradual and imperceptible transition.

The Upper Silurian strata of Britain vary from perhaps 3000 or 4000 feet in thickness up to 8000 or 10,000 feet. In North America the corresponding series, though also variable, is generally of much smaller thickness, and may be under 1000 feet. The general succession of the Upper Silurian deposits of North America is as follows:—

(1) *Medina Sandstone.*—This constitutes the base of the Upper Silurian, and consists of sandy strata, singularly devoid of life, and passing below in some localities into a conglomerate ("Oneida Conglomerate"), which is stated to contain pebbles derived from the older beds, and which would thus indicate an unconformity between the Upper and Lower Silurian.

(2) *Clinton Group.*—Above the Medina sandstone are beds of sandstone and shale, sometimes with calcareous bands, which constitute what is known as the "Clinton Group." The Medina and Clinton groups are undoubtedly the equivalent of the "May Hill Group" of Britain, as shown by the identity of their fossils.

(3) *Niagara Group.*—This group consists typically of a series of argillaceous beds ("Niagara Shale") capped by limestone ("Niagara Limestone"); and the name of the group is derived from the fact that it is over limestones of this age that the Niagara river is precipitated to form the great Falls. In places the Niagara group is wholly calcareous, and it is continued upwards into a series of marls and sand-

GENERALIZED SECTION OF THE UPPER SILURIAN STRATA OF WALES AND SHROPSHIRE.

Fig. 57.

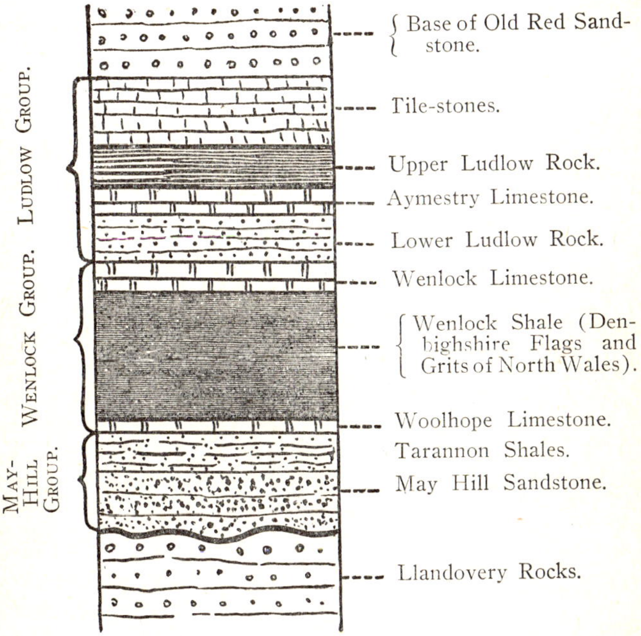

stones, with beds of salt and masses of gypsum (the "Salina Group"), or into a series of magnesian limestones ("Guelph Limestones"). The Niagara group, as a whole, corresponds unequivocally with the Wenlock group of Britain.

(4) *Lower Helderberg Group.*—The Upper Silurian period in North America was terminated by the deposition of a series of calcareous beds, which derive the name of "Lower Helderberg" from the Helderberg mountains, south of Albany, and which are divided into several zones, capable of recognition by their fossils, and known by local names (Tentaculite Limestone, Water-lime, Lower Pentamerus Limestone, Delthyris

Shaly Limestone, and Upper Pentamerus Limestone). As a whole, this series may be regarded as the equivalent of the Ludlow group of Britain, though it is difficult to establish any precise parallelism. The summit of the Lower Helderberg group is constituted by a coarse-grained sandstone (the "Oriskany Sandstone"), replete with organic remains, which have to a large extent a Silurian *facies*. Opinions differ as to whether this sandstone is to be regarded as the highest bed of the Upper Silurian or the base of the Devonian. We thus see that in America, as in Britain, no other line than an artificial one can be drawn between the Upper Silurian and the overlying Devonian.

As regards the *life* of the Upper Silurian period, we have, as before, a number of so-called "Fucoids," the true vegetable nature of which is in many instances beyond doubt. In addition to these, however, we meet for the first time, in deposits of this age, with the remains of genuine land-plants, though our knowledge of these is still too scanty to enable us to construct any detailed picture of the terrestrial vegetation of the period. Some of these remains indicate the existence of the remarkable genus *Lepidodendron*—a genus which played a part of great importance in the forests of the Devonian and Carboniferous periods, and which may be regarded as a gigantic and extinct type of the Club-mosses (*Lycopodiaceæ*). Near the summit of the Ludlow formation in Britain there have also been found beds charged with numerous small globular bodies, which Dr. Hooker has shown to be the seed-vessels or "sporangia" of Club-mosses. Principal Dawson further states that he has seen in the same formation fragments of wood with the structure of the singular Devonian Conifer known as *Prototaxites*. Lastly, the same distinguished observer has described from the Upper Silurian of North America the remains of the singular land-plants belonging to the genus *Psilophyton,* which will be referred to at greater length hereafter.

The *marine* life of the Upper Silurian is in the main constituted by types of animals similar to those characterizing the Lower Silurian, though for the most part belonging to different species. The *Protozoans* are represented principally by *Stromatopora* and *Ischadites,* along with a number of undoubted sponges (such as *Amphispongia, Astræospongia, Astylospongia,* and *Palæomanon*).

Amongst the *Cœlenterates*, we find the old group of *Graptolites* now verging on extinction. Individuals still remain numerous, but the variety of generic and specific types has now become greatly reduced. All the branching and complex forms of the Arenig, the twin-Graptolites and *Dicranograpti* of the Llandeilo, and the double-celled *Diplograpti* and *Climacograpti* of the Bala group, have now disappeared. In their place we have the singular *Retiolites,* with its curiously-reticulated skeleton; and several species of the single-celled genus *Monograptus,* of which a characteristic species (*M. priodon*) is here figured. If we remove from this group the plant-like *Dictyonemæ,* which are still present, and which survive into the Devonian, no known species of Graptolite has hitherto been detected in strata higher in geological position than the Ludlow. This, therefore, presents us with the first instance we have as yet met with of the total disappearance and extinction of a great and important series of organic forms.

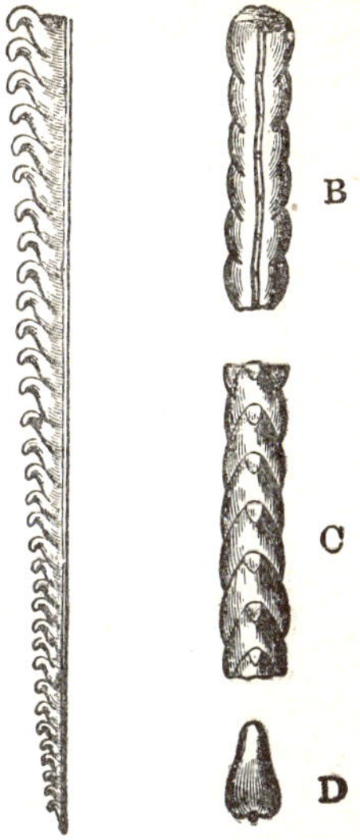

Fig. 58.—A, *Monograptus priodon,* slightly enlarged. B, Fragment of the same viewed from behind. C, Fragment of the same viewed in front, showing the mouths of the cellules. D, Cross-section of the same. From the Wenlock Group (Coniston Flags of the North of England.) (Original.)

The *Corals* are very numerously represented in the Upper Silurian rocks, some of the limestones (such as the Wenlock Limestone) being often largely composed of the skeletons of these animals. Almost all the known forms of this period belong to the two great divisions of the Rugose and Tabulate corals, the former being represented by species of *Zaphrentis, Omphyma, Cystiphyllum, Strombodes, Acervularia, Cyathophyllum,* &c.; whilst the later belong principally to the genera *Favosites, Chætetes, Halysites, Syringopora, Heliolites,* and *Plasmopora.* Amongst the *Rugosa,* the first appearance of the great and important genus *Cyathophyllum,* so characteristic of the Palæozoic period, is to be noted; and amongst the *Tabulata* we have similarly the first appearance, in force at any rate, of the widely-spread genus *Favosites*—the "Honeycomb-corals." The "Chain-corals" (*Halysites*), figured below (fig. 59), are also very common examples of the Tabulate corals during this period, though they occur likewise in the Lower Silurian.

Amongst the *Echinodermata,* all those orders which have hard parts capable of ready preservation are more or less

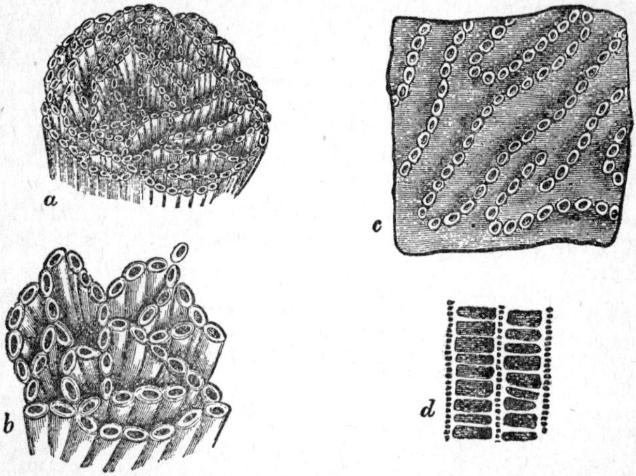

Fig. 59.—*a*, *Halysites catenularia*, small variety, of the natural size ; Fragment of a large variety of the same, of the natural size ; *c*, Fragment of limestone with the tubes of *Halysites agglomerata*, of the natural size ; *d*, Vertical section of two tubes of the same, showing the tabulæ, enlarged. Niagara Limestone (Wenlock), Canada. (Original.)

largely represented. We have no trace of the Holothurians or Sea-cucumbers; but this is not surprising, as the record of the past is throughout almost silent as to the former existence of these soft-bodied creatures, the scattered plates and spicules in their skin offering a very uncertain chance of preservation in the fossil condition. The Sea-urchins (*Echinoids*) are said to be represented by examples of the old genus *Palæchinus.* The Star-fishes (*Asteroids*) and the Brittle-stars (*Ophiuroids*) are, comparatively speaking, largely represented, the former by species of *Palasterina* (fig. 60), *Palæaster* (fig. 60), *Palæocoma* (fig. 60), *Petraster, Glyptaster,* and *Lepidaster*—and the latter by species of *Protaster* (fig. 61), *Palæodiscus, Acroura,* and *Eucladia.* The singular *Cystideans,* or "Globe Crinoids," with their globular or ovate, tesselated bodies (fig. 46, A, C, D,), are also not uncommon in the Upper Silurian; and if they do not become finally extinct here, they certainly survive the close of this period by but a very brief time. By far the most important, however, of the Upper Silurian Echinoderms, are the

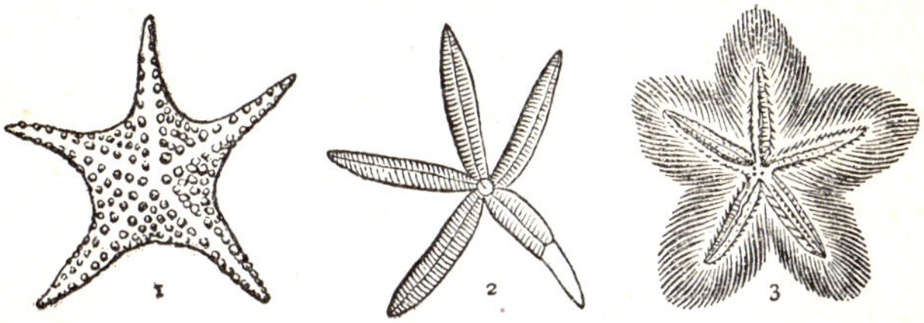

Fig. 60.—Upper Silurian Star-fishes. 1, *Palasterina primæva*. Lower Ludlow; 2, *Palæaster Ruthveni* Lower Ludlow; 3, *Palæocoma Colvini*, Lower Ludlow. (After Salter.)

Sea-lilies or *Crinoids*. The Limestones of this period are often largely composed of the fragmentary columns and detached plates of these creatures, and some of them (such as the Wen-

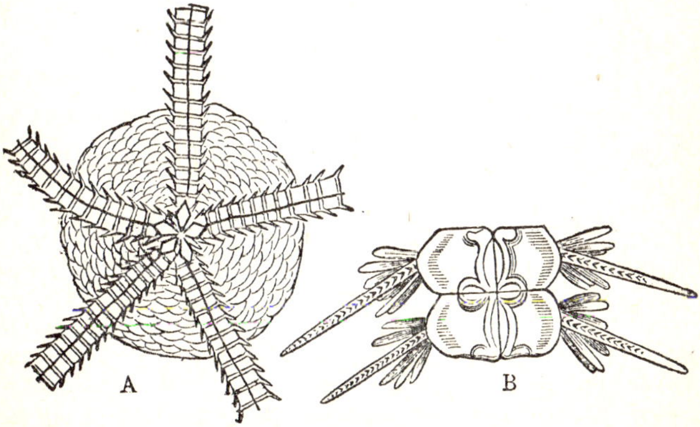

Fig. 61.—A, *Protaster Sedgwickii*, showing the disc and bases of the arms; B, Portion of an arm, greatly enlarged. Lower Ludlow. (After Salter.)

lock Limestone of Dudley) have yielded perhaps the most exquisitely-preserved examples of this group with which we are as yet acquainted. However varied in their forms, these beautiful organisms consists of a globular, ovate, or pear-shaped body (the "calyx"), supported upon a longer or shorter jointed stem (or "column"). The body is covered externally with an armour of closely-fitting calcareous plates (fig. 62), and its upper surface is protected by similar but smaller plates more loosely connected by a leathery integument. From the upper surface of the body, round its margin, springs a series of longer or shorter flexible processes, composed of innumerable calcareous joints or pieces, movably united with one another. The arms are typically five in number; but they generally subdivide at least once, sometimes twice, and they are furnished with similar but more slender lateral branches

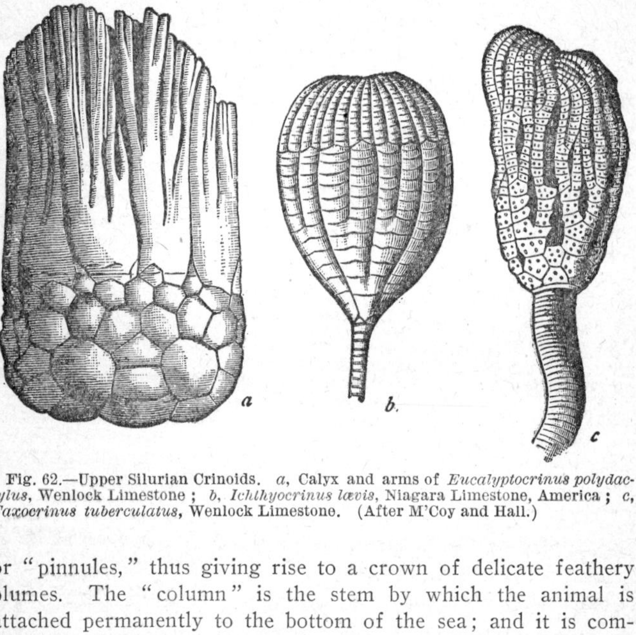

Fig. 62.—Upper Silurian Crinoids. *a*, Calyx and arms of *Eucalyptocrinus polydactylus*, Wenlock Limestone ; *b*, *Ichthyocrinus lævis*, Niagara Limestone, America ; *c*, *Taxocrinus tuberculatus*, Wenlock Limestone. (After M'Coy and Hall.)

or "pinnules," thus giving rise to a crown of delicate feathery plumes. The "column" is the stem by which the animal is attached permanently to the bottom of the sea; and it is composed of numerous separate plates, so jointed together that whilst the amount of movement between any two pieces must be very limited, the entire column acquires more or less flexibility, allowing the organism as a whole to wave backwards and forwards on its stalk. Into the exquisite *minutiæ* of structure by which the innumerable parts entering into the composition of a single Crinoid are adapted for their proper purpose in the economy of the animal, it is impossible to enter here. No period, as before said, has yielded examples of greater beauty than the Upper Silurian, the principal genera represented being *Cyathocrinus, Platycrinus, Marsupiocrinus, Taxocrinus,*

Eucalyptocrinus, Ichthyocrinus, Mariacrinus, Periechocrinus, Glyptocrinus, Crotalocrinus, and *Edriocrinus.*

The tracks and burrows of *Annelides* are as abundant in the Upper Silurian strata as in older deposits, and have just as commonly been regarded as plants. The most abundant forms are the cylindrical, twisted bodies (Planolites), which are so frequently found on the surfaces of sandy beds, and which have been described as the stems of sea-weeds. These fossils (fig. 63), however, can be nothing more, in most cases, than the filled-up burrows of marine worms resembling the living Lobe-worms. There are also various remains which belong to the group of the tube-inhabiting Annelides (*Tubicola*). Of this nature are the tubes of *Serpulites* and *Cornulites,* and the little spiral discs of *Spirorbis Lewisii.*

Amongst the *Articulates,* we still meet only with the remains of *Crustaceans.* Besides the little bivalved *Ostracoda*—which here are occasionally found of the size of beans—and various

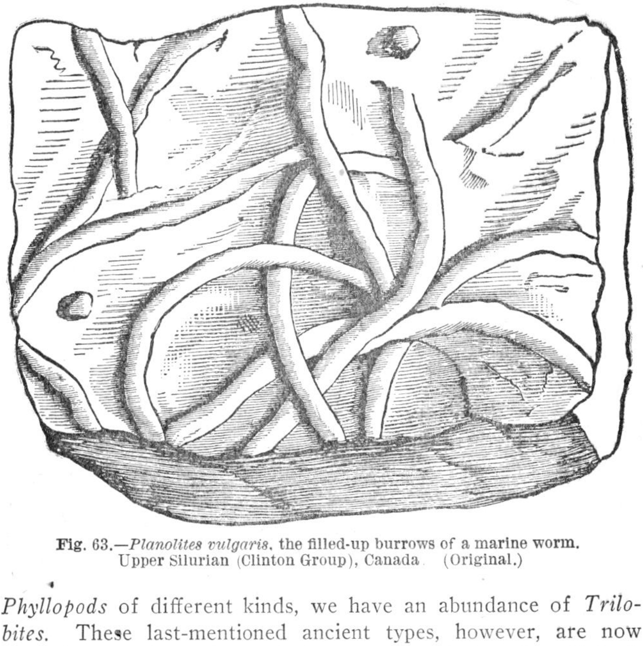

Fig. 63.—*Planolites vulgaris*, the filled-up burrows of a marine worm. Upper Silurian (Clinton Group), Canada. (Original.)

Phyllopods of different kinds, we have an abundance of *Trilobites.* These last-mentioned ancient types, however, are now beginning to show signs of decadence; and though still individually numerous, there is a great diminution in the number of generic types. Many of the old genera, which flourished so abundantly in Lower Silurian seas, have now died out; and the group is represented chiefly by species of *Cheirurus, Encrinurus, Harpes, Proetus, Lichas, Acidaspis, Illænus, Calymene, Homalonotus,* and *Phacops*—the last of these, one of the highest and most beautiful of the groups of Trilobites, attaining here its maximum of development. In the annexed illustration (fig. 64) some of the characteristic Upper Silurian Trilobites are represented—all, however, belonging to genera which have their commencement in the Lower Silurian period. In

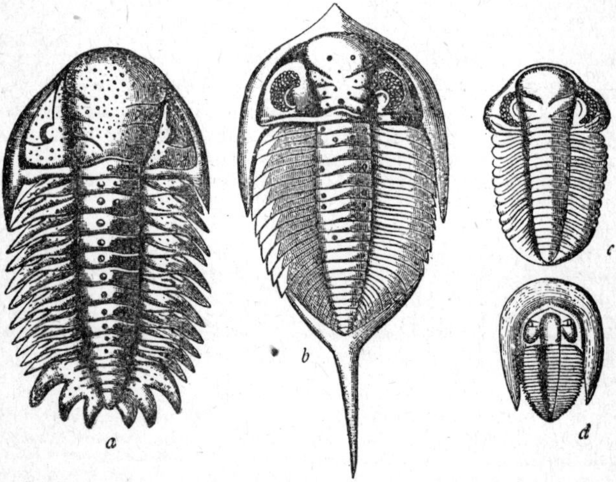

Fig. 64.—Upper Silurian Trilobites. *a, Cheirurus bimucronatus*, **Wenlock and Caradoc;** *b, Phacops longicaudatus*, **Wenlock, Britain and America;** *c, Phacops Downingiæ*, **Wenlock and Ludlow;** *d, Harpes ungula*, **Upper Silurian, Bohemia. (After Salter and Barrande.)**

addition to the above, the Ludlow rocks of Britain and the Lower Helderberg beds of North America have yielded the remains of certain singular Crustaceans belonging to the extinct order of the *Eurypterida.* Some of these wonderful forms are not remarkable for their size; but others, such as *Pterygotus Anglicus* (fig. 65), attain a length of six feet or more, and may fairly be considered as the giants of their class. The Eurypterids are most nearly allied to the existing King-crabs (*Limuli*), and have the anterior end of the body covered with a great head-shield, carrying two pairs of eyes, the one simple and the other compound. The feelers are converted into pincers, whilst the last pair of limbs have their bases covered with spiny teeth so as to act as jaws, and are flattened and widened out towards their extremities so as to officiate as swimming-paddles. The hinder extremity of the body is composed of thirteen rings, which have no legs attached to them; and the last segment of the tail is either a flattened plate or a narrow, sword-shaped spine. Fragments of the skeleton are easily recognized by the peculiar scale-like markings with which the surface is adorned, and which look not at all unlike the scales of a fish. The most famous locality for these great Crustaceans is Lesmahagow, in Lanarkshire, where many different species have been found. The true King-crabs (*Limuli*) of existing seas also appear to have been represented by at least one form (*Neolimulus*) in the Upper Silurian.

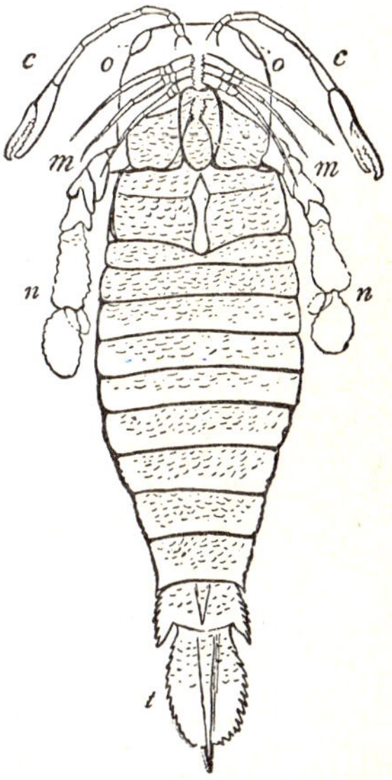

Fig. 65. — *Pterygotus Anglicus*, viewed from the under side, reduced in size and restored. *c c*, The feelers (antennæ), terminating in nipping-claws; *o o*, Eyes; *m m*, Three pairs of jointed limbs, with pointed extremities; *n n*, Swimming-paddles, the bases of which are spiny and act as jaws. Upper Silurian, Lanarkshire. (After Henry Woodward.)

Coming to the *Mollusca,* we note the occurrence of the same great groups as in the Lower Silurian. Amongst the Sea-mosses (*Polyzoa*), we have the ancient Lace-corals (*Fenestella* and *Retepora*), with the nearly-allied *Glauconome,* and species of *Ptilodictya* (fig. 66); whilst many forms often referred here may probably have to be transferred to the Corals, just as some so-called Corals will ultimately be removed to the present group.

The Brachiopods continued to flourish during the Upper Silurian period in immense numbers and under a greatly increased variety of forms. The three prominent Lower Silurian genera *Orthis, Strophomena,* and *Leptæna* are still well represented, though they have lost their former pre-eminence. Amongst the numerous types which have now come upon the scene for the first time, or which have now a special development, are *Spirifera* and *Pentamerus*. In the

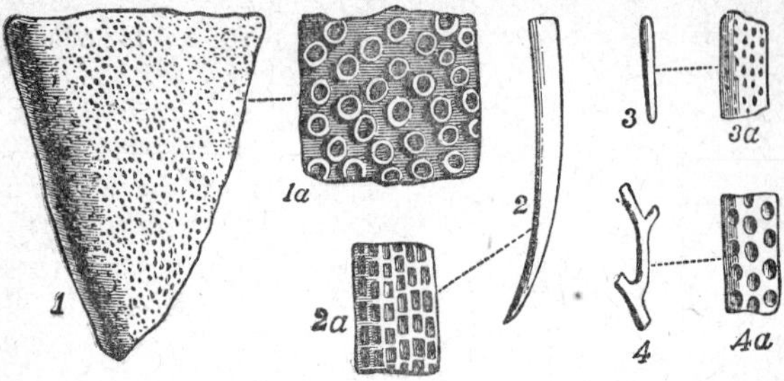

Fig. 66.—Upper Silurian Polyzoa. 1, Fan-shaped frond of *Rhinopora verrucosa;* 1*a*, Portion of the surface of the same, enlarged; 2 and 2*a*, *Phænopora ensiformis*, of the natural size and enlarged; 3 and 3*a*, *Helopora fragilis*, of the natural size and enlarged; 4 and 4*a*, *Ptilodictya raripora*, of the natural size and enlarged. The specimens are all from the Clinton Formation (May Hill Group) of Canada. (Original.)

first of these (fig. 69, *b, c*), one of the valves of the shell (the dorsal) is furnished in its interior with a pair of great calcareous spires, which served for the support of the long and

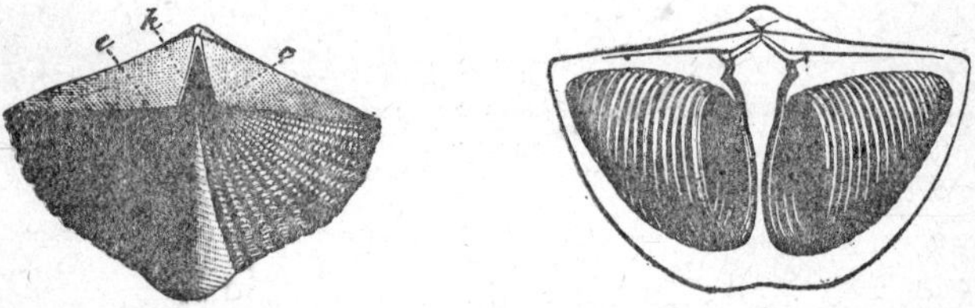

Fig. 67.—*Spirifera hysterica.* The right-hand figure shows the interior of the dorsal valve, with the calcareous spires for the support of the arms.

fringed fleshy processes or "arms" which were attached to the sides of the mouth.* In the genus *Pentamerus* (fig. 70) the shell is curiously subdivided in its interior by calcareous plates. The *Pentameri* commenced their existence at the very close of the Lower Silurian (Llandovery), and survived to the

* In all the Lamp-shells the mouth is provided with two long fleshy organs, which carry delicate filaments on their sides, and which are usually coiled into a spiral. These organs are known as the "arms," and it is from their presence that the name of "*Brachiopoda*" is derived (Gr. *brachion*, arm; *podes*, feet). In some cases the arms are merely coiled away within the shell, without any support; but in other cases they are carried upon a more or less elaborate shelly loop, often spoken of as the "carriage-spring apparatus." In the *Spirifers*, and in other ancient genera, this apparatus is coiled up into a complicated spiral (fig. 67). It is these "arms," with or without the supporting loops or spires, which serve as one of the special characters distinguishing the *Brachiopods* from the true Bivalves (*Lamellibranchiata*).

close of the Upper Silurian; but they are specially characteristic of the May Hill and Wenlock groups, both in Britain

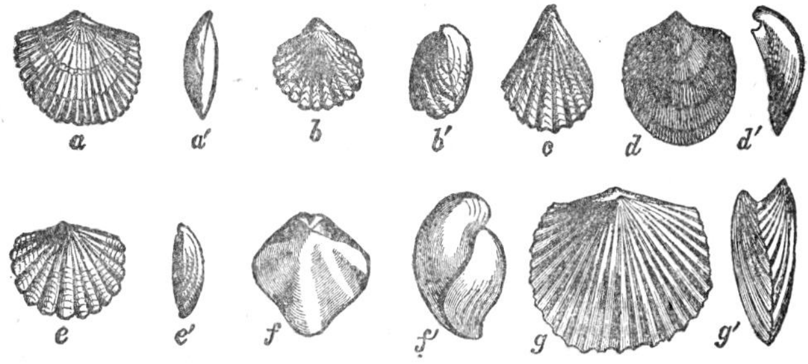

Fig. 68.—Upper Silurian Brachiopods. *a a'*, *Leptocœlia plano-convexa*, Clinton Group, America; *b b'*, *Rhynchonella neglecta*, Clinton Group, America; *c*, *Rhynchonella cuneata*, Niagara Group, America, and Wenlock Group, Britain; *d d'*, *Orthis elegantula*, Llandeilo to Ludlow, America and Europe; *e e'*, *Atrypa hemispherica*, Clinton Group, America, and Llandovery and May Hill Groups, Britain; *f f'*, *Atrypa congesta*, Clinton Group, America; *g g'*, *Orthis Davidsoni*, Clinton Group, America. (After Hall, Billings, and the Author.)

and in other regions. One species, *Pentamerus galeatus,* is common to Sweden, Britain, and America. Amongst the

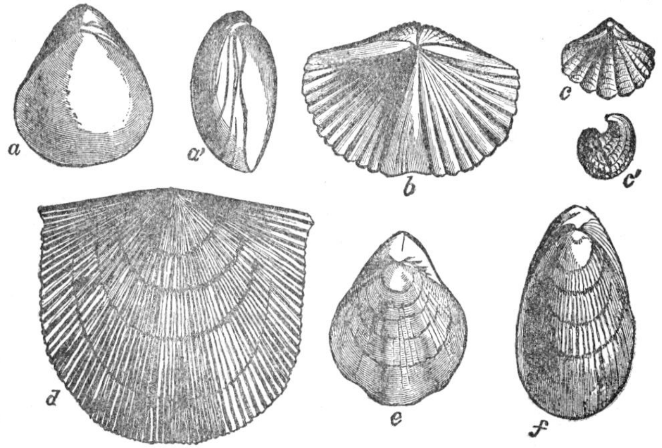

Fig. 69.—*a, a' Meristella intermedia*, Niagara Group, America; *b*, *Spirifera Niagarensis*, Niagara Group, America; *c c'*, *Spirifera crispa*, May Hill to Ludlow, Britain, and Niagara Group, America; *d*, *Strophomena* (*Streptorhynchus*) *subplana*, Niagara Group, America; *e*, *Meristella naviformis*, Niagara Group, America; *f*, *Meristella cylindrica*, Niagara Group, America. (After Hall, Billings, and the Author.)

remaining Upper Silurian Brachiopods are the extraordinary *Trimerellids*; the old and at the same time modern *Lingulæ*, *Discinæ*, and *Craniæ;* together with many species of *Atrypa* (fig. 68, *e*), *Leptocœlia* (fig. 68, *a*), *Rhynchonella* (fig. 68, *b, c*), *Meristella* (fig. 69, *a, e, f*), *Athyris, Retzia, Chonetes,* &c.

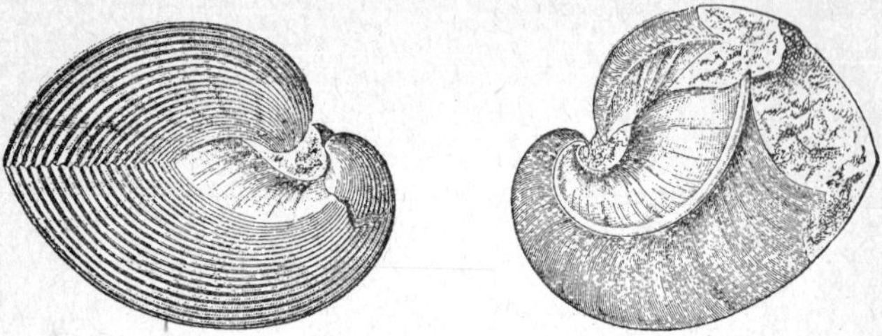

Fig. 70.—*Pentamerus Knightii*, Wenlock and Ludlow. The right-hand figure shows the internal partitions of the shell.

The higher groups of the *Mollusca* are also largely represented in the Upper Silurian. Apart from some singular types, such as the huge and thick-shelled *Megalomi* of the American

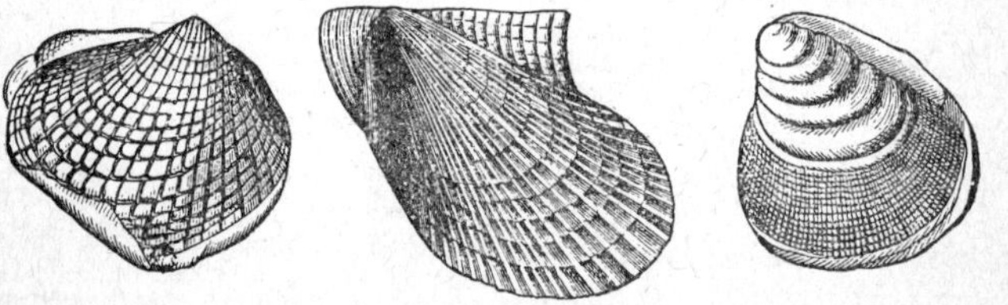

A B C

Fig. 71.—Upper Silurian Bivalves, A, *Cardiola interrupta*, Wenlock and Ludlow; B, *Pterinea subfalcata*, Wenlock; C, *Cardiola fibrosa*, Ludlow. (After Salter and M'Coy.)

Wenlock formation, the Bivalves (*Lamellibranchiata*) present little of special interest; for though sufficiently numerous, they are rarely well preserved, and their true affinities are often uncertain. Amongst the most characteristic genera of this period may be mentioned *Cardiola* (fig. 71, A and C) and *Pterinea* (fig. 71, B), though the latter survives to a much later date. The Univalves (*Gasteropoda*) are very numerous, and a few characteristic forms are here figured (fig. 72). Of these, no genus is perhaps more characteristic than *Euomphalus* (fig. 72, *b*), with its flat discoidal shell, coiled up into an oblique spiral, and deeply hollowed out on one side; but examples of this group are both of older and of more modern date. Another very extensive genus, especially in America, is *Platyceras* (fig. 72, *a* and *f*), with its thin fragile shell—often hardly coiled up at all—its minute spire, and its widely-expanded, often sinuated mouth. The British *Acroculiæ* should probably be placed here, and the group has with reason been regarded as allied to the Violet-snails (*Ianthina*) of the open Atlantic. The

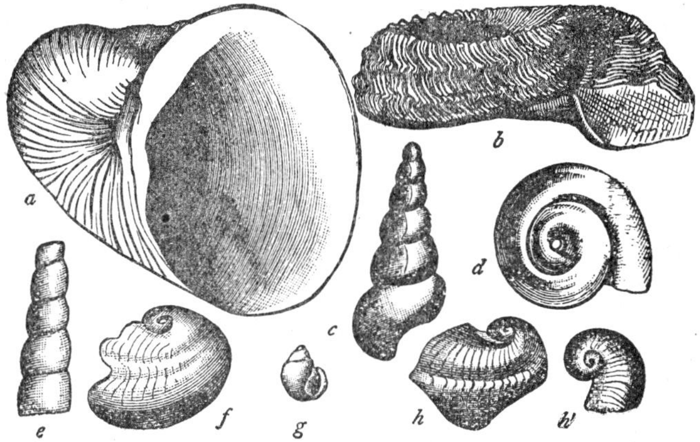

Fig. 72.—Upper Silurian Gasteropods. *a*, *Platyceras ventricosum*, Lower Helderberg, America; *b*, *Euomphalus discors*, Wenlock, Britain; *c*, *Holopella obsoleta*, Ludlow, Britain; *d*, *Platyschisma helicites*, Upper Ludlow, Britain; *e*, *Holopelia gracilior*, Wenlock, Britain; *f*, *Platyceras multisinuatum*, Lower Helderberg, America; *g*, *Holopea subconica*, Lower Helderberg, America; *h*, *h'* *Platyostoma Niagarense*, Niagara Group, America. (After Hall, M'Coy, and Salter.)

species of *Platyostoma* (fig. 72, *h*) also belong to the same family; and the entire group is continued throughout the Devonian into the Carboniferous. Amongst other well-known Upper Silurian Gasteropods are species of the genera *Holopea* (fig. 72, *g*), *Holopella* (fig. 72, *e*), *Platyschisma* (fig. 72, *d*), *Cyclonema, Pleurotomaria, Murchisonia, Trochonema,* &c. The oceanic Univalves (*Heteropods*) are represented mainly by species of *Bellerophon;* and the Winged Snails, or *Pteropods,* can still boast of the gigantic *Thecæ* and *Conulariæ,* which characterize yet older deposits. The commonest genus of *Pteropoda,* however, is *Tentaculites* (fig. 73), which clearly belongs here, though it has commonly been regarded as the tube of an Annelide. The shell in this group is a conical tube, usually adorned with prominent transverse rings, and often with finer transverse or longitudinal striæ as well; and many beds of the Upper Silurian exhibit myriads of such tubes scattered promiscuously over their surfaces.

Fig. 73.—*Tentaculites ornatus.* Upper Silurian of Europe and North America.

The last and highest group of the *Mollusca*—that of the *Cephalopoda*—is still represented only by *Tetrabranchiate* forms; but the abundance and variety of these is almost beyond belief. Many hundreds of different species are known, chiefly belonging to the straight *Orthoceratites,* but the slightly-curved *Cyrtoceras* is only little less common. There are also numerous forms of the genera *Phragmoceras, Ascoceras, Gyroceras, Lituites,* and *Nautilus.* Here, also, are the first-known species of the genus *Goniatites*—a group which attains considerable importance in later deposits, and which is to be regarded as the precursor of the *Ammonites* of the Secondary period.

Finally, we find ourselves for the first time called upon to consider the remains of undoubted vertebrate animals, in the form of *Fishes.* The oldest of these remains, so far as yet known, are found in the Lower Ludlow rocks, and they consist of the bony head-shields or bucklers of certain singular armored fishes belonging to the group of the *Ganoids,* represented at the present day by the Sturgeons, the Gar-pikes of North America, and a few other less familar forms. The principal Upper Silurian genus of these is *Pteraspis,* and the annexed illustration (fig. 74) will give some idea of the extraordinary form of the shield covering the head in these ancient fishes. The remarkable stratum near the top of the Ludlow formation known as the "bone-bed" has also yielded the remains of shark-like fishes. Some of these, for which the name of *Onchus* has been proposed, are in the form of compressed, slightly-curved spines (fig. 75, A), which would appear

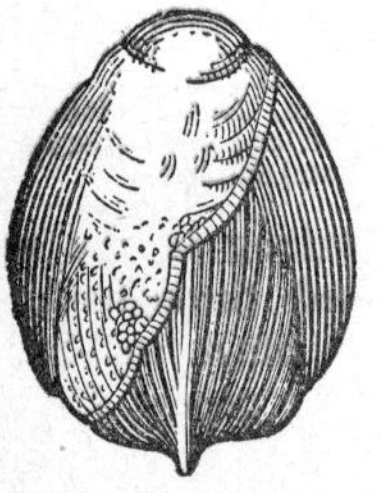

Fig. 74.—Head-shield of *Pteraspis Banksii,* **Ludlow rocks. (After Murchison.)**

Fig. 75.—A, Spine of *Onchus tenuistriatus* **; B, Shagreen-scales of** *Thelodus.* **Both from the "bone-bed" of the Upper Ludlow rocks. (After Murchison.)**

to be of the nature of the strong defensive spines implanted in front of certain of the fins in many living fishes. Besides these, have been found fragments of prickly skin or shagreen (*Sphagodus*), along with minute cushion-shaped bodies (*Thelo-*

dus, fig. 75, B), which are doubtless the bony scales of some fish resembling the modern Dog-fishes. As the above mentioned remains belong to two distinct, and at the same time highly-organized, groups of the fishes, it is hardly likely that we are really presented here with the first examples of this great class. On the contrary, whether the so-called "Conodonts" should prove to be the teeth of fishes or not, we are justified in expecting that unequivocal remains of this group of animals will still be found in the Lower Silurian. It is interesting, also, to note that the first appearance of fishes—the lowest class of vertebrate animals—so far as known to us at present, does not take place until after all the great sub-kingdoms of invertebrates have been long in existence; and there is no reason for thinking that future discoveries will materially affect the *relative* order of succession thus indicated.

LITERATURE.

From the vast and daily-increasing mass of Silurian literature, it is impossible to do more than select a small number of works which have a classical and historical interest to the English-speaking geologists, or which embody researches on special groups of Silurian animals—anything like an enumeration of all the works and papers on this subject being wholly out of the question. Apart, therefore, from numerous and in many cases extremely important memoirs, by various well-known observers, both at home and abroad, the following are some of the more weighty works to which the student may refer in investigating the physical characters and succession of the Silurian strata and their fossil contents:—

(1) 'Siluria.' Sir Roderick Murchison.
(2) 'Geology of Russia and Europe.' Murchison (with M. de Verneuil and Count von Keyserling).
(3) 'Bassin Siluren de Bohême Centrale.' Barrande.
(4) 'Introduction to the Catalogue of British Palæozoic Fossils in the Woodwardian Museum of Cambridge.' Sedgwick.
(5) 'Die Urwelt Russlands.' Eichwald.
(6) 'Report on the Geology of Londonderry, Tyrone,' &c. Portlock.
(7) "Geology of North Wales"—'Mem. Geol. Survey of Great Britain,' vol. iii. Ramsay.
(8) 'Geology of Canada.' 1863, Sir W. E. Logan; and the 'Reports of Progress of the Geological Survey' since 1863.
(9) 'Memoirs of the Geological Survey of Great Britain.'
(10) Reports of the Geological Surveys of the State of New York, Illinois, Ohio, Iowa, Michigan, Vermont, Wisconsin, Minnesota,' &c. By Emmons, Hall, Worthen, Meek, Newberry, Orton, Winchell, Dale Owen, &c.
(11) 'Thesaurus Siluricus.' Bigsby.

(12) 'British Palæozoic Fossils.' M'Coy.
(13) 'Synopsis of the Silurian Fossils of Ireland.' M'Coy
(14) "Appendix to the Geology of North Wales"—'Mem Geol. Survey,' vol. iii. Salter.
(15) 'Catalogue of the Cambrian and Silurian Fossils in the Woodwardian Museum of Cambridge.' Salter.
(16) 'Characteristic British Fossils.' Baily.
(17) 'Catalogue of British Fossils.' Morris.
(18) 'Palæozoic Fossils of Canada.' Billings.
(19) 'Decades of the Geological Survey of Canada.' Billings, Salter, Rupert Jones.
(20) 'Decades of the Geological Survey of Great Britain.' Salter, Edward Forbes.
(21) 'Palæontology of New York,' vols i.-iii. Hall.
(22) 'Palæontology of Illinois.' Meek and Worthen.
(23) 'Palæontology of Ohio.' Meek, Hall, Whitfield, Nicholson.
(24) 'Silurian Fauna of West Tennessee' (Silurische Fauna des Westlichen Tennessee). Ferdinand Rœmer.
(25) 'Reports on the State Cabinet of New York.' Hall.
(26) 'Lethæa Geognostica.' Bronn.
(27) 'Index Palæontologicus.' Bronn.
(28) 'Lethæa Rossica.' Eichwald.
(29) 'Lethæa Suecica.' Hisinger.
(30) 'Palæontologica Suecica.' Angelin.
(31) 'Petrefacta Germaniæ.' Goldfuss.
(32) 'Versteinerungen der Grauwacken-Formation in Sachsen.' Geinitz.
(33) 'Organization of Trilobites' (Ray Society). Burmeister.
(34) 'Monograph of the British Trilobites' (Palæontographical Society). Salter.
(35) 'Monograph of the British Merostomata' (Palæontographical Society). Henry Woodward.
(36) 'Monograph of British Brachiopoda' (Palæontolographical Society). Thomas Davidson.
(37) 'Graptolites of the Quebec Group.' James Hall.
(38) 'Monograph of the British Graptolitidæ.' Nicholson.
(39) 'Monographs on the Trilobites, Pteropods, Cephalopods, Graptolites,' &c. Extracted from the 'Système Silurien du Centre de la Bohême.' Barrande.
(40) 'Polypiers Fossiles des Terrains Paleozoiques.' and 'Monograph of the British Corals' (Palæontographical Society). Milne Edwards and Jules Haime.

CHAPTER XI.

THE DEVONIAN AND OLD RED SANDSTONE PERIOD.

Between the summit of the Ludlow formation and the strata which are universally admitted to belong to the Carboniferous series is a great system of deposits, to which the name of "Old

Red Sandstone" was originally applied, to distinguish them from certain arenaceous strata which lie above the coal ("New Red Sandstone"). The Old Red Sandstone, properly so-called, was originally described and investigated as occurring in Scotland and in South Wales and its borders; and similar strata occur in the south of Ireland. Subsequently it was discovered that Sediments of a different mineral nature, and containing different organic remains, intervened between the Silurian and the Carboniferous rocks on the continent of Europe, and strata with similar palæontological characters to these were found occupying a considerable area in Devonshire. The name of "Devonian" was applied to these deposits; and this title, by common usage, has come to be regarded as synonymous with the name of "Old Red Sandstone." Lastly, a magnificent series of deposits, containing marine fossils, and undoubtedly equivalent to the true "Devonian" of Devonshire, Rhenish Prussia, Belgium, and France, is found to intervene in North America between the summit of the Silurian and the base of the Carboniferous rocks.

Much difficulty has been felt in correlating the true "Devonian Rocks" with the typical "Old Red Sandstone"— this difficulty arising from the fact that though both formations are fossiliferous, the peculiar fossils of each have only been rarely and partially found associated together. The characteristic crustaceans and many of the characteristic fishes of the Old Red are wanting in the Devonian; whilst the corals and marine shells of the latter do not occur in the former. It is impossible here to enter into any discussion as to the merits of the controversy to which this difficulty has given origin. No one, however, can doubt the importance and reality of the Devonian series as an independent system of rocks to be intercalated in point of time between the Silurian and the Carboniferous. The want of agreement, both lithologically and palæontologically, between the Devonian and the Old Red, can be explained by supposing that these two formations, though wholly or in great part *contemporaneous,* and therefore strict equivalents, represent deposits in two different geographical areas, laid down under different conditions. On this view, the typical Devonian rocks of Europe, Britain, and North America are the deep-sea deposits of the Devonian period, or, at any rate, are genuine marine sediments formed far from land. On the other hand, the "Old Red Sandstone" of Britain and the corresponding "Gaspé Group" of Eastern

Canada represent the shallow-water shore-deposits of the same period. In fact, the former of these last-mentioned deposits contains no fossils which can be asserted positively to be *marine* (unless the Eurypterids be considered so); and it is even conceivable that it represents the sediments of an inland sea. Accepting this explanation in the meanwhile, we may very briefly consider the general succession of the deposits of this period in Scotland, in Devonshire, and in North America.

In Scotland the "Old Red" forms a great series of arenaceous and conglomeratic strata, attaining a thickness of many thousands of feet, and divisible into three groups. Of these, the *Lower Old Red Sandstone* repose with perfect conformity upon the highest beds of the Upper Silurian, the two formations being almost inseparably united by an intermediate series of "passage-beds." In mineral nature this group consists principally of massive conglomerates, sandstones, shales, and concretionary limestones; and its fossils consist chiefly of large crustaceans belonging to the family of the *Eurypterids*, fishes, and plants. The *Middle Old Red Sandstone* consists of flagstones, bituminous shales, and conglomerates, sometimes with irregular calcareous bands; and its fossils are principally fishes and plants. It may be wholly wanting, when the *Upper Old Red* seems to repose unconformably upon the lower division of the series. The *Upper Old Red Sandstone* consists of conglomerates and grits, along with a great series of red and yellow sandstones—the fossils, as before, being fishes and remains of plants. The Upper Old Red graduates upwards conformably into the Carboniferous series.

The Devonian rocks of Devonshire are likewise divisible into a lower, middle, and upper division. The *Lower Devonian* or *Lynton Group* consists of red and purple sandstones, with marine fossils, corresponding to the "Spirifer Sandstein" of Germany, and to the arenaceous deposits (Schoharie and Cauda-Galli Grits) at the base of the American Devonian. The *Middle Devonian* or *Ilfracombe Group* consists of sandstones and flags, with calcareous slates and crystalline limestones, containing many corals. It corresponds with the great "Eifel Limestone" of the Continent, and, in a general way, with the Corniferous Limestone and Hamilton Group of North America. The *Upper Devonian* or *Pilton Group*, lastly, consists of sandstones and calcareous shales which correspond with the "Clymenia Limestone" and Cyprindina Shales" of the Continent, and with the Chemung and Portage groups of North America. It seems quite possible, also, that the so-called "Carboniferous Slates" of Ireland correspond with this group, and that the former would be more properly regarded as forming the summit of the Devonian than the base of the Carboniferous.

In no country in the world, probably, is there a finer or more complete exposition of the strata intervening between the Silurian and Carboniferous deposits than in the United States. The following are the main subdivisions of the Devonian rocks in the State of New York, where the series may be regarded as being typically developed (fig. 67):—

(1) *Cauda-Galli Grit* and *Schoharie Grit.*—Considering the "Oriskany Sandstone" as the summit of the Upper Silurian, the base of the Devonian is constituted by the arenaceous deposits known by the above names, which rest quite conformably upon the Silurian, and which represent the Lower Devonian of Devonshire. The *Cauda-Galli Grit* is so-called from the abundance of a peculiar spiral fossil (*Spirophyton cauda-Galli*), which is of common occurrence in the Carboniferous rocks of Britain, and is supposed to be the remains of a sea-weed.

(2) The *Corniferous* or *Upper Helderberg Limestone.*—A series of limestones usually charged with considerable quantities of siliceous matter in the shape of hornstone or chert (Lat. *cornu,* horn). The thickness of this group rarely exceeds 3000 feet; but it is replete with fossils, more especially with the remains of corals. The Corniferous Limestone is the equivalent of the coral-bearing limestones of the Middle Devonian of Devonshire and the great "Eifel Limestone" of Germany.

(3) The *Hamilton Group*—consisting of shales at the base ("Marcellus shales"); flags, shales, and impure limestones ("Hamilton beds") in the middle; and again a series of shales ("Genesee Slates") at the top. The thickness of this group varies from 200 to 1200 feet, and it is richly charged with marine fossils.

(4) The *Portage Group.*—A great series of shales, flags and shaly sandstones, with few fossils.

(5) The *Chemung Group.*—Another great series of sandstones and shales, but with many fossils. The Portage and Chemung groups may be regarded as corresponding with the

Upper Devonian of Devonshire. The Chemung beds are succeeded by a great series of red sandstones and shales—the "*Catskill Group*"—which pass conformably upwards into the Carboniferous, and which may perhaps be regarded as the equivalent of the great sandstones of the Upper Old Red in Scotland.

Throughout the entire series of Devonian deposits in North America no unconformability or physical break of any kind has hitherto been detected; nor is there any marked interruption to the current of life, though each subdivision of the series has its own fossils. No completely natural line can thus be indicated, dividing the Devonian in this region from the Silurian on the one hand, and the Carboniferous on the other hand. At the same time, there is the most ample evidence, both stratigraphical and palæontological, as to the complete independence of the American Devonian series as a distinct life-system between the older Silurian and the later Carboniferous. The subjoined section (fig. 76) shows diagrammatically the general succession of the Devonian rocks of North America.

As regards the *life* of the Devonian period, we are now acquainted with a large and abundant terrestrial *flora*—this being the first time that we have met with a land vegetation capable of reconstruction in any fulness. By the researches of Gœppert, Unger, Dawson, Carruthers, and other botanists, a knowledge has been acquired of a large number of Devonian plants, only a few of which can be noticed here. As might have been anticipated, the greater number of the vegetable remains of this period have been obtained from such shallow-water deposits as the Old Red Sandstone proper and the Gaspè series of North America, and few traces of plant-life occur in the strictly marine sediments. Apart from numerous remains, mostly of a problematical nature, referred to the comprehensive group of the Sea-weeds, a large number of Ferns have now been recognized, some being of the ordinary plant-like type (*Pecopteris, Neuropteris, Alethopteris, Sphenopteris,* &c.), whilst others belong to the gigantic group of the "Tree-ferns" (*Psaronius, Caulopteris,* &c.) Besides these there is an abundant development of the singular extinct types of the *Lepidodendroids*, the *Sigillarioids*, and the *Calamites*, all of which attained their maximum in the Carboniferous. Of these, the *Lepidodendra* may be regarded as gigantic, tree-like Club-mosses

GENERALIZED SECTION OF THE DEVONIAN ROCKS OF NORTH AMERICA.

Fig. 76.

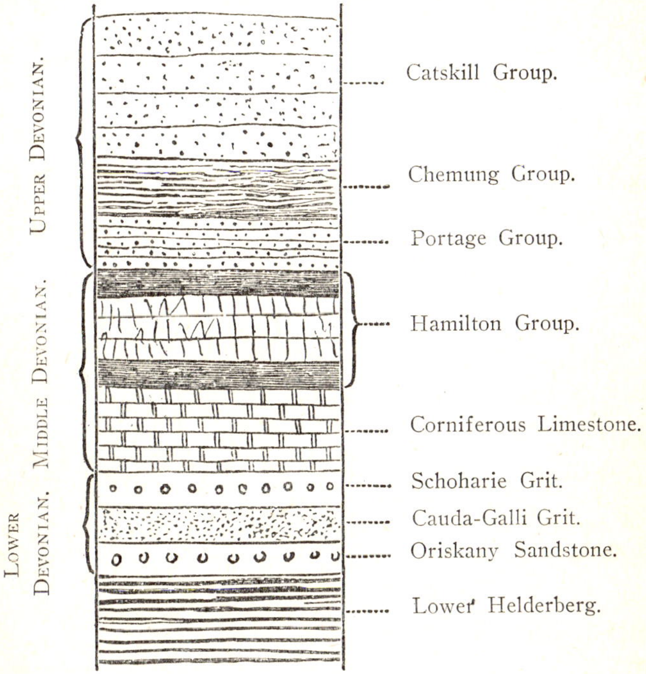

(*Lycopodiaceæ*); the *Calamites* are equally gigantic Horse-tails (*Equisetaceæ*); and the *Sigillarioids*, equally huge in size, in some respects hold a position intermediate between the Club-mosses and the Pines (*Conifers*). The Devonian rocks have also yielded traces of many other plants (such as *Annularia, Asterophyllites, Cardiocarpon*, &c.), which acquire a greater predominance in the Carboniferous period, and which will be spoken of in discussing the structure of the plants of the Coal-measures. Upon the whole, the one plant which may be considered as specially characteristic of the Devonian (though not confined to this series) is the *Psilophyton* (fig. 77) of Dr. Dawson. These singular plants have slender branching stems, with sparse needle-shaped leaves, the young stems being at first coiled up, crosier-fashion, like the young fronds of ferns, whilst the old branches carry numerous spore-cases. The stems and branches seem to have attained a height of two or three feet; and they sprang from prostrate "root-stocks" or creeping stems. Upon the whole, Principal Dawson is disposed to regard *Psilophyton* as a "generalized type" of plants intermediate between the Ferns and the Club-mosses. Lastly, the Devonian deposits have yielded the remains of the first actual *trees* with which we are as yet acquainted. About the nature of some of these (*Ormoxylon* and *Dadoxylon*) no doubt can be entertained, since their trunks not only show the concentric rings of growth characteristic of exogenous trees in general, but their woody tissue exhibits under the microscope the "discs" which are characteristic of the wood of the Pines and Firs (*see* fig. 2). The singular genus *Prototaxites,* however, which occurs in an older portion of the Devonian series than the above, is not in an absolutely unchallenged position. By Principal Dawson it is regarded as the trunk of an ancient *Conifer*—the most ancient known; but Mr. Carruthers regards it as more probably the stem of a gigantic sea-weed. The trunks of *Prototaxites* (fig. 78, A) vary from one to three feet in diameter, and exhibit concentric rings of growth; but its woody fibres have not hitherto been clearly demonstrated to possess discs. Before leaving the Devonian vegetation, it may be mentioned that the hornstone or chert so abundant in the Corniferous limestone of North America has been shown to contain the remains of various

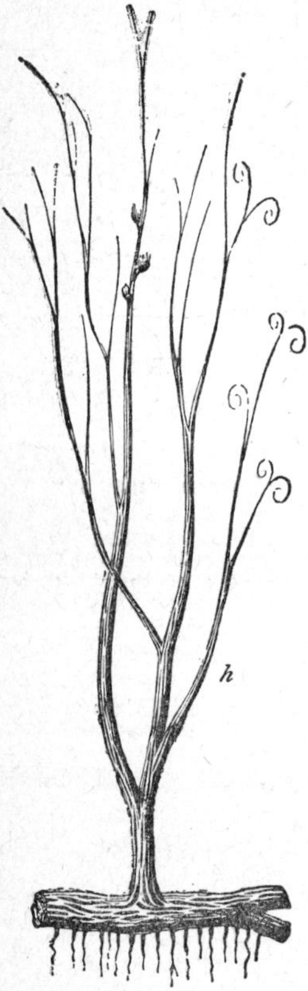

Fig. 77.—Restoration of *Psilophyton princeps* Devonian, Canada. (After Dawson.)

microscopic plants (*Diatoms and Desmids*). We find also in the same siliceous material the singular spherical bodies, with radiating spines, which occur so abundantly in the chalk flints, and which are termed *Xanthidia*. These may be regarded as probably the spore-cases of the minute plants known as *Desmidiæ*.

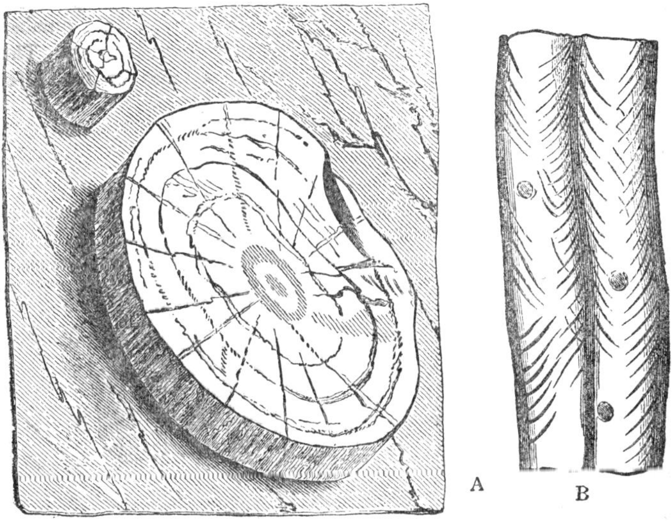

A B

Fig. 78.—A, Trunk of *Prototaxites Logani*, eighteen inches in diameter, as seen in the cliff near L'Anse Brehaut, Gaspé ; B, Two wood-cells showing spiral fibres and obscure pores, highly magnified. Lower Devonian, Canada. (After Dawson.)

The Devonian *Protozoans* have still to be fully investigated. True Sponges (such as *Astræospongia, Sphærospongia*, &c.) are not unknown; but by far the commonest representatives of this sub-kingdom in the Devonian strata are *Stromatopora* and its allies. These singular organisms (fig. 79) are not only very abundant in some of the Devonian limestones—both in the Old World and the New—but they often attain very large dimensions. However much they may differ in minor details, the general structure of these bodies is that of numerous, concentrically-arranged, thin, calcareous laminæ, separated by narrow interspaces, which in turn are crossed by numerous delicate vertical pillars, giving the whole mass a cellular structure, and dividing it into innumerable minute quadrangular compartments. Many of the Devonian *Stromato-*

poræ also exhibit on their surface the rounded openings of canals, which can hardly have served any other purpose than that of permitting the sea-water to gain ready access to every part of the organism.

No true *Graptolites* have ever been detected in strata of Devonian age; and the whole of this group has become extinguished—unless we refer here the still surviving *Dictyonemæ*. The *Cœlenterates,* however, are represented by a vast number of *Corals,* of beautiful forms and very varied types. The marbles of Devonshire, the Devonian limestones of the Eifel

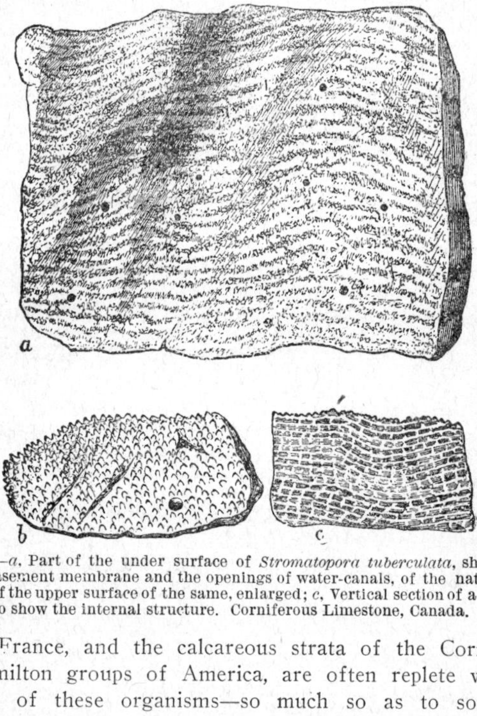

Fig 79.—*a*, Part of the under surface of *Stromatopora tuberculata*, showing the wrinkled basement membrane and the openings of water-canals, of the natural size; *b*, Portion of the upper surface of the same, enlarged; *c*, Vertical section of a fragment, magnified to show the internal structure. Corniferous Limestone, Canada. (Original.)

and of France, and the calcareous strata of the Corniferous and Hamilton groups of America, are often replete with the skeletons of these organisms—so much so as to sometimes entitle the rock to be considered as representing an ancient coral-reef. In some instances the Corals have preserved their primitive calcareous composition; and if they are embedded in soft shales, they may weather out of the rock in almost all their original perfection. In other cases, as in the marbles of Devonshire, the matrix is so compact and crystalline that the included corals can only be satisfactorily studied by means of polished sections. In other cases, again, the corals have been more or less completely converted into flint, as in the Cornifer-

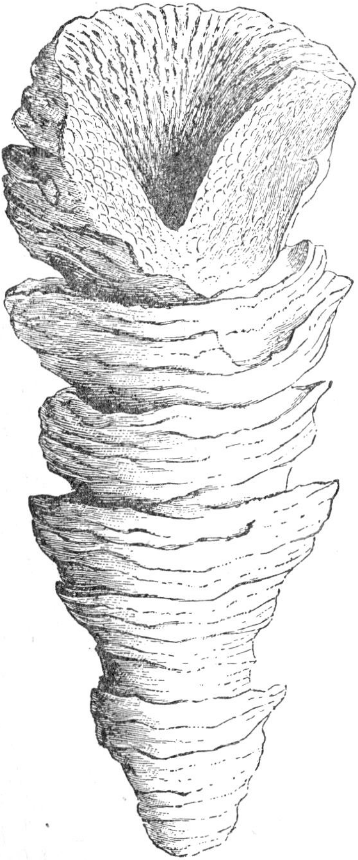

Fig. 80. — *Cystiphyllum vesiculosum*, showing a succession of cups produced by budding from the original coral. Of the natural size. Devonian, America and Europe. (Original.)

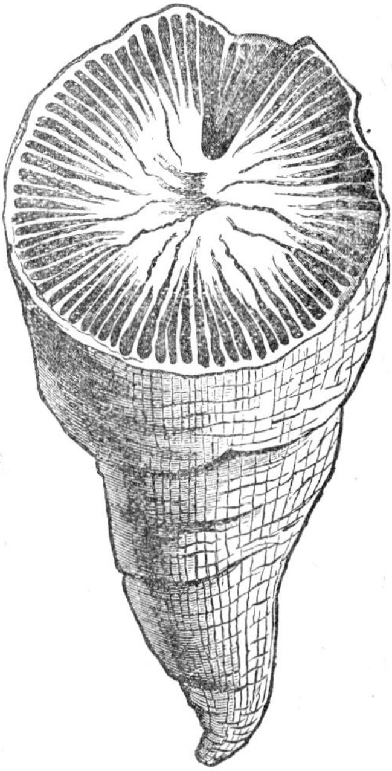

Fig. 81.—*Zaphrentis cornicula*, of the natural size. Devonian, America. (Original.)

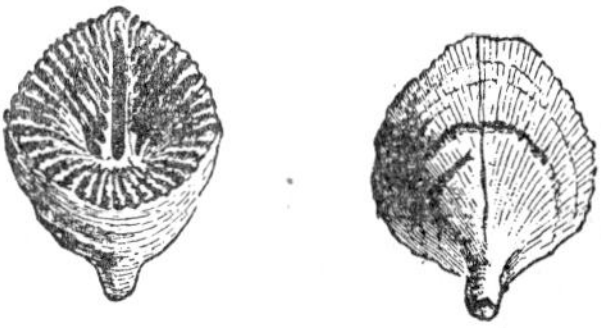

Fig. 82.—*Heliophyllum exiguum*, viewed from in front and behind. Of the natural size. Devonian, Canada. (Original.)

ous limestone of North America. When this is the case, they often come, by the action of the weather, to stand out from the enclosing rock in the boldest relief, exhibiting to the observer the most minute details of their organization. As before, the principal representatives of the Corals are still referable to the groups of the *Rugosa* and *Tabulata*. Amongst the Rugose group we find a vast number of simple "cup-corals," generally known by the quarrymen as "horns," from their shape. Of the many forms of these, the species of *Cyathophyllum, Heliophyllum* (fig. 82), *Zaphrentis* (fig. 81), and *Cystiphyllum* (fig. 80), are perhaps those most abundantly represented—none of these genera, however, except *Heliophyllum,* being peculiar to the

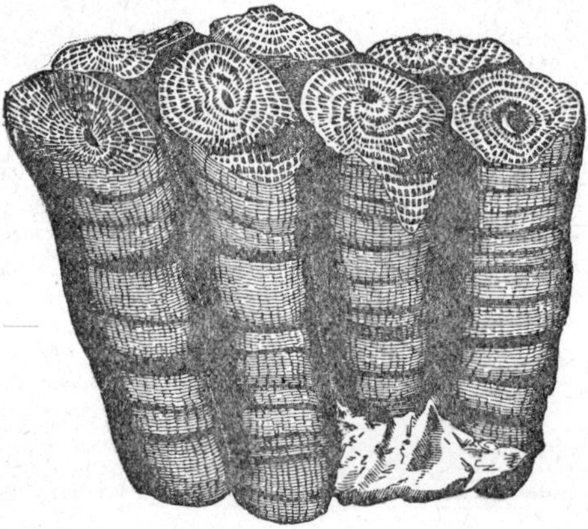

Fig. 83.—Portion of a mass of *Crepidophyllum Archiaci*, of the natural size. Hamilton Formation, Canada. (After Billings.)

Devonian period. There are also numerous compound Rugose corals, such as species of *Eridophyllum, Diphyphyllum, Syringopora, Phillipsastræa,* and some of the forms of *Cyathophyllum* and *Crepidophyllum* (fig. 83). Some of these compound corals attain a very large size, and form of themselves regular beds, which have an analogy, at any rate, with existing coral-reefs, though there are grounds for believing that these ancient types differed from the modern reef-builders in being inhabitants of deep water. The "Tabulate Corals" are hardly less abundant in the Devonian rocks than the *Rugosa;* and being invariably compound, they hardly yield to the latter in the dimensions of the aggregations which they sometimes form.

The commonest, and at the same time the largest, of these are the "honeycomb corals," forming the genus *Favosites* (figs. 84, 85), which derive both their vernacular and their technical names from their great likeness to masses of petrified honeycomb. The most abundant species are *Favosites Gothlandica* and *F. hemispherica,* both here figured, which form masses sometimes not less than two or three feet in diameter. Whilst *Favosites* has acquired a popular name by its honeycombed appearance, the resemblance of *Michelinia* to a fossilized wasp's nest with the comb exposed is hardly less striking, and has earned for it a similar recognition from the

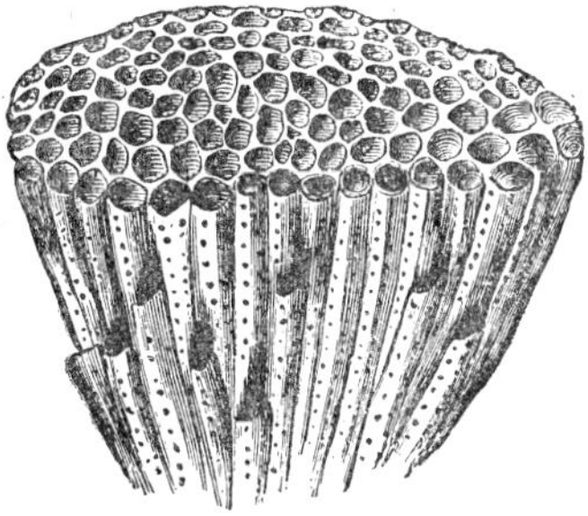

Fig. 84.—Portion of a mass of *Favosites Gothlandica*, of the natural size. Upper Silurian and Devonian of Europe and America. (Original.)

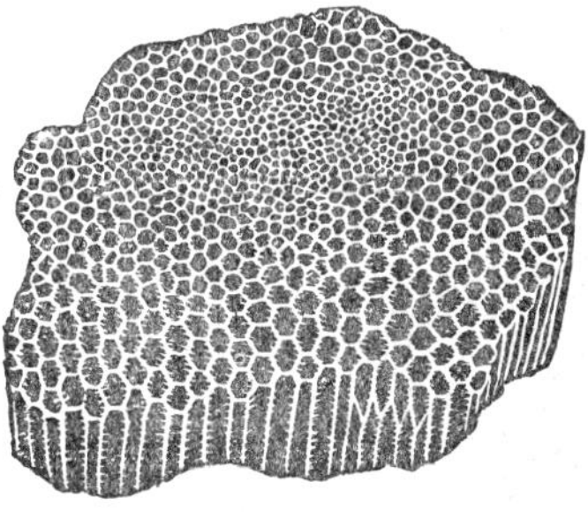

Fig. 85. Fragment of *Favosites hemispherica*, of the natural size. Upper Silurian and Devonian of America. (After Billings.)

non-scientific public. In addition to these, there are numerous branching or plant-like Tabulate Corals, often of the most graceful form, which are distinctive of the Devonian in all parts of the world.

The *Echinoderms* of the Devonian period call for little special notice. Many of the Devonian limestones are "crinoidal;" and the *Crinoids* are the most abundant and widely-distributed representatives of their class in the deposits of this period.

The *Cystideans,* with doubtful exceptions, have not been recognized in the Devonian; and their place is taken by the allied group of the "Pentremites," which will be further spoken of as occurring in the Carboniferous rocks. On the other hand, the Star-fishes, Brittle-fishes, and Sea-urchins are all continued by types more or less closely allied to those of the preceding Upper Silurian.

Of the remains of Ringed-worms (*Annelides*), the most numerous and the most interesting are the calcareous envelopes of some small tube-inhabiting species. No one who has visited the seaside can have failed to notice the little spiral tubes of the existing *Spirorbis* growing attached to shells, or covering the fronds of the commoner Sea-weeds (especially *Fucus serratus*). These tubes are inhabited by a small Annelide, and structures of a similar character occur not uncommonly from the Upper Silurian upwards. In the Devonian rocks, *Spirorbis* is an extremely common fossil, growing in hundreds attached to the outer surface of corals and shells, and appearing in many specific forms (figs. 86 and 87); but almost all the known examples are of small size, and are liable to escape a cursory examination.

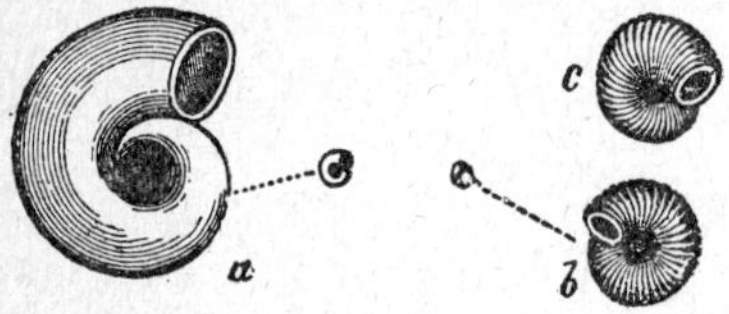

Fig. 86.—*a*, *Spirorbis omphalodes* natural size and enlarged, Devonian, Europe and America; *b*, *Spirorbis Arkonensis*, of the natural size and enlarged; *c*, The same, with the tube twisted in the reverse direction. Devonian, America. (Original.)

Fig. 87.—*a b*, *Spirorbis laxus*, enlarged, Upper Silurian, America; *c*, *Spirorbis spinulifera*, of the natural size and enlarged, Devonian, Canada. (After Hall and the Author.)

The *Crustaceans* of the Devonian are principally *Eurypterids* and *Trilobites*. Some of the former attain gigantic dimensions, and the quarrymen in the Scotch Old Red give them the name of "seraphim," from their singular scale-like ornamentation. The *Trilobites*, though still sufficiently abundant in some localities, have undergone a yet further diminution since the close of the Upper Silurian. In both America and Europe quite a number of generic types have survived from the Silurian, but few or no new ones make their appearance during this period in either the Old World or the New. The *species*, however, are distinct; and the principal forms belong to the genera *Phacops* (fig. 88, *a, c, d*), *Homalonotus* (fig. 88, *b*), *Proetus*, and *Bronteus*. The species figured opposite under the name of *Phacops latifrons* (fig. 88, *a*), has an almost world-wide distribution, being found in the Devonian of Britain, Belgium, France, Germany, Russia, Spain, and South America; whilst its place is taken in North America by the closely-allied *Phacops rana*. In addition to the *Trilobites*, the Devonian deposits have yielded the remains of a number of the minute *Ostracoda*, such as *Entomis* ("*Cypridina*"), *Leperditia*, &c., which sometimes occur in vast numbers, as in the so-called "*Cypridina* Slates" of the German

Devonian. There are also a few forms of *Phyllopods* (*Estheria*). Taken as a whole, the Crustacean fauna of the Devonian period presents many alliances with that of the

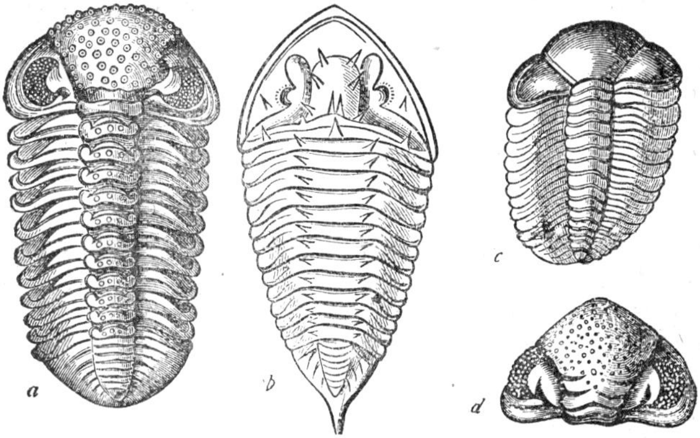

Fig. 88.—Devonian Trilobites. *a*, *Phacops latifrons*, Devonian of Britain, the Continent of Europe, and South America ; *b*, *Homalonotus armatus*, Europe ; *c*, *Phacops* (*Trimerocephalus*) *lævis*, Europe; *d*, Head-shield of *Phacops* (*Portlockia*) *granulatus*, Europe. (After Salter and Burmeister.)

Upper Silurian, but has only slight relationships with that of the Lower Carboniferous.

Besides *Crustaceans*, we meet here for the first time with the remains of *air-breathing Articulates*, in the shape of *Insects*. So far, these have only been obtained from the Devonian rocks of North America, and they indicate the existence of at least four generic types, all more or less allied to the existing May-flies (*Ephemeridæ*). One of these interesting primitive insects, namely, *Platephemera antiqua* (fig. 89), appears to have measured five inches in expanse of wing; and another (*Xenoneura antiquorum*) has attached to its wing the remains of a " stridulating-organ " similar to that possessed by the modern Grasshoppers—the instrument, as Principal Dawson remarks, of " the first music of living things that Geology as yet reveals to us."

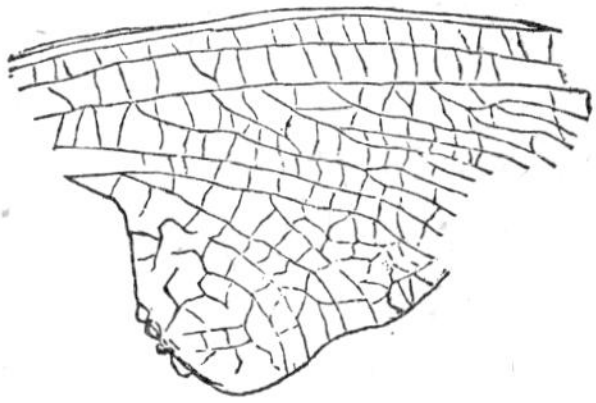

Fig. 89.—Wing of *Platephemera antiqua*, Devonian, America. (After Dawson.)

Amongst the *Mollusca*, the Devonian rocks have yielded a

great number of the remains of Sea-mosses (*Polyzoa*). Some of these belong to the ancient type *Ptilodictya,* which seems to

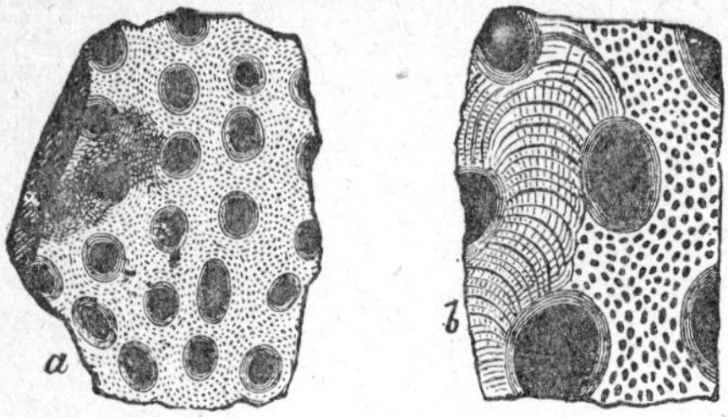

Fig. 90.—Fragment of *Clathopora intertexta,* of the natural size and enlarged. Devonian, Canada. (Original.)

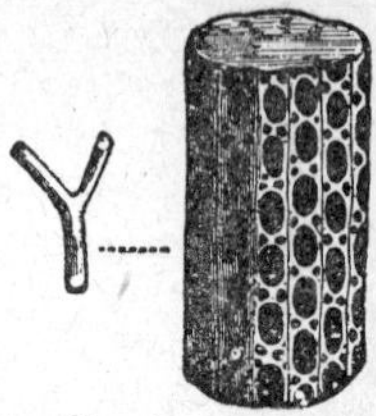

Fig. 91.—Fragment of *Ceriopora Hamiltonensis,* of the natural size and enlarged. Devonian, Canada. (Original.)

disappear here, or to the allied *Clathropora* (fig. 90), with its fenestrated and reticulated fronds. We meet also with the graceful and delicate stems of *Ceriopora* (fig. 91).

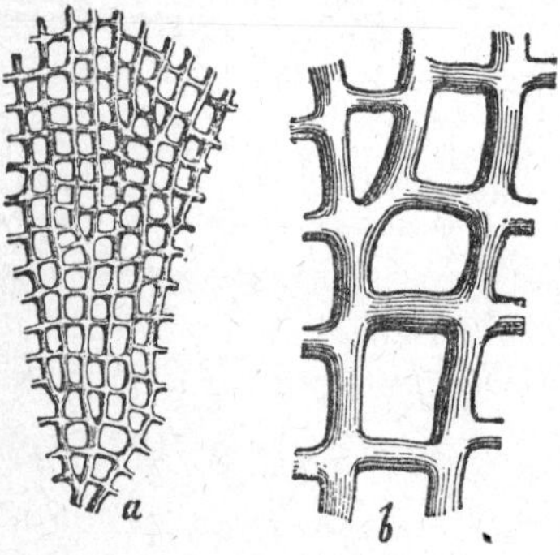

Fig. 92.—Fragment of *Fenestella magnifica,* of the natural size and enlarged. Devonian, Canada. (Original.)

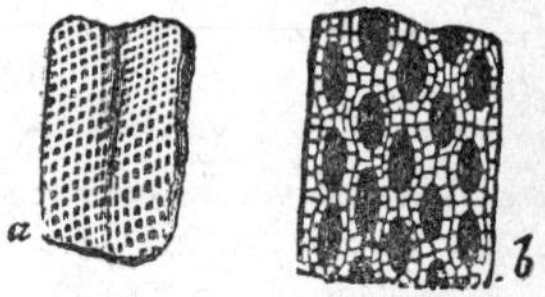

Fig. 93.—Fragment of *Retepora Phillipsi,* of the natural size and enlarged. Devonian, Canada. (Original.)

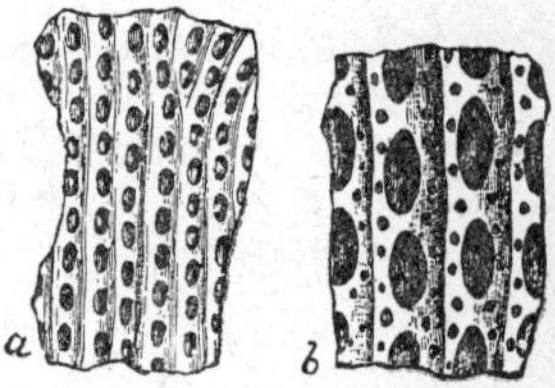

Fig. 94.—Fragment of *Fenestella cribrosa,* of the natural size and enlarged. Devonian, Canada. (Original.)

The majority of the Devonian *Polyzoa* belong, however, to the great and important Palæozoic group of the Lace-corals (*Fenestella,* figs. 92 and 94, *Retepora,* fig. 93, *Polypora,* and their allies). In all these forms there is a horny skeleton, of a fan-like or funnel-shaped form, which grew attached by its base to some foreign body. The frond consists of slightly-diverging or nearly parallel branches, which are either united by delicate cross-bars, or which bend alternately from side to side, and become directly united with one another at short intervals—in either case giving origin to numerous oval or oblong perforations, which communicate to the whole plant-like colony a characteristic netted and lace-like appearance. On one of its surfaces—sometimes the internal, sometimes the external—the frond carries a number of minute chambers or "cells," which are generally borne in rows on the branches, and of which each originally contained a minute animal.

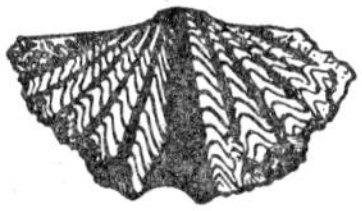

Fig. 95. — *Spirifera sculptilis*. Devonian, Canada. (After Billings.)

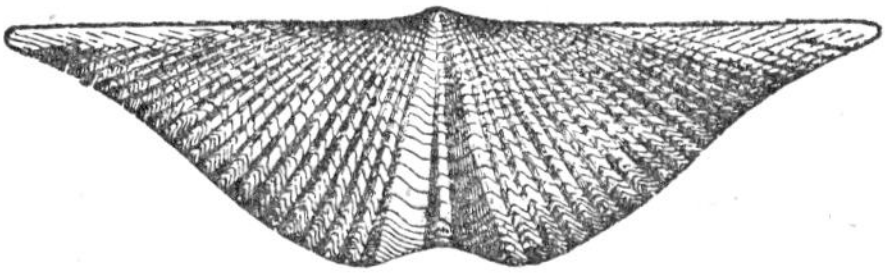

Fig. 96.—*Spirifera mucronata*. Devonian, America. (After Billings.)

The *Brachiopods* still continue to be represented in great force through all the Devonian deposits, though not occurring in the true Old Red Sandstone. Besides such old types as *Orthis, Strophomena, Lingula, Athyris,* and *Rhynchonella,* we find some entirely new ones; whilst various types which only commenced their existence in the Upper Silurian, now undergo a great expansion and development. This last is especially the case with the two families of the *Spiriferidæ* and the *Productidæ*. The *Spirifers,* in particular, are especially characteristic of the Devonian, both in the Old and New Worlds—some of the most typical forms, such as *Spirifera mucronata* (fig. 96), having the shell "winged," or with the lateral angles prolonged to such an extent as to have earned for them the popular name of "fossil-butterflies." The closely-allied *Spirifera disjuncta* occurs in Britain, France, Spain, Belgium, Germany, Russia, and China. The family of the *Productidæ* commenced to exist in the Upper Silurian, in the genus *Chonetes;* and we shall hereafter find it culminating in the Carboniferous in many forms of the great genus *Producta** itself. In the Devonian period, there is an intermediate state of things, the genus *Chonetes* being continued in new and varied types, and the Carboniferous *Productæ* being represented by many forms of the allied group *Productella*. Amongst other well-known Devonian Brachiopods may be mentioned the two long-lived and persistent types *Atrypa reticularis* (fig. 97) and *Strophomena rhomboidalis* (fig. 98). The former of these commences in the Upper Silurian, but is more abundantly developed in the Devonian, having a geographical range that is nothing less than

Fig. 97.—*Atrypa reticularis*. Upper Silurian and Devonian of Europe and America. (After Billings.)

world-wide; whilst the latter commences in the Lower Silurian, and, with an almost equally cosmopolitan range, survives into the Carboniferous period.

The Bivalves (*Lamellibranchiata*) of the Devonian call for

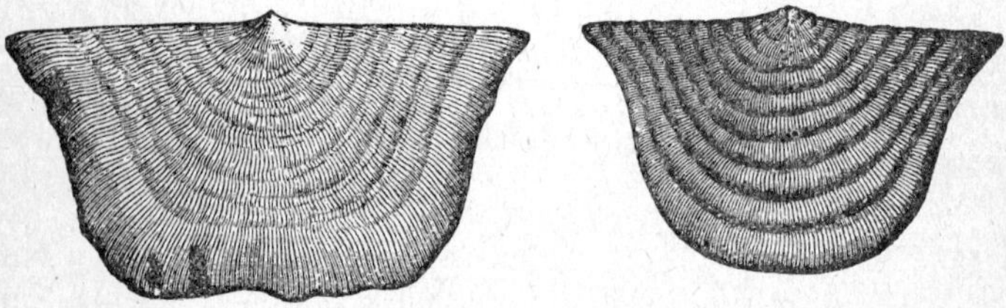

Fig. 98.—*Strophomena rhomboidalis*. Lower Silurian, Upper Silurian, and Devonian of Europe and America.

no special comment, the genera *Pterinea* and *Megalodon* being, perhaps, the most noticeable. The Univalves (*Gasteropods*), also, need not be discussed in detail, though many interesting

* The name of this genus is often written *Productus*, just as *Spirifera* is often given in the masculine gender as *Spirifer* (the name originally given to it). The masculine termination to these names is, however, grammatically incorrect, as the feminine noun *cochlea* (shell) is in these cases *understood*.

forms of this group are known. The type most abundantly represented, especially in America, is *Platyceras* (fig. 99), comprising thin, wide-mouthed shells, probably most nearly allied to the existing "Bonnet-limpets," and sometimes attaining very considerable dimensions. We may also note the continuance of the genus *Euomphalus,* with its discoidal spiral shell. Amongst the *Heteropods,* the survival of *Bellerophon* is to be recorded; and in the "Winged-snails," or *Pteropods,* we find new forms of the old genera *Tentaculites* and *Conularia* (fig. 100). The latter, with its fragile, conical, and often beautifully ornamented shell, is especially noticeable.

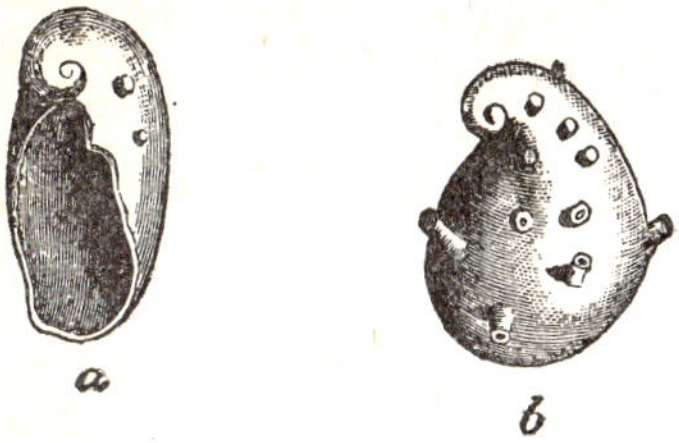

Fig. 99.—Different views of *Platyceras dumosum,* of the natural size. Devonian, Canada. (Original.)

The remains of *Cephalopoda* are far from uncommon in the Devonian deposits, all the known forms being still Tetrabranchiate. Besides the ancient types *Orthoceras* and *Cyrtoceras,* we have now a predominance of the spirally-coiled chambered shells of *Goniatites* and *Clymenia.* In the former of these the shell is shaped like that of the *Nautilus;* but the partitions between the chambers ("septa") are more or less lobed, folded, or angulated, and the "siphuncle" runs along the *back* or convex side of the shell—these being characters which approximate *Goniatites* to the true Ammonites of the later rocks. In *Clymenia,* on the other hand, whilst the shell (fig. 101) is coiled into a flat spiral, and the partitions or septa are simple or only slightly lobed, there is still this difference, as compared with the *Nautilus,* that the tube of the siphuncle is placed on the *inner* or concave side of the shell. The species of *Clymenia* are exclusively Devonian in their range; and some of the limestones of this period in Germany are so richly charged with fossils of this genus as to have received the name of "Clymenien-kalk."

Fig. 100.—*Conularia ornata,* of the natural size. Devonian, Europe.

The sub-kingdom of the *Vertebrates* is still represented by *Fishes* only; but these are so abundant, and belong to such varied types, that the Devonian period has been appropriately called the "Age of Fishes." Amongst the existing fishes there are three great groups which are of special geological importance, as being more or less extensively represented in past time. These groups are: (1) The *Bony Fishes* (*Teleostei*), comprising most existing fishes, in which the skeleton is more or less completely converted into bone; the tail is symmetrically lobed or divided into equal moieties; and the scales are usually thin, horny, flexible plates, which overlap one another to a greater

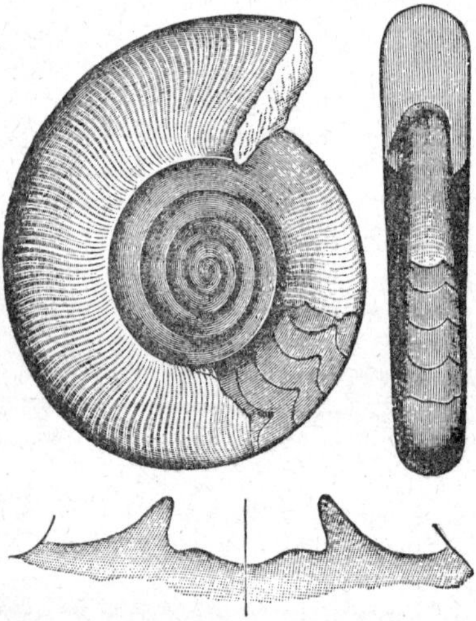

Fig. 101.—*Clymenia Sedgwickii.* Devonian, Europe.

or less extent. (2) The *Ganoid Fishes* (*Ganoidei*), comprising the modern Gar-pikes, Sturgeons, &c., in which the skeleton usually more or less completely retains its primitive soft and cartilaginous condition; the tail is generally markedly unsymmetrical, being divided into two unequal lobes; and the scales (when present) have the form of plates of bone, usually covered by a layer of shining enamel. These scales may overlap; or they may be rhomboidal plates, placed edge to edge in oblique rows; or they have the form of large-sized bony plates, which are commonly united in the region of the head to form a regular buckler. (3) The *Placoid Fishes*, or *Elasmobranchii*, comprising the Sharks, Rays, and *Chimæræ* of the present day, in which the skeleton is cartilaginous; the tail is unsymmetrically lobed; and the scales have the form of detached bony plates of variable size, scattered in the integument.

It is to the two last of these groups that the Devonian fishes belong, and they are more specially referable to the *Ganoids*. The order of the Ganoid fishes at the present day comprises by some seven or eight genera, the species of which principally or exclusively inhabit fresh waters, and all of which are confined to the northern hemisphere. As compared, therefore, with the Bony fishes, which constitute the great majority of existing forms, the Ganoids form but an extremely small and limited group. It was far otherwise, however, in Devonian times. At this period, the bony fishes are not known to have come into existence at all, and the Ganoids held almost undisputed possession of the waters. To what extent the Devonian Ganoids were confined to fresh waters remains yet to be proved; and that many of them lived in the sea is certain. It was formerly supposed that the Old Red Sandstone of Scotland and Ireland, with its abundant fish-remains, might perhaps be a fresh-water deposit, since the habitat of its fishes is uncertain, and it contains no indubitable marine fossils. It has been now shown, however, that the marine Devonian strata of Devonshire and the continent of Europe contain some of the most characteristic of the Old Red Sandstone fishes of Scotland; whilst the undoubted marine deposit of the Corniferous limestone of North America contains numerous shark-like and Ganoid fishes, including such a characteristic Old Red genus as *Coccosteus*. There can be little doubt, therefore, but that the majority of the Devonian fishes were truly marine in their habits, though it is probable that many of them lived in shallow water, in the immediate neighborhood of the shore, or in estuaries.

The Devonian Ganoids belong to a number of groups; and it is only possible to notice a few of the most important forms here. The modern group of the Sturgeons is represented, more or less remotely, by a few Devonian fishes—such as *Asterosteus;* and the great *Macropetalichthys* of the Corniferous limestone of North America is believed by Newberry to belong to this group. In this fish (fig. 102, *b*) the skull was of large size, its outer surface being covered with a tuberculated enamel; and, as in the existing Sturgeons, the mouth seems to have been wholly destitute of teeth. Somewhat allied, also, to the Sturgeons, is a singular group of armored fishes, which is highly characteristic of the Devonian of Britain and Europe, and less so of that of America. In these curious forms the head and front extremity of the body were protected by a buckler composed of large enameled plates, more or less firmly united to one another; whilst the hinder end of the body was naked, or was protected with small scales. Some forms of

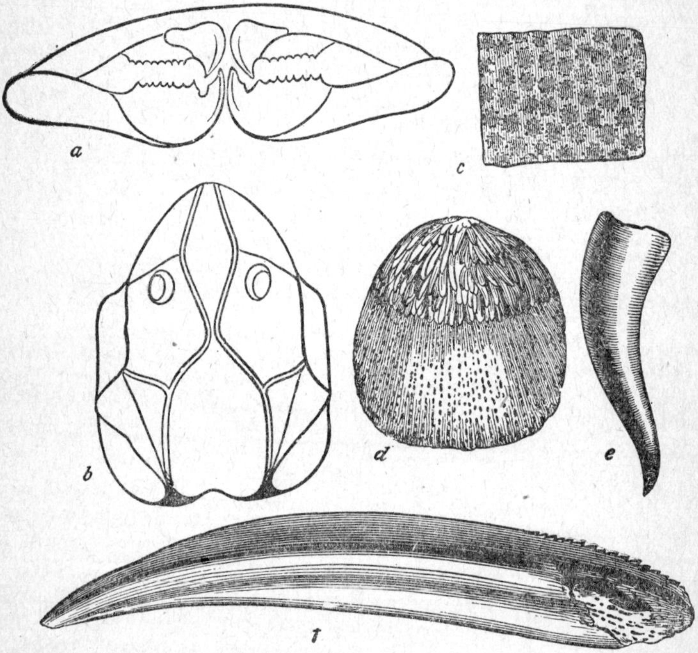

Fig. 102.—Fishes of the Devonian rocks of America. *a*, Diagram of the jaws and teeth of *Dinichthys Hertzeri*, viewed from the front, and greatly reduced; *b*, Diagram of the skull of *Macropetalichthys Sullivanti*, reduced in size; *c*, A portion of the enamelled surface of the skull of the same, magnified; *d*, One of the scales of *Onychodus sigmoides*, of the natural size; *e*, One of the front teeth of the lower jaw of the same, of the natural size; *f*, Fin-spine of *Machæracanthus major*, a shark-like fish reduced in size. (After Newberry.)

this group—such as *Pteraspis* and *Coccosteus*—date from the Upper Silurian; but they attain their maximum in the Devonian, and none of them are known to pass upwards into the overlying Carboniferous rocks. Amongst the most characteristic forms of this group may be mentioned *Cephalaspis* (fig. 103) and *Pterichthys* (fig. 104). In the former of these the head-shield is of a crescentic shape, having its hinder angles produced backwards into long "horns," giving it the shape of a "saddler's knife." No teeth have been discovered; but the body was covered with small ganoid scales, and there was an unsymmetrical tail-fin. In *Pterichthys*—which, like the preceding, was first brought to light by the labors of Hugh Miller—the whole of the head and the front part of the body were defended by a buckler of firmly-united enamelled plates, whilst the rest of the body was covered with small scales. The form of the "pectoral fins" was quite unique—these having the shape of two long, curved spines, somewhat like wings, covered by finely-tuberculated ganoid plates. All the preceding forms of this group are of small size; but few fishes, living or extinct, could rival the proportions of the great *Dinichthys,* referred to

Fig. 103.—*Cephalaspis Lyellii.* Old Red Sandstone, Scotland. (After Page.)

this family by Newberry. In this huge fish (fig. 102, *a*) the head alone is over three feet in length, and the body is supposed to have been twenty-five or thirty feet long. The head was protected by a massive cuirass of bony plates firmly articulated together, but the hinder end of the body seems to have been simply enveloped in a leathery skin. The teeth are of the most formidable description, consisting in both jaws of serrated dental plates behind, and in front of enormous conical tusks (fig. 102, *a*). Though immensely larger, the teeth of *Dinichthys* present a curious resemblance to those of the existing Mud-fishes (*Lepidosiren*).

In another great group of Devonian Ganoids, we meet with fishes more or less closely allied to the living *Polypteri* (fig. 105) of the Nile and Senegal. In this group (fig. 106) the pectoral fins consist of a central scaly lobe carrying the fin-rays on both sides, the scales being sometimes rounded and overlapping (fig. 106), or more commonly rhomboidal and placed edge to edge (fig. 105, A). Numerous forms of these

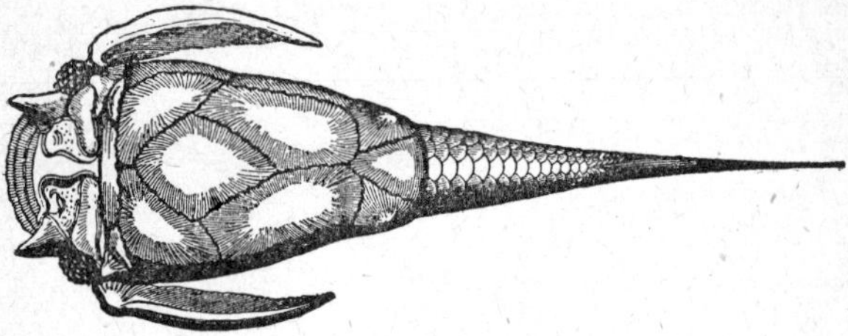

Fig. 104.—*Pterichthys cornutus.* Old Red Sandstone, Scotland. (After Agassiz.)

"Fringe-finned" Ganoids occur in the Devonian strata, such as *Holoptychius, Glyptolæmus, Osteolepis, Phaneropleuron,* &c. To this group is also to be ascribed the huge *Onychodus* (fig. 102, *d* and *e*), with its large, rounded, overlapping scales, an inch in diameter, and its powerful pointed teeth. It is to be remembered, however, that some of these "Fringe-finned" Ganoids are probably referable to the small but singular group of the "Mud-fishes" (*Dipnoi*), represented at the present day by the singular *Lepidosiren* of South America and Africa, and the *Ceratodus* of the rivers of Queensland.

Leaving the Ganoid fishes, it still remains to be noticed that the Devonian deposits have yielded the remains of a number of fishes more or less closely allied to the existing Sharks, Rays, and *Chimæræ* (the *Elasmobranchii*). The majority of the forms here alluded to are allied not to the true Sharks and Dog-fishes, but to the more peaceable "Port Jackson Sharks," with their blunt teeth, adapted for crushing the shells of Molluscs. The collective name of "Cestracionts" is applied to these; and we have evidence of their past existence in the Devonian seas both by their teeth, and by the defensive spines which were implanted in front of a greater or less number of the fins. These are bony spines, often variously grooved, serrated, or ornamented, with hollow bases, implanted in the integument, and capable of being erected or depressed at will. Many of these "fin-spines" have been preserved to us in the fossil condition, and the Devonian rocks have yielded examples belonging to many genera. As some of the true Sharks and

Dog-fishes, some of the Ganoids, and even some Bony fishes, possess similar defences, it is often a matter of some uncertainty to what group a given spine is to be referred. One of these spines, belonging to the genus *Machæracanthus,* from the Devonian rocks of America, has been figured in a previous illustration (fig. 102, *f*).

In conclusion, a very few words may be said as to the validity of the Devonian series as an independent system of rocks, preserving in its successive strata the record of an independent system of life. Some high authorities have been inclined to the view that the Devonian formation has in nature no actual existence, but that it is made up partly of beds which should be referred to the summit of the Upper Silurian,

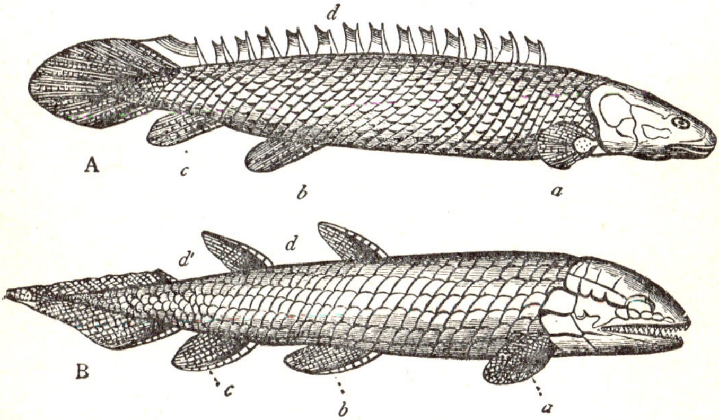

Fig. 105.—A, *Polypterus*, a recent Ganoid fish ; B, *Osteolepis*, a Devonian Ganoid ; *a*, *a*, Pectoral fins, showing the fin-rays arranged round a central lobe.

and partly of beds which properly belong to the base of the Carboniferous. This view seems to have been arrived at in consequence of a too exclusive study of the Devonian series of the British Isles, where the physical succession is not wholly clear, and where there is a striking discrepancy between the organic remains of those two members of the series which are known as the "Old Red Sandstone" and the "Devonian" rocks proper. This discrepancy, however, is not complete; and, as we have seen, can be readily explained on the supposition that the one group of rocks presents us with the shallow water and littoral deposits of the period, while in the other we are introduced to the deep-sea accumulations of the same period. Nor can the problem at issue be solved by an appeal to the phenomena of the British area alone, be the testimony of these what it may. As a matter of fact, there is at present no sufficient ground for believing that there is any irreconcilable discordance between the succession of rocks and of life in Britain during the period which elapsed between the deposition of the Upper Ludlow and the formation of the

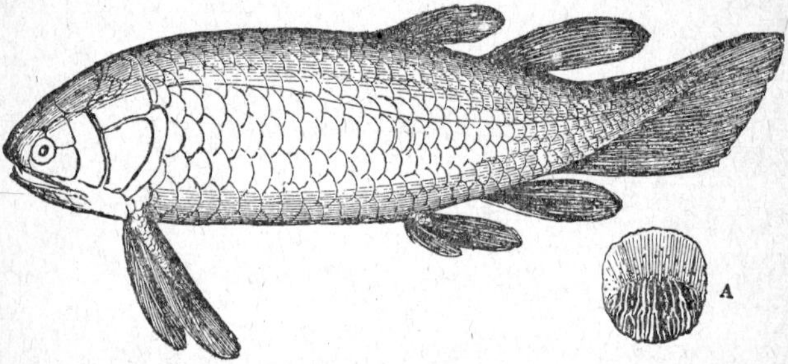

A

Fig. 106.—*Holoptychius nobilissimus*, **restored. Old Red Sandstone, Scotland. A, Scale of the same.**

Carboniferous Limestone, and the order of the same phenomena during the same period in other regions. Some of the Devonian types of life, as is the case with all great formations, have descended unchanged from older types; others pass upwards unchanged to the succeeding period: but the fauna and flora of the Devonian period are, as a whole, quite distinct from those of the preceding Silurian or the succeeding Carboniferous; and they correspond to an equally distinct rock-system, which in point of time holds an intermediate position between the two great groups just mentioned. As before remarked, this conclusion may be regarded as sufficiently proved even by the phenomena of a British area; but it may be said to be rendered a certainty by the study of the Devonian deposits of the continent of Europe—or, still more, by the investigation of the vast, for the most part uninterrupted and continuous series of sediments which commenced to be laid down in North America at the beginning of the Upper Silurian, and did not cease till, at any rate, the close of the Carboniferous.

LITERATURE.

The following list comprises the more important works and memoirs to which the student of Devonian rocks and fossils may refer:—

(1) 'Siluria.' Sir Roderick Murchison.
(2) 'Geology of Russia in Europe.' Murchison (together with De Verneuil and Count von Keyserling).
(3) "Classification of the Older Rocks of Devon and Cornwall"—'Proc. Geol. Soc.,' vol. iii., 1839. Sedgwick and Murchison.
(4) "On the Physical Structure of Devonshire;" and on the "Classification of the Older Stratified Rocks of Devonshire and Cornwall"—'Trans. Geol. Soc.,' vol. v., 1840. Sedgwick and Murchison.
(5) "On the Distribution and Classification of the Older or Palæozoic Rocks of North Germany and Belgium"—'Geol. Trans.,' 2d ser., vol. vi., 1842. Sedgwick and Murchison.
(6) 'Report on the Geology of Cornwall, Devon, and West Somerset.' De la Beche.
(7) 'Memoirs of the Geological Survey of Ireland and Scotland.' Jukes and Geikie.
(8) "On the Carboniferous Slate (or Devonian Rocks) and the Old Red Sandstone of South Ireland and North Devon"—'Quart. Journ. Geol. Soc.,' vol. xxii. Jukes.
(9) "On the Physical Structure of West Somerset and North Devon;" and on the "Palæontological Value of Devonian Fossils"—'Quart. Journ. Geol. Soc.,' vol. iii. Etheridge.
(10) "On the connection of the Lower, Middle, and Upper Old Red Sandstone of Scotland"—'Trans. Edin. Geol. Soc.,' vol. i. part ii. Powrie.
(11) 'The Old Red Sandstone,' 'The Testimony of the Rocks,' and 'Footprints of the Creator.' Hugh Miller.
(12) "Report on the 4th Geological District"—'Geology of New York.' vol. iv. James Hall.
(13) 'Geology of Canada,' 1863. Sir W. E. Logan.
(14) 'Acadian Geology.' Dawson.
(15) 'Manual of Geology.' Dana.
(16) 'Geological Survey of Ohio,' vol. i.
(17) 'Geological Survey of Illinois,' vol i.
(18) 'Palæozoic Fossils of Cornwall, Devon, and West Somerset.' Phillips.
(19) 'Recherches sur les Poissons Fossiles.' Agassiz.
(20) 'Poissons de l'Old Red.' Agassiz.
(21) "On the Classification of Devonian Fishes"—'Mem. Geol. Survey of Great Britain.' Decade X. Huxley.
(22) 'Monograph of the Fishes of the Old Red Sandstone of Britain' (Palæontographical Society). Powrie and Lankester.
(23) 'Fishes of the Devonian System, Palæontology of Ohio.' Newberry.

(24) 'Monograph of British Trilobites' (Palæontographical Society). Salter.
(25) 'Monograph of British Merostomata' (Palæontographical Society). Henry Woodward.
(26) 'Monograph of British Brachiopoda' (Palæontographical Society). Davidson.
(27) 'Monograph of British Fossil Corals' Palæontographical Society). Milne-Edwards and Haime.
(28) 'Polypiers Foss. des Terrains Paléozoiques.' Milne-Edwards and Jules Haime.
(29) "Devonian Fossils of Canada West"—'Canadian Journal,' new ser., vols. iv.-vi Billings.
(30) 'Palæontology of New York,' vol. iv. James Hall.
(31) 'Thirteenth, Fifteenth, and Twenty-third Annual Reports on the State Cabinet.' James Hall.
(32) 'Palæozoic Fossils of Canada,' vol. ii. Billings.
(33) 'Reports on the Palæontology of the Province of Ontario for 1874 and 1875.' Nicholson.
(34) "The Fossil Plants of the Devonian and Upper Silurian Formations of Canada"—'Geol. Survey of Canada.' Dawson.
(35) 'Petrefacta Germaniæ.' Goldfuss.
(36) 'Versteinerungen der Grauwacken-formation.' &c. Geinitz.
(37) 'Beitrag zur Palæontologie des Thüringer-Waldes.' Richter and Unger.
(38) 'Ueber die Placodermen des Devonischen Systems.' Pander.
(39) 'Die Gattungen der Fossilen Pflanzen.' Gœppert.
(40) 'Genera et Species Plantarum Fossilium.' Unger.

CHAPTER XII.

THE CARBONIFEROUS PERIOD.

Overlying the Devonian formation is the great and important series of the *Carboniferous Rocks*, so called because workable beds of coal are more commonly and more largely developed in this formation than in any other. Workable coal-seams, however, occur in various other formations (Jurassic, Cretaceous, Tertiary), so that coal is not an exclusively Carboniferous product; whilst even in the Coal-measures themselves the coal bears but a very small proportion to the total thickness of strata, occurring only in comparatively thin beds intercalated in a great series of sandstones, shales, and other genuine aqueous sediments.

Stratigraphically, the Carboniferous rocks usually repose conformably upon the highest Devonian beds, so that the line of demarcation between the Carboniferous and Devonian formations is principally a palæontological one, founded on the observed differences in the fossils of the two groups. On the other hand, the close of the Carboniferous period seems to have been generally, though not universally, signalized by movements of the crust of the earth, so that the succeeding Permian beds often lie unconformably upon the Carboniferous sediments.

Strata of Carboniferous age have been discovered in almost every large land-area which has been sufficiently investigated; but they are especially largely developed in Britain, in various parts of the continent of Europe, and in North America. Their general composition, however, is, comparatively speaking, so uniform, that it will suffice to take a comprehensive view of the formation without considering any one area in detail, though in each region the subdivisions of the formation are known by distinctive local names. Taking such a comprehensive view, it is found that the Carboniferous series is generally divisible into a *Lower* and essentially calcareous group (the "Sub-Carboniferous" or "Carboniferous Limestone"); a *Middle* and principally arenaceous group (the "Millstone Grit"); and an *Upper* group, of alternating shales and sandstones, with workable seams of coal (the "Coal-measures").

I. The *Carboniferous, Sub-Carboniferous,* or *Mountain Limestone Series* constitutes the general base of the Carboniferous system. As typically developed in Britain, the Carboniferous Limestone is essentially a calcareous formation, sometimes consisting of a mass of nearly pure limestone from 1000 to 2000 feet in thickness, or at other times of successive great beds of limestone with subordinate sandstones and shales. In the north of England the base of the series consists of pebbly conglomerates and coarse sandstones; and in Scotland generally, the group is composed of massive sandstones with a comparatively feeble development of the calcareous element. In Ireland, again, the base of the Carboniferous Limestone is usually considered to be formed by a locally-developed group of grits and shales (the "Coomhola Grits" and "Carboniferous Slate"), which attain the thickness of about 5000 feet, and contain an intermixture of Devonian with Carboniferous types of fossils. Seeing that the Devonian formation is generally conformable to the Carboniferous, we need feel no surprise at this intermixture of forms; nor does it appear to be of great moment whether these strata be referred to the former or to the latter series. Perhaps the most satisfactory course is to regard the Coomhola Grits and Carboniferous Slates as "passage-beds" between the Devonian and Carboniferous; but any view that may be taken as to the position of these beds, really leaves unaffected the integrity of the Devonian series as a distinct life-system, which, on the whole, is more closely allied to the Silurian than to the Carboniferous. In North America, lastly, the Sub-Carboniferous series is never purely calcareous, though in the interior of the continent it becomes mainly so. In other regions, however, it consists principally of shales and sandstones, with subordinate beds of limestone, and sometimes with thin beds of coal or deposits of clay-ironstone.

II. *The Millstone Grit.*—The highest beds of the Carboniferous Limestone series are succeeded, generally with perfect conformity, by a series of arenaceous beds, usually known as the *Millstone Grit.* As typically developed in Britain, this group consists of hard quartzose sandstones, often so large-grained and coarse in texture as to properly constitute fine conglomerates. In other cases there are regular conglomerates, sometimes with shales, limestones, and thin beds of coal—the thickness of the whole series, when well developed, varying from 1000 to 5000 feet. In North America, the Millstone Grit rarely reaches 1000 feet in thickness; and, like its British equivalent, consists of coarse sandstones and grits, sometimes with regular conglomerates. Whilst the Carboniferous Limestone was undoubtedly deposited in a tranquil ocean of considerable depth, the coarse mechanical sediments of the Millstone Grit indicate the progressive shallowing of the Carboniferous seas, and the consequent supervention of shore-conditions.

III. *The Coal-measures.*—The Coal-measures properly so called rest conformably upon the Millstone Grit, and usually consist of a vast series of sandstones, shales, grits, and coals, sometimes with beds of limestone, attaining in some regions a total thickness of from 7000 to nearly 14,000 feet. Beds of workable coal are by no means unknown in some areas in the inferior group of the Sub-Carboniferous; but the general statement is true, that coal is mostly obtained from the true Coal-measures—the largest known, and at present most productive coal-fields of the world being in Great Britain, North America, and Belgium. Wherever they are found, with limited exceptions, the Coal-measures present a singular *general* uniformity of mineral composition. They consist, namely, of an indefinite alternation of beds of sandstone, shale, and coal, sometimes with bands of clay-ironstone or beds of limestone, repeated in no constant order, but sometimes attaining the enormous aggregate thickness of 14,000 feet, or little short of 3 miles. The beds of coal differ in number and thickness in different areas, but they seldom or never exceed one-fiftieth part of the total bulk of the formation in thickness. The characters of the coal itself, and the way in which the coal-beds were deposited, will be briefly alluded to in speaking of the vegetable life of the period. In Britain, and in the Old World generally, the Coal-measures are composed partly of genuine terrestrial deposits—such as the Coal—and partly of sediments accumulated in the fresh or brackish waters of vast lagoons, estuaries, and marshes. The fossils of the Coal-measures in these regions are therefore necessarily the remains either of terrestrial plants and animals, or of such forms of life as inhabit fresh or brackish waters, the occurrence of strata with marine fossils being quite a local and occasional phenomenon. In various parts of North America, on the other hand, the Coal-measures, in addition to sandstones, shales, coal-seams, and bands of clay-ironstone, commonly include beds of limestone, charged with marine remains, and indicating marine conditions. The subjoined section (fig. 107) gives, in a generalized form, the succession of the Carboniferous strata in such a British area as the north of England, where the series is developed in a typical form.

As regards the *life* of the Carboniferous period, we naturally find, as has been previously noticed, great differences in different parts of the entire series, corresponding to the different mode of origin of the beds. Speaking generally, the Lower Carboniferous (or the Sub-Carboniferous) is characterized by the remains of marine animals; whilst the Upper Carboniferous (or Coal-measures) is characterized by the remains of plants and terrestrial animals. In all those cases, however, in which marine beds are found in the series of the Coal-measures, as is common in America, then we find that the fossils agree in their general characters with those of the older marine deposits of the period.

Owing to the fact that coal is simply compressed and otherwise altered vegetable matter, and that it is of the highest economic value to man, the Coal-measures have been more thoroughly explored than any other group of strata of equiva-

GENERALIZED SECTION OF THE CARBONIFEROUS STRATA OF THE NORTH OF ENGLAND.

Fig. 107.

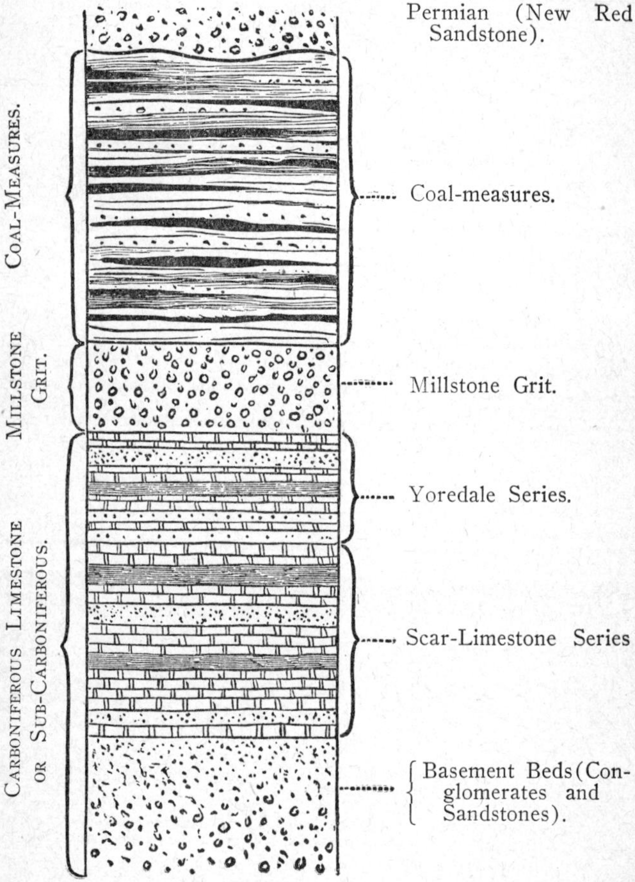

lent thickness in the entire geological series. Hence we have already a very extensive acquaintance with the *plants* of the

Carboniferous period; and our knowledge on this subject is daily undergoing increase. It is not to be supposed, however, that the remains of plants are found solely in the Coal-measures; for though most abundant towards the summit, they are found in less numbers in all parts of the series. Wherever found, they belong to the same great types of vegetation; but, before reviewing these, a few words must be said as to the orgin and mode of formation of *coal.*

The coal-beds, as before mentioned, occur interstratified with shales, sandstones, and sometimes limestones; and there may, within the limits of a single coal-field, be as many as 80 of 100 of such beds, placed one above the other at different levels, and varying in thickness from a few inches up to 20 or 30 feet. As a general rule, each bed of coal rests upon a bed of shale or clay, which is termed the "under-clay," and in which are found numerous roots of plants; whilst the strata immediately on the top of the coal may be shaly or sandy, but in either case are generally charged with the leaves and stems of plants, and often have upright trunks passing vertically through them. When we add to this that the coal itself is, chemically, nearly wholly composed of carbon, and that its microscopic structure shows it to be composed almost entirely of fragments of stems, leaves, bark, seeds, and vegetable *debris* derived from *land-plants,* we are readily enabled to understand how the coal was formed. The "*under-clay*" immediately beneath the coal-bed represents an old land-surface—sometimes, perhaps, the bottom of a swamp or marsh, covered with a luxuriant vegetation; the *coal-bed* itself represents the slow accumulation, through long periods, of the leaves, seeds, fruits, stems, and fallen trunks of this vegetation, now hardened and compressed into a fraction of its original bulk by the pressure of the superincumbent rocks; and the strata of sand or shale above the coal-bed—the so-called "roof" of the coal—represent sediments quietly deposited as the land, after a long period of repose, commenced to sink beneath the sea. On this view, the rank and long-continued vegetation which gave rise to each coal-bed was ultimately terminated by a slow depression of the surface on which the plants grew. The land-surface then became covered by the water, and aqueous sediments were accumulated to a greater or less thickness upon the dense mass of decaying vegetation below, enveloping any trunks of trees which might still be in an erect position, and preserving between their layers the leaves and branches of plants brought down from the neighboring land by streams, or blown into the water by the wind. Finally, there set in a slow movement of elevation,—the old land again reappeared above the water; a new and equally luxuriant vegetation flourished upon the new land-surface; and another coal-bed was accumulated, to be preserved ultimately in a similar fashion. Some few beds of coal may have been formed by drifted vegetable matter brought down into the ocean by rivers, and deposited directly on the bottom of the sea; but in the majority of cases the coal is undeniably the result of the slow growth and decay of plants *in situ*; and as the plants of the coal are not *marine* plants, it is necessary to adopt some such theory as the above to account for the formation of coal-seams. By this theory, as is obvious, we are compelled to suppose that the vast alluvial and marshy flats upon which the coal-plants grew were liable to constantly-recurring oscillations of level, the successive land-surfaces represented by the successive coal-beds of any coal-field being thus successively buried beneath accumulations of mud or sand. We have no need, however, to suppose that these oscillations affected large areas at the same time; and geology teaches us that local elevations and depressions of the land have been matters of constant occurrence throughout the whole of past time.

All the varieties of coal (bituminous coal, anthracite, cannel-coal, &c.) show a more or less distinct "lamination"—that is to say, they are more or less obviously composed of successive thin layers, differing slightly in color and texture. All the varieties of coal, also, consist chemically of *carbon,* with varying proportions of certain gaseous constituents and a small amount of incombustible mineral or "ash." By cutting thin and transparent slices of coal, we are further enabled, by means of the microscope, to ascertain precisely not only that the carbon of the coal is derived from vegetables, but also, in many cases, what kinds of plants, and what parts of these, enter into the formation of coal. When examined in this way, all coals are found to consist more or less entirely of vegetable matter; but there is considerable difference in different coals as to the exact nature of this. By Professor Huxley it has been shown that many of the English coals consist largely of accumulations of rounded discoidal sacs or bags, which are unquestionably the seed-vessels or "spore-cases" of certain of the commoner coal-plants (such as the *Lepidodendra*). The best bituminous coals seem to be most largely composed of these spore-cases; whilst inferior kinds possess a progressively increasing amount of the dull carbonaceous substance which is known as "mineral charcoal," and which is undoubtedly composed of "the stems and leaves of plants reduced to little more than their carbon." On the other hand, Principal Dawson finds that the American coals only occasionally exhibit spore-cases to any extent, but consist principally of the cells, vessels, and fibres of the bark, integumentary coverings, and woody portions of the Carboniferous plants.

The number of plants already known to have existed during the Carboniferous period is so great, that nothing more can be done here than to notice briefly the typical and characteristic *groups* of these—such as the Ferns, the Calamites, the Lepidodendroids, the Sigillarioids, and the Conifers.

In accordance with M. Brongniart's generalization, that the Palæozoic period is, botanically speaking, the "Age of Acrogens," we find the Carboniferous plants to be still mainly referable to the Flowerless or "Cryptogamous" division of the vegetable kingdom. The flowering or "Phanerogamous" plants, which form the bulk of our existing vegetation, are hardly known, with certainty, to have existed at all in the Carboniferous era, except as represented by trees related to the existing Pines and Firs, and possibly by the Cycads or "false palms."* Amongst the "Cryptogams," there is no more striking or beautiful group of Carboniferous plants than the *Ferns*. Remains of these are found all through the Carboniferous, but in exceptional numbers in the Coal-measures, and include both herbaceous forms like the majority of existing species, and arborescent forms resembling the living Tree-ferns of New Zealand. Amongst the latter, together with some new types, are examples of the genera *Psaronius* and *Caulopteris*, both of which date from the Devonian. The simply herbaceous ferns are extremely numerous, and belong to such widely-distributed and largely-represented genera as *Neuropteris*, *Odontopteris* (fig. 108), *Alethopteris*, *Pecopteris*, *Sphenopteris*, *Hymenophyllites*, &c.

The fossils known as *Calamites* (fig. 109) are very common in the Carboniferous deposits, and have given occasion to an abundance of research and speculation. They present themselves as prostrate and flattened striated stems, or as similar

* Whilst the vegetation of the Coal-period was mainly a terrestrial one, aquatic plants are not unknown. Sea-weeds (such as the *Spirophyton cauda-Galli*) are common in some of the marine strata; whilst coal, according to the researches of the Abbé Castracane, is asserted commonly to contain the siliceous envelope of Diatoms.

uncompressed stems growing in an erect position, and sometimes attaining a length of twenty feet or more. Externally, the stems are longitudinally ribbed, with transverse joints at regular intervals, these joints giving origin to a whorl of branchlets, which may or may not give origin to similar whorls of smaller branchlets still. The stems, further, were hollow, with transverse partitions at the joints, and having neither true wood nor bark, but only a thin external fibrous shell. There can be little doubt but that the *Calamites* are properly regarded as colossal representatives of the little Horse-tails (*Equisetaceæ*) of the present day. They agree with these not only in the general details of their organization, but also in the fact that the fruit was a species of cone, bearing "spore-cases" under scales. According to Principal Dawson, the *Calamites* "grew in dense

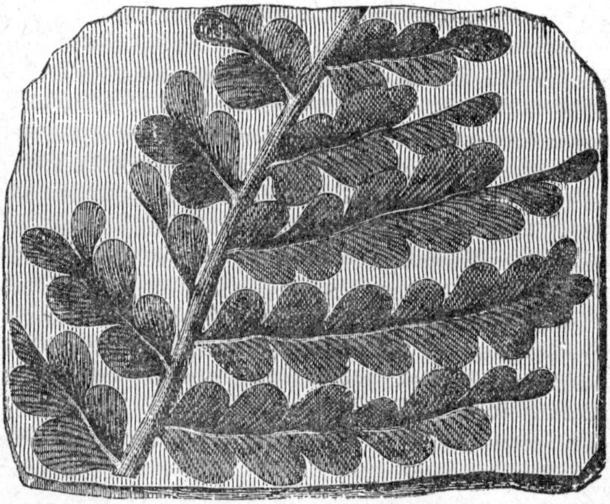

Fig. 108.—*Odontopteris Schlotheimii.* Carboniferous, Europe and North America.

brakes on the sandy and muddy flats, subject to inundation, or perhaps even in water; and they had the power of budding out from the base of the stem, so as to form clumps of plants, and also of securing their foothold by numerous cord-like roots proceeding from various heights on the lower part of the stem."

The *Lepidodendroids,* represented mainly by the genus *Lepidodendron* itself (fig. 110), were large tree-like plants, which attain their maximum in the Carboniferous period, but which appear to commence in the Upper Silurian, are well represented in the Devonian, and survive in a diminished form into the Permian. The trunks of the larger species of *Lepido-*

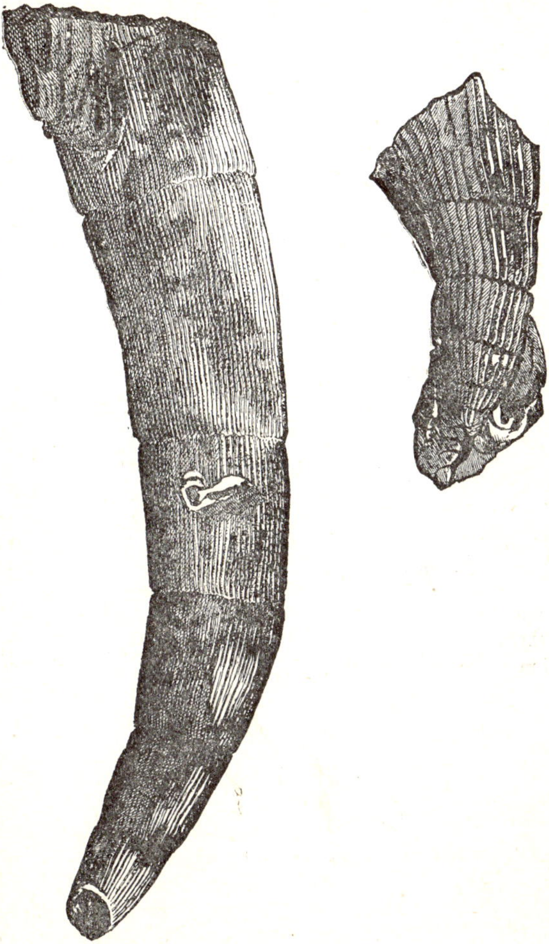

Fig. 109.—*Calamites cannæformis*. Carboniferous Rocks, Europe and North America.

dendron at times reach a length of fifty feet and upwards, giving off branches in a regular bifurcating manner. The bark is marked with numerous rhombic or oval scars, arranged in quincunx order, and indicating the points where the long, needle-shaped leaves were formerly attached. The fruit consisted of cones or spikes, carried at the ends of the branches, and consisting of a central axis surrounded by overlapping scales, each of which supports a "spore-case" or seed-vessel. These cones have commonly been described under the name of *Lepidostrobi.* In the structure of the trunk there is nothing comparable to what is found in existing trees, there being a thick bark surrounding a zone principally composed of "scalariform" vessels, this in turn enclosing a large central pith. In their general appearance the *Lipidodendra* bring to mind the existing Araucarian Pines; but they are true "Cryptogams," and are to be regarded as a gigantic extinct type of the modern Club-mosses (*Lycopodiaceæ*). They are amongst the commonest and most characteristic of the Carboniferous plants; and the majority of the "spore-cases" so commonly found in the coal appear to have been derived from the cones of Lepidodendroids.

The so-called *Sigillarioids,* represented mainly by *Sigillaria* itself (fig. 111), were no less abundant and characteristic of the Carboniferous forests than the *Lepidodendra.* They commence their existence, so far as known, in the Devonian period, but they attain their maximum in the Carboniferous; and—unlike the Lepidodendroids—they are not known to occur in the Permian period. They are comparatively gigantic in size, often attaining a height of from thirty to fifty feet or more; but though abundant and well preserved, great divergence of opinion prevails as to their true affinities. The *name* of Sigillarioids (Lat. *sigilla,* little seals or images) is derived from the fact that the bark is marked with seal-like impressions or leaf-scars (fig. 111).

Externally, the trunks of *Sigillaria* present strong longitudinal ridges, with vertical alternating rows of oval leaf-scars indicating the points where the leaves were originally attached. The trunk was furnished with a large central pith, a thick outer bark, and an intermediate woody zone—composed, according to Dawson, partly of the disc-bearing fibres so characteristic of Conifers; but, according to Carruthers, entirely made up of the "scalariform" vessels characteristic of Cryptogams. The size of the pith was very great, and the bark seems to have been the most durable portion of the trunk. Thus we have evidence that in many cases the stumps and "stools" of *Sigillariæ,* standing upright in the old Carboniferous swamps, were completely hollowed out by internal decay, till nothing but an exterior shell of bark was left. Often these hollow stumps became

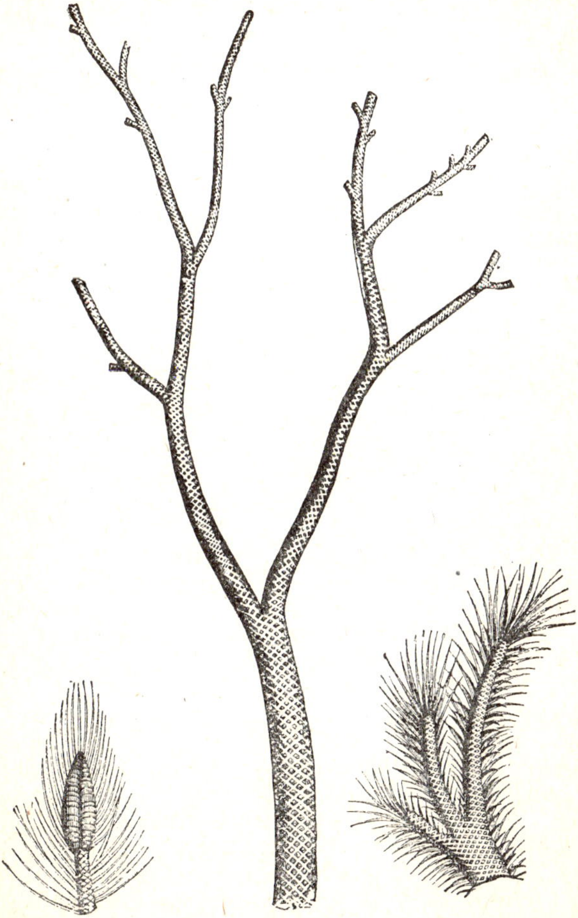

Fig. 110.—*Lepidodendron Sternbergii*. Carboniferous, Europe. The central figure represents a portion of the trunk with its branches, much reduced in size. The right-hand figure is a portion of a branch with the leaves partially attached to it; and the left-hand figure represents the end of a branch bearing a cone of fructification.

ultimately filled up with sediment, sometimes enclosing the remains of galley-worms, land-snails or Amphibians, which formerly found in the cavity of the trunk a congenial home; and from the sandstone or shale now filling such trunks some of the most interesting fossils of the Coal-period have been obtained. There is little certainty as to either the leaves or fruits of *Sigillaria,* and there is equally little certainty as to the true botanical position of these plants. By Principal Dawson they are regarded as being probably flowering plants allied to the existing "false palms" or "*Cycads;*" but the high authority of Mr. Carruthers is to be quoted in support of the belief that they are Cryptogamic, and most nearly allied to the Club-mosses.

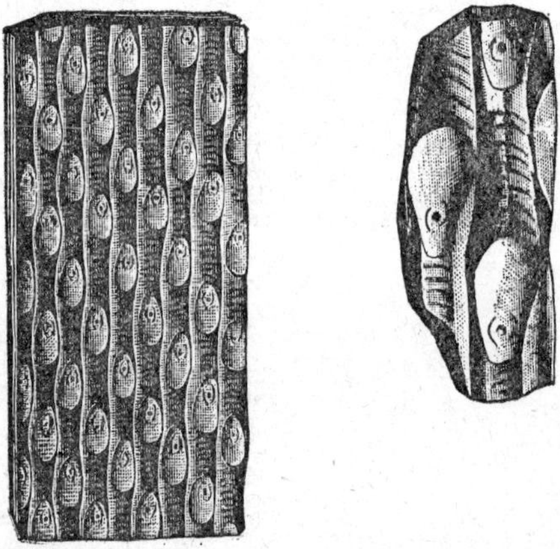

Fig. 111.—Fragment of the external surface of *Sigillaria Græseri*, showing the ribs and leaf-scars. The left-hand figure represents a small portion enlarged. Carboniferous, Europe.

Leaving the botanical position of *Sigillaria* thus undecided, we find that it is now almost universally conceded that the fossils originally described under the name of *Stigmaria* are the *roots* of *Sigillaria,* the actual connection between the two having been in numerous instances demonstrated in an unmistakable manner. The *Stigmariæ* (fig. 112) ordinarily present themselves in the form of long, compressed or rounded fragments, the external surface of which is covered with rounded pits or shallow tubercles, each of which has a little pit or depression in its center. From each of these pits there proceeds, in perfect examples, a long cylindrical rootlet; but in many cases these have altogether disappeared. In their internal structure, *Stigmaria* exhibits a central pith surrounded by a sheath of scalariform vessels, the whole enclosed in a cellular envelope. The *Stigmariæ* are generally found ramifying in the "under-clay," which forms the floor of a bed of coal, and which represents the ancient soil upon which the *Sigillariæ* grew.

The Lepidodendroids and Sigillarioids, though the first were certainly, and the second possibly, Cryptogamic or flowerless plants, must have constituted the main mass of the forests of the Coal period; but we are not without evidence of the existence at the same time of genuine "trees," in the technical sense of this term—namely, flowering plants with large woody

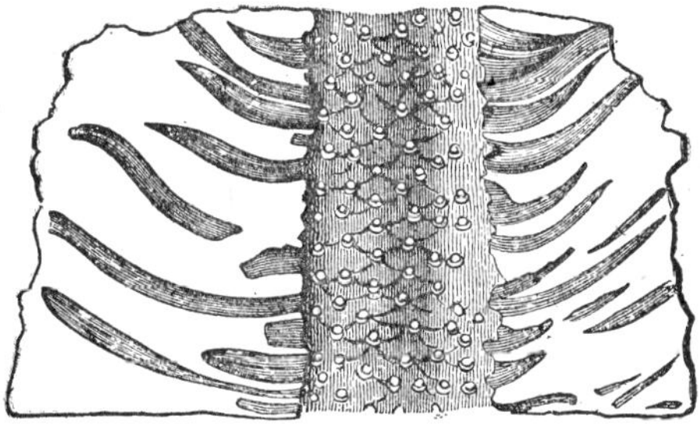

Fig. 112.—*Stigmaria ficoides*. Quarter natural size. Carboniferous.

stems. So far as is certainly known, all the true trees of the Carboniferous formation were *Conifers,* allied to the existing Pines and Firs. They are recognized by the great size and concentric woody rings of their prostrate, rarely erect trunks, and by the presence of disc-bearing fibres in their wood, as demonstrated by the microscope; and the principal genera which have been recognized are *Dadoxylon, Palæoxylon, Araucarioxylon,* and *Pinites.* Their fruit is not known with absolute certainty, unless it be represented, as often conjectured, by *Trigonocarpon* (fig. 113). The fruits known under this name are nut-like, often of considerable size, and commonly three- or six-angled. They probably originally possessed a fleshy envelope; and if truly referable to the *Conifers,* they would indicate that these ancient evergreens produced berries instead of cones, and thus resembled the modern Yews rather than the Pines. It seems further, that the great group of the *Cycads,* which are nearly allied to the *Conifers,* and

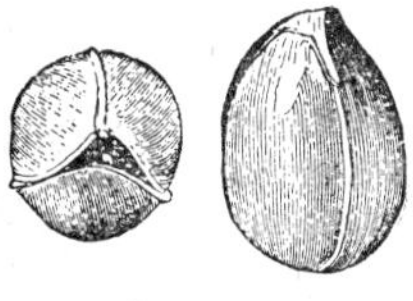

Fig. 113. — *Trigonocarpon ovatum*. Coal-measures, Britain. (After Lindley and Hutton.)

which attained such a striking prominence in the Secondary period, probably commenced its existence during the Coal period; but these anticipatory forms are comparatively few in number, and for the most part of somewhat dubious affinities.

CHAPTER XIII.

THE CARBONIFEROUS PERIOD—Continued.

ANIMAL LIFE OF THE CARBONIFEROUS.

We have seen that there exists a great difference as to the mode of origin of the Carboniferous sediments, some being purely marine, whilst others are terrestrial; and others, again, have been formed in inland swamps and morasses, or in brackish-water lagoons, creeks, or estuaries. A corresponding difference exists necessarily in the animal remains of these deposits, and in many regions this difference is extremely well marked and striking. The great marine limestones which characterize the lower portion of the Carboniferous series in Britain, Europe, and the eastern portion of America, and the calcareous beds which are found high up in the Carboniferous in the western States of America, may, and do, often contain the remains of drifted plants; but they are essentially characterized by marine fossils; and moreover, they can be demonstrated by the microscope to be almost wholly composed of the remains of animals which formerly inhabited the ocean. On the other hand, the animal remains of the beds accompanying the coal are typically the remains of air-breathing, terrestrial, amphibious, or aerial animals, together with those which inhabit fresh or brackish waters. Marine fossils may be found in the Coal-measures, but they are invariably confined to special horizons in the strata, and they indicate temporary depressions of the land beneath the sea. Whilst the distinction here mentioned is one which cannot fail to strike the observer, it is convenient to consider the animal life of the Carboniferous as a whole: and it is simply necessary, in so doing, to remember that the marine fossils are in general derived from the inferior portion of the system; whilst the air-breathing, fresh-water, and brackish-water forms are almost exclusively derived from the superior portion of the same.

The Carboniferous *Protozoans* consist mainly of *Foraminifera* and *Sponges.* The latter are still very insufficiently known, but the former are very abundant, and belong to very varied types. Thin slices of the limestones of the period, when examined by the microscope, very commonly exhibit the shells of *Foraminifera* in greater or less plenty. Some limestones, indeed, are made up of little else than these minute and elegant shells, often belonging to types, such as the Textularians and Rotalians, differing little or not at all from those now in existence. This is the case, for example, with the Carboniferous Limestone of Spergen Hill in Indiana (fig. 114), which is almost wholly made up of the spiral shells of a species of *Endothyra.* In the same way, though to a less extent the black Carboniferous marbles of Ireland, (and the similar marbles of Yorkshire, the limestones of the west of England and of Derbyshire, and the great "Scar Limestones" of the north of England, contain great numbers of Foraminiferous shells; whilst similar organisms commonly occur in the shale-beds associated with the limestones throughout the Lower Carboniferous series. One of the most interesting of the British Carboniferous forms is the *Saccammina* of Mr. Henry Brady, which is sometimes present in considerable numbers in the limestones of Northumberland, Cumberland, and the west of Scotland, and which is conspicuous for the comparatively large size of its spheroidal or pear-shaped shell (reaching from an eighth to a fifth of an inch in size). More widely distributed are the generally spindle-shaped shells of *Fusulina* (fig. 115), which occur in vast numbers in the Carboniferous Limestone of Russia, Armenia, the Southern Alps, and Spain, similar forms occurring in equal profusion in the higher

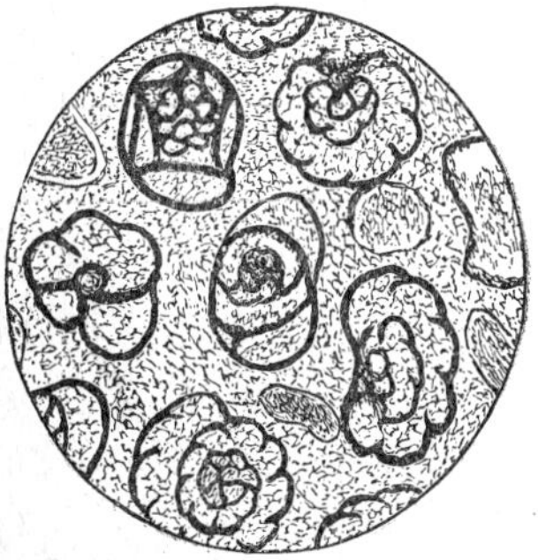

Fig. 114.—Transparent slice of Carboniferous Limestone, from Spergen Hill, Indiana, U. S., showing numerous shells of *Endothyra* (*Rotalia*), *Baileyi*, slightly enlarged. (Original.)

limestones which are found in the Coal-measures of the United States, in Ohio, Illinois, Indiana, Missouri, &c. Mr. Henry Brady, lastly, has shown that we have in the *Nummulina pristina* of the Carboniferous Limestone of Namur a genuine *Nummulite,* precursor of the great important family of the Tertiary Nummulites.

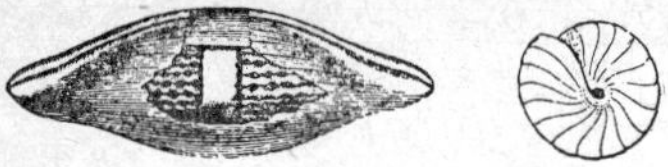

Fig. 115.—*Fusulina cylindrica*, Carboniferous Limestone, Russia.

The sub-kingdom of the *Cœlenterates,* so far as certainly known, is represented only by *Corals;* * but the remains of these are so abundant in many of the limestones of the Carboniferous formation as to constitute a feature little or not at all less conspicuous than that afforded by the Crinoids. As is the case in the preceding period, the Corals belong, almost exclusively, to the groups of the *Rugosa* and *Tabulata;* and there is a general and striking resemblance and relationship between the coral-fauna of the Devonian as a whole, and that of the Carboniferous. Nevertheless, there is an equally decided and striking amount of difference between these successive faunas, due to the fact that the great majority of the Carboniferous *species* are new; whilst some of the most characteristic Devonian *genera* have nearly or quite disappeared, and several new genera now make their appearance for the first time. Thus, the characteristic Devonian types *Heliophyllum, Pachyphyllum, Chonophyllum, Acervularia, Spongophyllum, Smithia, Endophyllum,* and *Cystiphyllum,* have now disappeared; and the great masses of *Favosites* which are such a striking feature in the Devonian limestones, are represented but by one or two degenerate and puny successors. On the other hand, we meet in the Carboniferous rocks not only with entirely new genera—such as *Axophyllum, Lophophyllum,* and *Londsdaleia*—but we have an enormous expansion of certain types which had just begun to exist in the preceding period. This is especially well seen in the case of the genus *Lithostrotion* (fig. 116, *b*), which more than any other may be considered as the predominant Carboniferous group of Corals. All the species of *Lithostrotion* are compound, consisting either of bundles of

* A singular fossil has been described by Professor Martin Duncan and Mr. Jenkins from the Carboniferous rocks under the name of *Palæocoryne*, and has been referred to the Hydroid Zoophytes (*Corynida*). Doubt, however, has been thrown by other observers on the correctness of this reference.

loosely-approximated cylindrical stems, or of similar "corallites" closely aggregated together into astræiform colonies, and rendered polygonal by mutual pressure. This genus has a historical interest, as having been noticed as early as in the year 1699 by Edward Lhwyd; and it is geologically important from its wide distribution in the Carboniferous rocks of both the Old and New Worlds. Many species are known, and whole beds of limestone are often found to be composed of little else than the skeletons of these ancient corals, still standing upright as they grew. Hardly less characteristic of the Carboniferous than the above is the great group of simple "cup-corals," of which *Clisiophyllum* is the central type. Amongst types which commenced in the Silurian and Devonian, but which are still well represented here, may be mentioned *Syringopora* (fig. 116, *e*), with its colonies of delicate cylindrical tubes united at intervals by cross-bars; *Zaphrentis* (fig. 116, *d*), with its cup-shaped skeleton and the well-marked depression (or "fossula") on one side of the calice; *Amplexus* (fig. 116, *c*), with its cylindrical, often irregularly swollen coral and short septa; *Cyathophyllum* (fig. 116, *a*), sometimes simple, sometimes forming great masses of star-like corallites; and *Chætetes,* with its branched stems, and its minute, "tabulate" tubes (fig. 116, *f*). The above, together with other and hardly less characteristic forms, combine to constitute a coral-fauna which is not only in itself perfectly distinctive, but which is of especial interest, from the fact that almost all the varied types of which it is composed disappeared utterly before the close of the Carboniferous period. In the first marine sediments of a calcareous nature which succeeded to the Coal-measures (the magnesian limestones of the Permian), the great group of the *Rugose corals,* which flourished so largely throughout the Silurian, Devonian, and Carboniferous periods, is found to have all but disappeared, and it is never again represented save sporadically and by isolated forms.

Amongst the *Echinoderms,* by far the most important forms are the Sea-lilies and the Sea-urchins—the former from their great abundance, and the latter from their singular structure; but the little group of the "Pentremites" also requires to be noticed. The Sea-lilies are so abundant in the Carboniferous rocks, that it has been proposed to call the earlier portion of the period the "Age of Crinoids." Vast masses of the limestones of the period are "crinoidal," being more or less extensively composed of the broken columns, and detached plates and joints of Sea-lilies, whilst perfect "heads" may be exceedingly rare and difficult to procure. In North America the remains of Crinoids are even more abundant at this horizon than

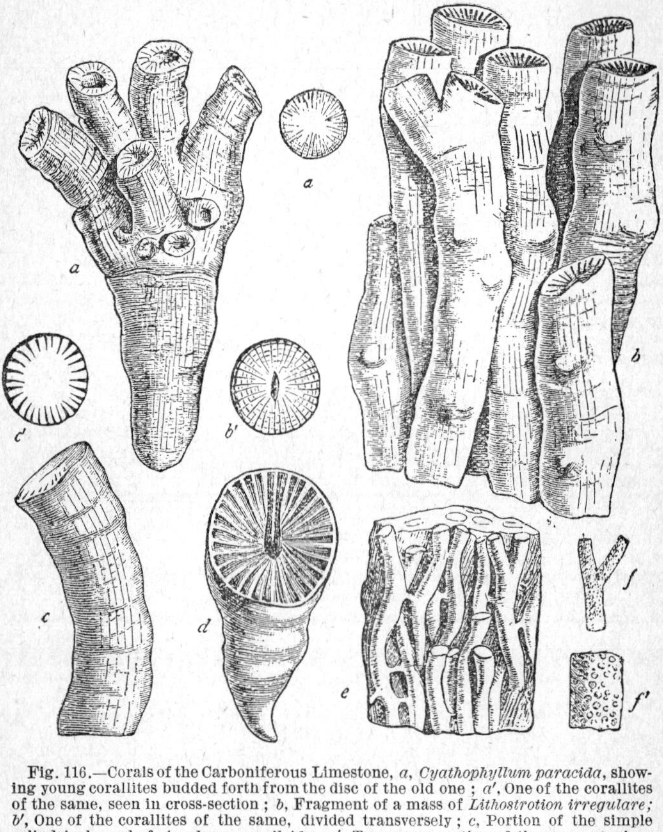

Fig. 116.—Corals of the Carboniferous Limestone, *a*, *Cyathophyllum paracida*, showing young corallites budded forth from the disc of the old one ; *a'*, One of the corallites of the same, seen in cross-section ; *b*, Fragment of a mass of *Lithostrotion irregulare; b'*, One of the corallites of the same, divided transversely ; *c*, Portion of the simple cylindrical coral of *Amplexus coralloides; c'*, Transverse section of the same species ; *d*, *Zaphrentis vermicularis*, showing the depression or "fossula" on one side of the cup ; *e*, Fragment of a mass of *Syringopora ramulosa; f*. Fragment of *Chætetes tumidus; f'*, Portion of the surface of the same, enlarged. From the Carboniferous Limestone of Britain and Belgium. (After Thomson, De Koninck, Milne-Edwards and Haime, and the Author.)

in Britain, and the specimens found seem to be commonly more perfect. The commonest of the Carboniferous Crinoids belong to the genera *Cyathocrinus, Actinocrinus, Platycrinus,* (fig. 117), *Poteriocrinus, Zeacrinus,* and *Forbesiocrinus.* Closely allied to the Crinoids, or forming a kind of transition between these and the Cystideans, is the little group of the "Pentremites," or *Blastoids,* (fig. 118). This group is first known to have commenced its existence in the Upper Silurian, and it increased considerably in numbers in the Devonian; but it was in the seas of the Carboniferous period that it attained its maximum, and no certain representative of the family has been detected in any later deposits. The "Pentremites" resemble the Crinoids in having a cup-shaped body (fig. 118, A) enclosed by closely-fitting calcareous plates, and supported on a short stem or "column," composed of numerous calcareous pieces flexibly articulated together. They differ from the Crinoids, however, in the fact that the upper surface of the body does not support the crown of branched feathery "arms," which are so characteristic of the latter. On the contrary, the summit of the cup is closed up in the fashion of a flower-bud, whence the

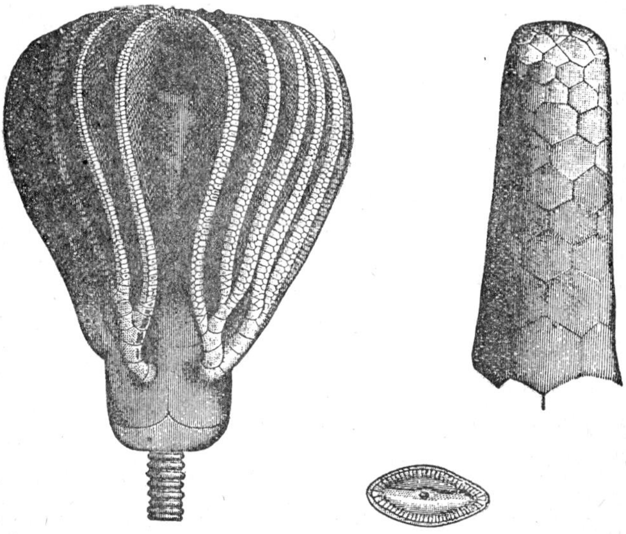

Fig. 117.—*Platycrinus tricontadactylus*, Lower Carboniferous. The left-hand figure shows the calyx, arms, and upper part of the stem; and the figure next this shows the surface of one of the joints of the column. The right-hand figure shows the proboscis. (After M'Coy.)

technical name of *Blastoidea* applied to the group (Gr. *blastos,* a bud; *eidos,* form). From the top of the cup radiate five broad, transversely-striated areas (fig. 118, C), each with a longitudinal groove down its middle; and along each side of each of these grooves there seems to have been attached a row of short jointed calcareous filaments or "pinnules."

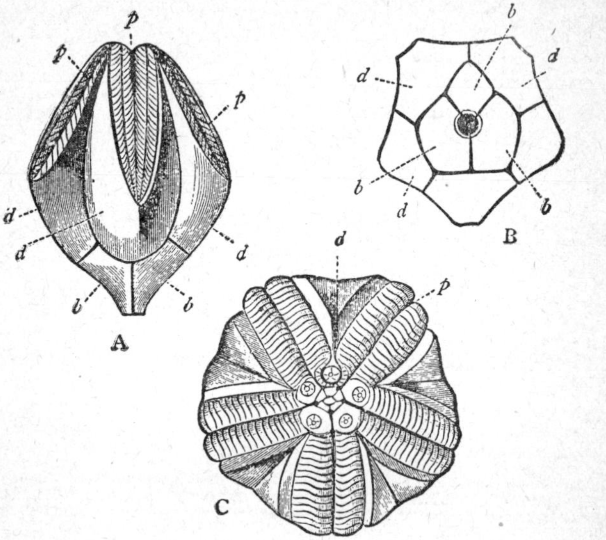

Fig. 118.—A, *Pentremites pyriformis*, side-view of the body ("calyx"); The same viewed from below, showing the arrangement of the plates; C, Body of *Pentremites conoideus*, viewed from above. Carboniferous.

A few Star-fishes and Brittle-stars are known to occur in the Carboniferous rocks; but the only other Echinoderms of this period which need be noticed are the Sea-urchins (*Echinoids*). Detached plates and spines of these are far from rare in the Carboniferous deposits; but anything like perfect specimens are exceedingly scarce. The Carboniferous Sea-urchins agree with those of the present day in having the body enclosed in a shell, formed by an enormous number of calcareous plates articulated together. The shell may be regarded as, typically, nearly spherical in shape, with the mouth in the center of the base, and the excretory opening or vent at the summit. In both the ancient forms and the recent ones, the plates of the shell are arranged in ten zones which generally radiate from the summit to the center of the base. In five of these zones—termed the "ambulacral areas"—the plates are perforated by minute apertures or "pores," through which the animal can protrude the little water-tubes ("tube-feet") by which its locomotion is carried on. In the other five zones—the so-called "inter-ambulacral areas"—the plates are of larger size, and are not perforated by any apertures. In all the modern Sea-urchins each of these ten zones, whether perforate or imperforate, is composed of two rows of plates; and there are thus twenty rows of plates in all. In the Palæozoic Sea-urchins, on the other hand, the "ambulacral areas" are often like those of recent forms, in consisting of *two* rows of perforated plates (fig. 119); but the "inter-ambulacral areas" are always quite peculiar in consisting each of three, four, five, or more rows of large imperforate plates, whilst there are sometimes four or ten rows of plates in the "ambulacral areas" also: so that there are many more than twenty rows of plates in the entire shell. Some of the Palæozoic Sea-urchins, also, exhibit a very peculiar singularity of structure which is only known to exist in a

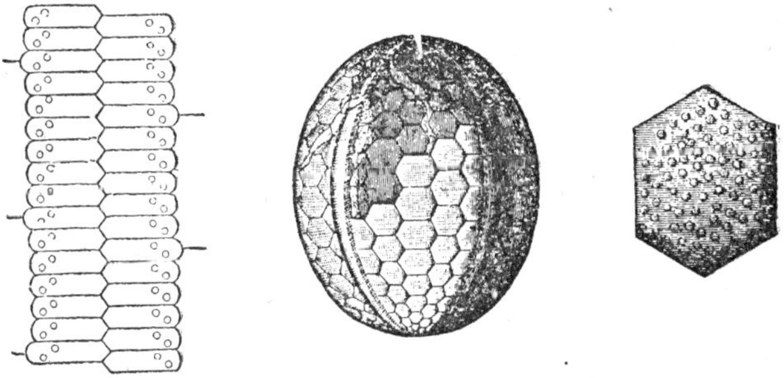

Fig. 119.—*Palæchinus ellipticus*, one of the Carboniferous Sea-urchins. The left-hand figure shows one of the "ambulacral areas" enlarged, exhibiting the perforated plates. The right-hand figure exhibits a single plate from one of the "inter-ambulacral areas." (After M'Coy.)

very few recently-discovered modern forms (viz., *Calveria* and *Phormosoma*). The plates of the inter-ambulacral areas, namely, overlap one another in an imbricating manner, so as to communicate a certain amount of flexibility to the shell; whereas in the ordinary living forms these plates are firmly articulated together by their edges, and the shell forms a rigid immovable box. The Carboniferous Sea-urchins which exhibit this extraordinary peculiarity belong to the genera *Lepidechinus* and *Lepidesthes,* and it seems tolerably certain that a similar flexibility of the shell existed to a less degree in the much more abundant genus *Archæocidaris.* The Carboniferous Sea-urchins, like the modern ones, possessed movable spines of greater or less length, articulated to the exterior of the shell; and these structures are of very common occurrence in a detached condition. The most abundant genera are *Archæocidaris* and *Palæchinus;* but the characteristic American forms belong principally to *Melonites, Oligoporus,* and *Lepidechinus.*

Amongst the *Annelides* it is only necessary to notice the little spiral tubes of *Spirorbis Carbonarius* (fig. 120), which are

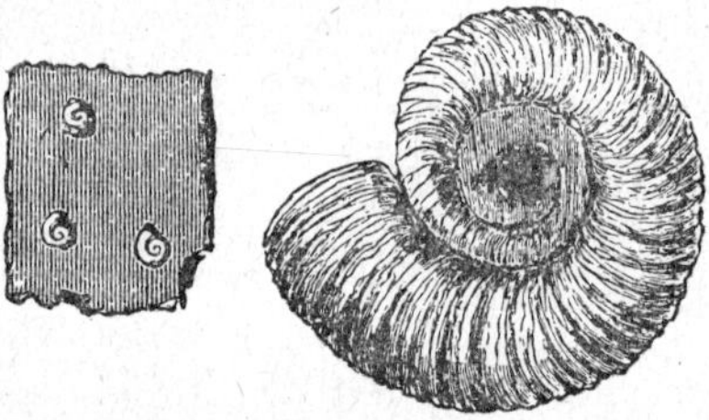

Fig. 120.—*Spirorbis* (*Microconchus*) *Carbonarius*, of the natural size, attached to a fossil plant, and magnified. Carboniferous, Britain and North America. (After Dawson.)

commonly found attached to the leaves or stems of the Coal-plants. This fact shows that though the modern species of *Spirorbis* are inhabitants of the sea, these old representatives of the genus must have been capable of living in the brackish waters of lagoons and estuaries.

The *Crustaceans* of the Carboniferous rocks are numerous, and belong partly to structural types with which we are already familar, and partly to higher groups which come into existence here for the first time. The gigantic *Eurypterids* of the Upper Silurian and Devonian are but feebly represented, and make their final exit here from the scene of life. Their place, however, is taken by peculiar forms belonging to the allied group of the *Xiphosura,* represented at the present day by the King-crabs or "Horse-shoe Crabs" (*Limulus*). Characteristic forms of this group appear in the Coal-measures both of Europe and America; and though constituting three distinct genera (*Prestwichia, Belinurus,* and *Euproöps*), they are all nearly related to one another. The best known of them, perhaps, is the

Prestwichia rotundata of Coalbrookdale, here figured (fig. 121). The ancient and formerly powerful order of the *Trilobites* also undergoes its final extinction here, not surviving the deposition of the Carboniferous Limestone series in Europe, but extending its range in America into the Coal-measures. All the known Carboniferous forms are small in size and degraded in point of structure, and they are referable to but three genera (*Phillipsia, Griffithides,* and *Brachymetopus*), belonging to a single family. The *Phillipsia seminifera* here figured (fig. 122, *a*,) is a characteristic species in the old World. The water-fleas (*Ostracoaa*) are extremely abundant in the Carboniferous rocks, whole strata being often made up of little else than the little bivalved shells of these Crustaceans. Many of them are extremely small, averaging about the size of a millet-seed; but a few forms, such as *Entomoconchus Scouleri* (fig. 122, *c*), may attain a length of from one to three quarters of an inch. The old group of the *Phyllopods* is likewise still represented in some abundance, partly by tailed forms of a shrimp-like appearance, such as *Dithyrocaris* (fig. 122, *d*), and partly by the curious striated *Estheriæ* and their allies, which present a curious resemblance to the true Bivalve Molluscs (fig. 122, *b*). Lastly, we meet for the first time in the Carboniferous rocks with the remains of the highest of all the groups of *Crustaceans*—namely, the so-called "Decapods," in which there are five pairs of walking-limbs, and the hinder end of the body ("abdomen") is composed of separate rings, whilst the anterior end is covered by a head-shield or "carapace." All the Carboniferous Decapods hitherto discovered resemble the existing Lobsters, Prawns, and Shrimps (the *Macrura*), in having a long and well-developed abdomen terminated by an expanded tail-fin.

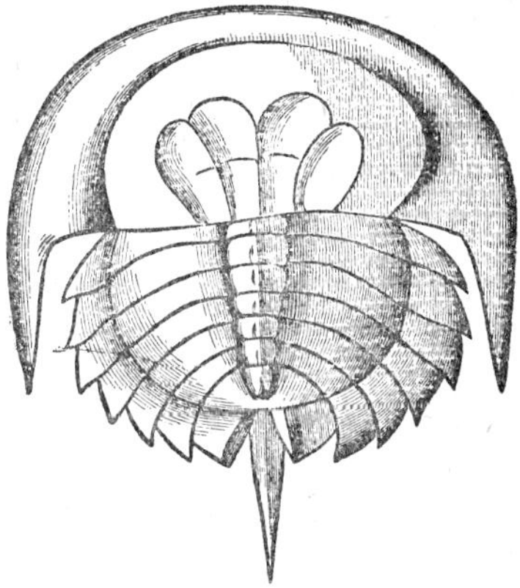

Fig. 121. — *Prestwichia rotundata*, a Limuloid Crustacean. Coal-measures, Britain. (After Henry Woodward.)

The *Palæocaris typus* (fig. 122, *e*) and the *Anthrapalæmon gracilis* (fig. 122, *f*), from the Coal-measures of Illinois, are two of the best understood and most perfectly preserved of the few known representatives of the "Long-tailed" Decapods in the Carboniferous series. The group of the Crabs or "Short-tailed" Decapods (*Brachyura*), in which the abdomen is short, not terminated by a tail-fin, and tucked away out of sight beneath the body, is at present not known to be represented at all in the Carboniferous deposits.

In addition to the water-inhabiting group of the Crustaceans, we find the articulate animals to be represented by members

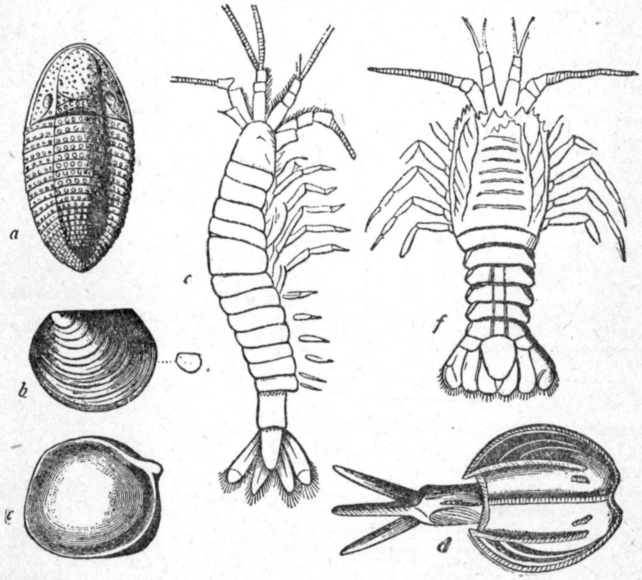

Fig. 122.—Crustaceans of the Carboniferous Rocks. *a*, *Phillipsia seminifera*, of the natural size—Mountain Limestone. Europe ; *b*, One valve of the shell of *Estheria tenella*, of the natural size and enlarged—Coal-measures, Europe ; *c*, Bivalved shell of *Entomoconchus Scouleri*, of the natural size—Mountain Limestone, Europe; *d*, *Dithyrocaris Scouleri*, reduced in size—Mountain Limestone, Ireland ; *e*, *Palæocaris typus*, slightly enlarged—Coal-measures, North America ; *f*, *Anthrapalæmon gracilis*, of the natural size—Coal-measures, North America. (After De Koninck, M'Coy, Rupert Jones, and Meek and Worthen.)

belonging to the air-breathing classes of the *Arachnida*, *Myriapoda*, and *Insecta*. The remains of these, as might have been expected, are not known to occur in the marine limestones of the Carboniferous series, but are exclusively found in beds associated with the Coal, which have been deposited in lagoons, estuaries, or marshes, in the immediate vicinity of the land, and which actually represent an old land-surface. The *Arachnids* are at present the oldest known of their class, and are represented both by true Spiders and Scorpions. Remains of the latter (fig. 123) have been found both in the Old and New Worlds, and indicate the existence in the Carboniferous period of Scorpions differing but very little from existing forms. The group of the *Myriapoda,* including the recent Centipedes and Galley-worms, is likewise represented in the Carboniferous strata, but by forms in many respects very unlike any that are known to exist at the present day. The most interesting of these were obtained by Principal Dawson, along with the bones of Amphibians and the shells of Land-snails, in the sediment filling the hollow trunks of *Sigillaria,* and they belong to the genera

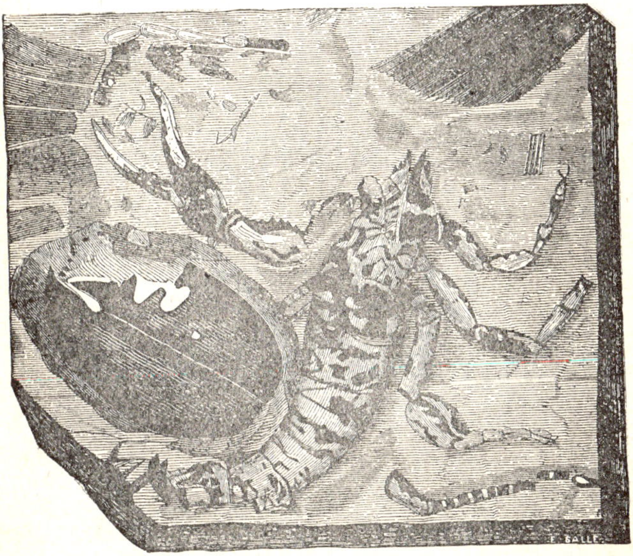

Fig. 123.—*Cyclophthalmus senior.* A fossil Scorpion from the Coal-measures of Bohemia.

Xylobius (fig. 124) and *Archiulus.* Lastly, the true *insects* are represented by various forms of Beetles (*Coleoptera*), *Orthoptera* (such as Cockroaches), and *Neuropterous* insects resembling those which we have seen to have existed towards the close of the Devonian period. One of the most remarkable of the latter is a huge May-fly (*Haplophlebium Barnesi,* fig. 125), with netted wings attaining an expanse of fully seven inches, and therefore much exceeding any existing Ephemerid in point of size.

The lower groups of the *Mollusca* are abundantly represented

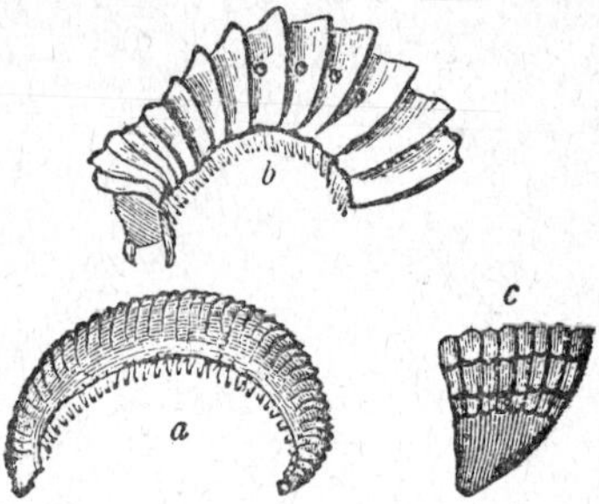

Fig. 124.—*Xylobius Sigillariæ*, a Carboniferous Myriapod. *a*, A specimen, of the natural size ; *b*, Anterior portion of the same, enlarged ; *c*, Posterior portion, enlarged. From the Coal-measures of Nova Scotia. (After Dawson.)

in the marine strata of the Carboniferous series by *Polyzoans* and *Brachiopods*. Amongst the former, although a variety of

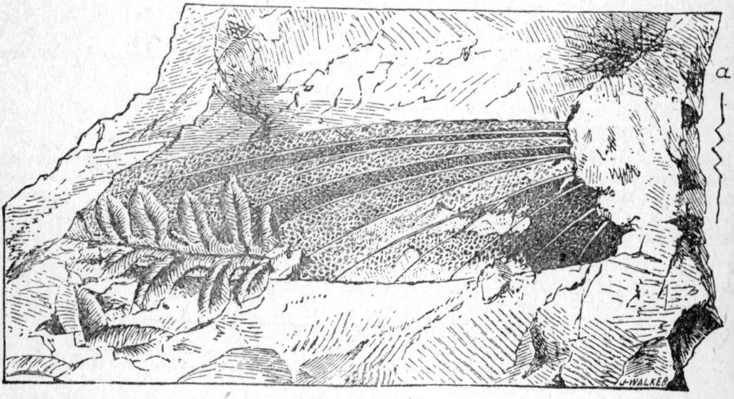

Fig. 125.—*Haplophlebium Barnesi*, a Carboniferous insect, from the Coal-measures of Nova Scotia. (After Dawson.)

other types are known, the majority still belong to the old group of the "Lace-corals" (*Fenestellidæ*), some of the characteristic forms of which are here figured (fig. 126). The graceful netted fronds of *Fenestella, Retepora,* and *Polypora* (fig. 126 *a*) are highly characteristic, as are the slender toothed branches of *Glauconome* (fig. 126, *b*). A more singular form, however, is the curious *Archimedes* (fig. 126, *c*), which is so characteristic of the Carboniferous formation of North America. In this remarkable type, the colony consists of a succession of funnel-

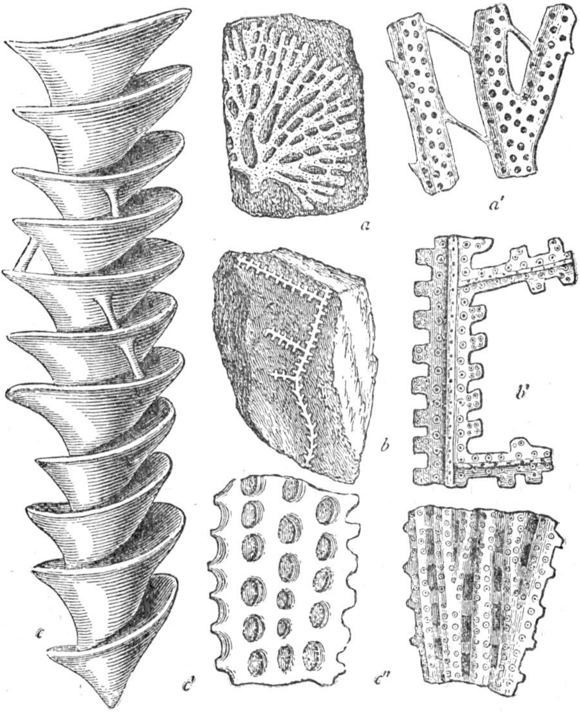

Fig. 126.—Carboniferous *Polyzoa*. *a*, Fragment of *Polypora dendroides*, of the natural size, Ireland; *a'* Small portion of the same, enlarged to show the cells; *b*, *Glauconome pulcherrima*, a fragment, of the natural size, Ireland; *b'*, Portion of the same, enlarged; *c*, The central screw-like axis of *Archimedes Wortheni*, of the natural size—Carboniferous, America; *c'*, Portion of the exterior of the frond of the same, enlarged; *c''*, Portion of the interior of the frond of the same showing the mouths of the cells, enlarged. (After M'Coy and Hall.)

shaped fronds, essentially similar to *Fenestella* in their structure, springing in a continuous spiral from a strong screw-like vertical axis. The outside of the fronds is simply striated; but the branches exhibit on the interior the mouths of the little cells in which the semi-independent beings composing the colony originally lived.

The *Brachiopods* are extremely abundant, and for the most part belong to types which are exclusively or principally Palæozoic in their range. The old genera *Strophomena, Orthis* (fig. 127, *c*), *Athyris* (fig. 127, *e*), *Rhynchonella* (fig. 127, *g*), and *Spirifera* (fig. 127, *h*), are still well represented—the latter, in particular, existing under numerous specific forms, conspicuous by their abundance and sometimes by their size. Along with these ancient groups, we have representatives—for the first time in any plenty—of the great genus *Terebratula* (fig. 127, *d*), which underwent a great expansion during later periods, and still exists at the present day. The most characteristic Carboniferous Brachiopods, however, belong to the family of the *Productidæ,* of which the principal genus is *Producta* itself. This family commenced its existence in the Upper Silurian with the genus *Chonetes,* distinguished by its spinose hinge-margin. This genus lived through the Devonian, and flourished in the Carboniferous (fig. 127, *f*). The genus *Producta* itself, represented in the Devonian by the nearly allied *Productella,* appeared first in the Carboniferous, at any rate in force, and survived into the Permian; but no member of this extensive family has yet been shown to have over-lived the Palæozoic period. The *Productæ* of the Carboniferous are not only exceedingly abundant, but they have in many instances a most extensive geographical range, and some species attain what may fairly be considered gigantic dimensions. The shell (fig. 127, *a* and *b*) is generally more or less semicircular, with a straight hinge-margin, and having its lateral angles produced into larger or smaller ears (hence its generic name—"*cochlea producta*"). One valve (the ventral) is usually strongly convex, whilst the other (the dorsal) is flat or concave, the surface of both being adorned with radiating ribs, and with hollow tubular spines, often of great length. The valves are not locked together by teeth, and there is no sign in the fully-grown shell of an opening in or between the valves for the emission of a muscular stalk for the attachment of the shell to foreign objects. It is probable, therefore, that the *Productæ,* unlike the ordinary Lamp-shells, lived an independent existence, their long spines apparently serving to anchor them firmly in the mud or ooze of the sea-bottom; but Mr. Robert Etheridge, jun., has recently shown that in one species the spines were actually employed as organs of adhesion, whereby the shell was permanently attached to some extraneous object, such as the stem of a Crinoid. The two species here figured are interesting for their extraordinarily extensive geographical range—*Producta semireticulata* (fig. 127, *a*) being found in the Carboniferous rocks of Britain, the continent of Europe, Central Asia, China, India, Australia, Spitzbergen, and North and South America; whilst *P. longispina* (fig. 127, *b*) has a distribution little if at all less wide.

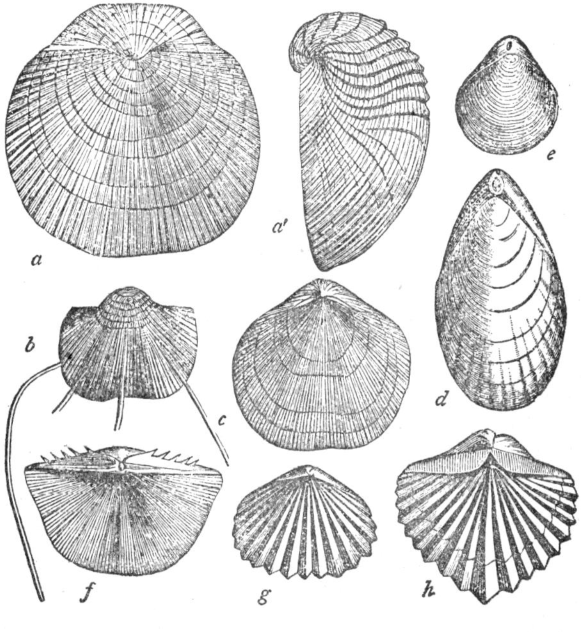

Fig. 127 —Carboniferous *Brachiopoda* *a*, *Producta semireticulata*, showing the slightly concave dorsal valve; *a'* Side view of the same, showing the convex ventral valve; *b*, *Producta longispina*; *c*, *Orthis resupinata*; *d*, *Terebratula hastata*; *e*, *Athyris subtilita*; *f*, *Chonetes Hardrensis*, *g*. *Rhynchonella pleurodon*; *h*, *Spirifera trigonalis*. Most of these forms are widely distributed in the Carboniferous Limestone of Britain, Europe, America, &c. All the figures are of the natural size. (After Davidson, De Koninck, and Meek.)

The higher *Mollusca* are abundantly represented in the Carboniferous rocks by Bivalves (*Lamellibranchs*), Univalves (*Gasteropoda*), Winged-snails (*Pteropoda*), and *Cephalopods*. Amongst the Bivalves we may note the great abundance of

Scallops (*Aviculopecten* and other allied forms), together with numerous other types—some of ancient origin, others represented here for the first time. Amongst the Gasteropods, we find the characteristically Palæozoic genera *Macrocheilus* and *Loxonema*, the almost exclusively Palæozoic *Euomphalus*, and the persistent genus *Pleurotomaria;* whilst the free-swimming Univalves (*Heteropoda*) are represented by *Bellerophon* and *Porcellia*, and the *Pteropoda* by the old genus *Conularia*. With regard to the Carboniferous Univalves, it is also of interest to note here the first appearance of true air-breathing or terrestrial Molluscs, as discovered by Dawson and Bradley in the Coal-measures of Nova Scotia and Illinois. Some of these (*Conulus priscus*) are true Land-snails, resembling the existing *Zonites;* whilst others (*Pupa vetusta*, fig. 128) appear to be generically inseparable from the "Chrysalis-shells" (*Pupa*) of the present day. All the known forms—three in number—are of small size, and appear to have been local in their distribution or in their preservation. More important, however, than any of the preceding, are the *Cephalopoda*, represented, as before, exclusively by the chambered shells of the Tetrabranchiates. The older and simpler type of these, with simple plain septa, and mostly a central siphuncle, is represented by the straight conical shells of the ancient genus *Orthoceras*, and the bow-shaped shells of the equally ancient *Cyrtoceras*—some of the former attaining a great size. The spirally-curved discoidal shells of the persistent genus *Nautilus* are also not unknown, and some of these likewise exhibit very considerable dimensions. Lastly, the more complex family of the *Ammonitidæ*, with lobed or angulated septa, and a dorsally-placed siphuncle (situated on the convex side of the curved shells), now for the first time commences to acquire a considerable prominence. The principal representative of this group is the genus *Gonia-*

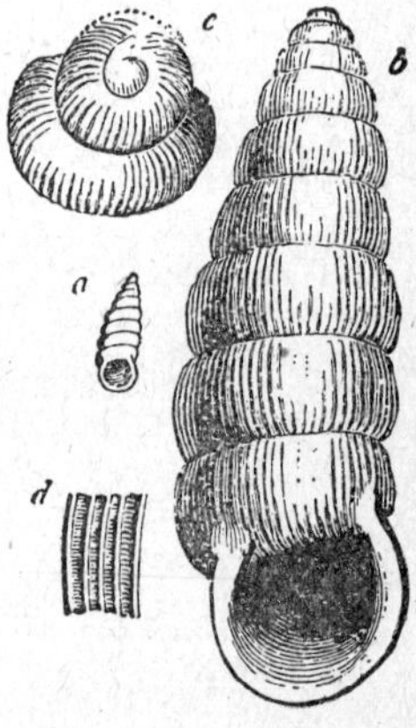

Fig. 128.—*Pupa* (*Dendropupa*) *vetusta*, a Carboniferous Land-snail from the Coal-measures of Nova Scotia. *a*, The shell, of the natural size ; *b*, The same, magnified ; *c*, Apex of the shell, enlarged ; *d*, Portion of the surface, enlarged. (After Dawson.)

tites (fig. 129), which commenced its existence in the Upper Silurian, is well represented in the Devonian, and attains its maximum here. In this genus, the shell is spirally curved, the septa are strongly lobed or angulated, though not elaborately frilled as in the Ammonites, and the siphuncle is dorsal. In addition to *Goniatites,* the shells of true *Ammonites,* so characteristic of the Secondary period, have been described by Dr. Waagen as occurring in the Carboniferous rocks of India.

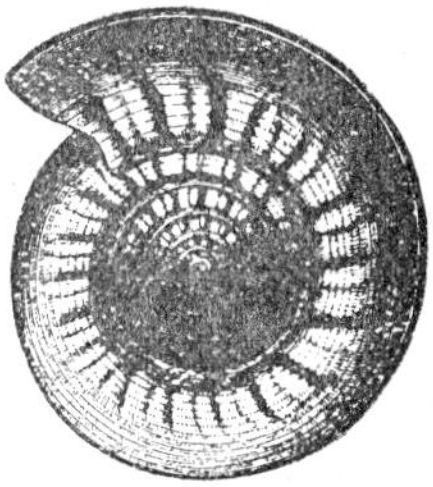

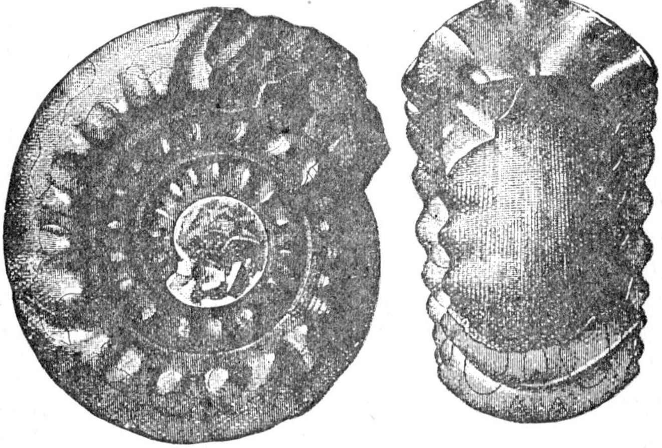

Fig. 129.—*Goniatites* (*Aganides*) *Jossæ.* Carboniferous Limestone.

Coming finally to the *Vertebrata,* we have in the first place to very briefly consider the Carboniferous *fishes.* These are numerous; but, with the exception of the still dubious "Conodonts," belong wholly to the groups of the *Ganoids* and the *Placoids* (including under the former head remains which perhaps are truly referable to the group of the *Dipnoi* or Mudfishes). Amongst the *Ganoids,* the singular buckler-headed fishes of the Upper Silurian and Devonian (*Cephalaspidæ*) have apparently disappeared; and the principal types of the Carboniferous belong to the groups respectively represented at the present day by the Gar pike (*Lepidosteus*) of the North American lakes, and the *Polypterus* of the rivers of Africa. Of the former, the genera *Palæoniscus* and *Amblypterus* (fig. 130), with their small rhomboidal and enamelled scales, and their strongly unsymmetrical tails, are perhaps the most abundant. Of the latter, the most important are species belonging to

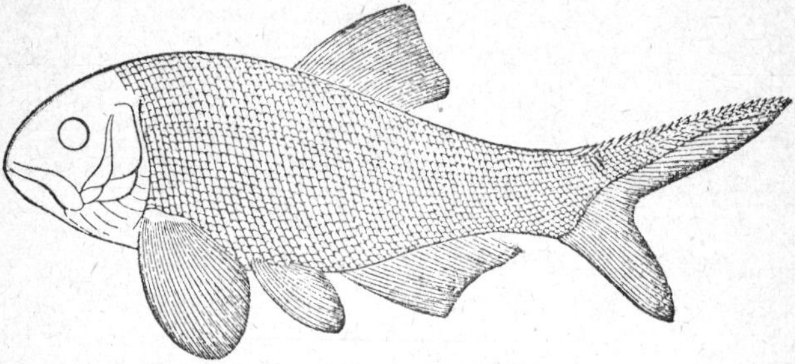

Fig. 130.—*Amblypterus macropterus*. Carboniferous.

the genera *Megalichthys* and *Rhizodus*, comprising large fishes, with rhomboidal scales, unsymmetrical ("heterocercal") tails, and powerful conical teeth. These fishes are sometimes said to be "sauroid," from their presenting some Reptilian features in their organization, and they must have been the scourges of the Carboniferous seas. The remains of *Placoid* fishes in the Carboniferous strata are very numerous, but consist wholly of teeth and fin-spines, referable to forms more or less closely allied to our existing Port Jackson Sharks, Dog-fishes, and Rays. The teeth are of very various shapes and sizes,—some with sharp, cutting edges (*Petalodus, Cladodus,* &c.); others in the form of broad crushing plates, adapted, like the teeth of the existing Port Jackson Shark (*Cestracion Philippi*), for breaking down the hard shells of Molluscs and Crustaceans. Amongst the many kinds of these latter, the teeth of *Psammodus* and *Cochliodus* (fig. 131) may be mentioned as specially characteristic. The fin-spines are mostly similar to those so common in the Devonian deposits, consisting of hollow defensive spines implanted in front of the pectoral or other fins, usually slightly curved, often superficially ribbed or sculptured, and not uncommonly serrated or toothed. The genera *Ctenacanthus, Gyracanthus, Homacanthus,* &c., have been founded for the reception of these defensive weapons, some of which indicate fishes of great size and predaceous habits.

In the Devonian rocks we meet with no other remains of

Vertebrated animals save fishes only; but the Carboniferous deposits have yielded remains of the higher group of the *Amphibians*. This class, comprising our existing Frogs, Toads, and Newts, stands to some extent in a position midway between the class of the fishes and that of the true reptiles, being distinguished from the latter by the fact that its members invariably possess gills in their early condition, if not throughout life; whilst they are separated from the former by always possessing true lungs when adult, and by the fact that the limbs (when present at all) are never in the form of fins. The Amphibians, therefore, are all water-breathers when young, and have respiratory organs adapted for an aquatic mode of life; whereas, when grown up, they develop lungs, and with these the capacity for breathing air directly. Some of them, like the Frogs and Newts, lose their gills altogether on attaining the adult condition; but others, such as the living *Proteus* and *Menobranchus*, retain their gills even after acquiring their lungs, and are thus fitted indifferently for an aquatic or terrestrial existence. The name of "Amphibia," though applied to the whole class, is thus not precisely appropriate except to these last-mentioned forms (Gr. *amphi*, both; *bios*, life). The Amphibians also differ amongst themselves according as to whether they keep permanently the long tail which they all possess when young (as do the Newts and Salamanders), or lose this appendage when grown up (as do the Frogs and Toads). Most of them have naked skins, but a few living and many extinct forms have hard structures in the shape of scales developed in the integument. All of them have well-ossified skeletons, though some fossil types are partially deficient in this respect; and all of them which possess limbs at all have these appendages supported by bones essentially similar to those found in the limbs of the higher Vertebrates. All the Carboniferous Amphibians belong to a group which has now wholly passed away—namely, that of the *Labyrinthodonts*. In the marine strata which form the base of the Carboniferous series these creatures have only

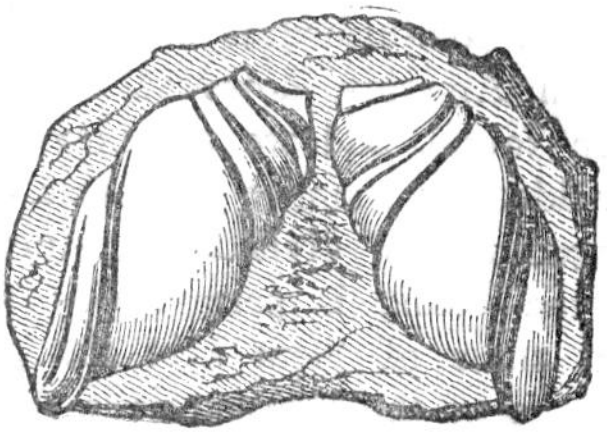

Fig. 131.—Teeth of *Cochliodus contortus*. Carboniferous Limestone, Britain.

been recognized by their curious hand-shaped footprints, similar in character to those which occur in the Triassic rocks, and which will be subsequently spoken of under the name of *Cheirotherium*. In the Coal-measures of Britain, the continent of Europe, and North America, however, many bones of these animals have been found, and we are now tolerably well acquainted with a considerable number of forms. All of them seem to have belonged to the division of Amphibians in which the long tail of the young is permanently retained; and there is evidence that some of them kept the gills also throughout life. The skull is of the characteristic Amphibian type (fig. 132, *a*), with two occipital condyles, and having its surface singularly pitted and sculptured; and the vertebræ are hollowed out at both ends. The lower surface of the body was defended by an armor of singular integumentary shields or scales (fig. 132, *c*); and an extremely characteristic feature (from which the entire

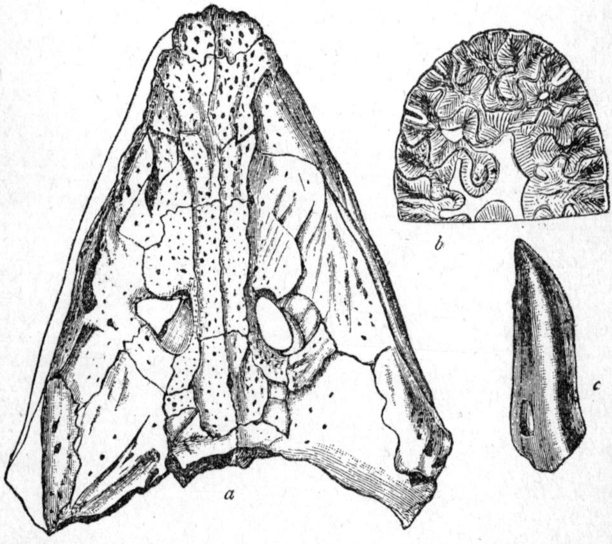

Fig. 132.—*a*, Upper surface of the skull of *Anthracosaurus Russelli*, one-sixth of the natural size; *b*, Part of one of the teeth cut across, and highly magnified to show the characteristic labyrinthine structure; *c*, One of the integumentary shields or scales, one-half of the natural size. Coal-measures, Northumberland. (After Atthey.)

group derives its name) is, that the walls of the teeth are deeply folded, so as to give rise to an extraordinary "labyrinthine" pattern when they are cut across (fig. 132, *b*). Many of the

Carboniferous Labyrinthodonts are of no great size, some of them very small, but others attain comparatively gigantic dimensions, though all fall short in this respect of the huge examples of this group which occur in the Trias. One of the largest, and at the same time most characteristic, forms of the Carboniferous series, is the genus *Anthracosaurus,* the skull of which is here figured.

No remains of the true Reptiles, Birds, or Quadrupeds have as yet been certainly detected in the Carboniferous deposits in any part of the world. It should, however, be mentioned, that Professor Marsh, one of the highest authorities on the subject, has described from the Coal-formation of Nova Scotia certain vertebræ which he believes to have belonged to a marine reptile (*Eosaurus Acadianus*), allied to the great *Ichthyosauri* of the Lias. Up to this time no confirmation of this determination has been obtained by the discovery of other and more unquestionable remains, and it therefore remains doubtful whether these bones of *Eosaurus* may not really belong to large Labyrinthodonts.

Literature.

The following list contains some of the more important of the original sources of information to which the student of Carboniferous rocks and fossils may refer:—

(1) 'Geology of Yorkshire,' vol. ii.; 'The Mountain Limestone District.' John Phillips.
(2) 'Siluria.' Sir Roderick Murchison.
(3) 'Memoirs of the Geological Survey of Great Britain and Ireland.'
(4) 'Geological Report on Londonderry,' &c. Portlock.
(5) 'Acadian Geology.' Dawson.
(6) 'Geology of Iowa.' vol. i. James Hall.
(7) 'Reports of the Geological Survey of Illinois' (Geology and Palæontology). Meek, Worthen, &c.
(8) 'Reports of the Geological Survey of Ohio' (Geology and Palæontology). Newberry, Cope, Meek, Hall, &c.
(9) 'Description des Animaux fossiles qui se trouvent dans le Terrain Carbonifère de la Belgique,' 1843; with subsequent monographs on the genera *Productus* and *Chonetes,* on *Crinoids,* on *Corals,* &c. De Koninck.
(10) 'Synopsis of the Carboniferous Fossils of Ireland.' M'Coy.
(11) 'British Palæozoic Fossils.' M'Coy.
(12) 'Figures of Characteristic British Fossils.' Baily.
(13) 'Catalogue of British Fossils.' Morris.
(14) 'Monograph of the Carboniferous Brachiopoda of Britain (Palæontographical Society). Davidson.
(15) 'Monograph of the British Carboniferous Corals' Palæontographical Society). Milne-Edwards and Haime.

(16) 'Monograph of the Carboniferous Bivalve Entomostraca of Britain' (Palæontographical Society). Rupert Jones, Kirkby, and George S. Brady.
(17) 'Monograph of the Carboniferous Foraminifera of Britain' (Palæontographical Society). H. B. Brady.
(18) "On the Carboniferous Fossils of the West of Scotland"—'Trans. Geol. Soc.,' of Glasgow, vol. iii., Supplement. Young and Armstrong.
(19) 'Poissons Fossiles.' Agassiz.
(20) "Report on the Labyrinthodonts of the Coal-measures"—'British Association Report,' 1873. L. C. Miall.
(21) 'Introduction to the Study of Palæontological Botany.' John Hutton Balfour.
(22) 'Traité de Paléontologie Végétale.' Schimper.
(23) 'Fossil Flora.' Lindley and Hutton.
(24) 'Histoire des Végétaux Fossiles.' Brongniart.
(25) 'On Calamites and Calamodendron' (Monographs of the Palæontographical Society). Binney.
(26) 'On the Structure of Fossil Plants found in the Carboniferous Strata' (Palæontographical Society). Binney.

Also numerous memoirs by Huxley, Davidson, Martin Duncan, Professor Young, John Young, R. Etheridge, jun.; Baily, Carruthers, Dawson, Binney, Williamson, Hooker, Jukes, Geikie, Rupert Jones, Salter, and many other British and Foreign observers.

CHAPTER XIV.

THE PERMIAN PERIOD.

The Permian formation closes the long series of the Palæozoic deposits, and may in some respects be considered as a kind of appendix to the Carboniferous system, to which it cannot be compared in importance, either as regards the actual bulk of its sediments or the interest and variety of its life-record. Consisting, as it does, largely of red rocks—sandstones and marls—for the most part singularly destitute of organic remains, the Permian rocks have been regarded as a lacustrine or fluviatile deposit; but the presence of well-developed limestones with indubitable marine remains entirely negatives this view. It is, however, not improbable that we are presented in the Permian formation, as known to us at present, with a series of sediments laid down in inland seas of great extent, due to the subsidence over large areas of the vast land-surfaces of the Coal-measures. This view, at any rate, would explain some of the more puzzling physical characters of the formation, and would not be definitely negatived by any of its fossils.

A large portion of the Permian series, as already remarked,

consists of sandstones and marls, deeply reddened by peroxide of iron, and often accompanied by beds of gypsum or deposits of salt. In strata of this nature few or no fossils are found; but their shallow-water origin is sufficiently proved by the presence of the footprints of terrestrial animals, accompanied in some cases by well-defined "ripple-marks." Along with these are occasionally found massive breccias, holding larger or smaller blocks derived from the older formations; and these have been supposed to represent an old "boulder-clay," and thus to indicate the prevalence of an arctic climate. Beds of this nature must also have been deposited in shallow water. In all regions, however, where the Permian formation is well developed, one of its most characteristic members is a Magnesian limestone, often highly and fantastically concretionary, but containing numerous remains of genuine marine animals, and clearly indicating that it was deposited beneath a moderate depth of salt water.

It is not necessary to consider here whether this formation can be retained as a distinct division of the geological series. The name of *Permian* was given to it by Sir Roderick Murchison, from the province of Perm in Russia, where rocks of this age are extensively developed. Formerly these rocks were grouped with the succeeding formation of the Trias under the common name of "New Red Sandstone." This name was given them because they contain a good deal of red sandstone, and because they are superior to the Carboniferous rocks, while the Old Red Sandstone is inferior. Nowadays, however, the term "New Red Sandstone" is rarely employed, unless it be for red sandstones and associated rocks, which are seen to overlie the Coal-measures, but which contain no fossils by which their exact age may be made out. Under these circumstances, it is sometimes convenient to employ the term "New Red Sandstone." The New Red, however, of the older geologists, is now broken up into the two formations of the Permian and Triassic rocks—the former being usually considered as the top of the Palæozoic series, and the latter constituting the base of the Mesozoic.

In many instances, the Permian rocks are seen to repose unconformably upon the underlying Carboniferous, from which they can in addition be readily separated by their lithological characters. In other instances, however, the Coal-measures terminate upwards in red rocks, not distinguishable by their mineral characters from Permian; and in other cases no physical discordance between the Carboniferous and Permian strata can be detected. As a general rule, also, the Permian rocks appear to pass upwards conformably into the Trias. The division, therefore, between the Permian and Triassic rocks, and consequently between the Palæozoic and Mesozoic series, is not founded upon any conspicuous or universal physical break, but upon the difference in life which is observed in comparing the marine animals of the Carboniferous and Permian with those of the Trias. It is to be observed, however, that this difference can be solely due to the fact that the Magnesian Limestone of the Permian series presents us with only a small, and not a typical, portion of the marine deposits which must have been accumulated in some area at present unknown to us during the period which elapsed between the formation of the great marine limestones of the Lower Carboniferous and the open-sea and likewise calcareous sediments of the Middle Trias.

The Permian rocks exhibit their most typical features in Russia and Germany, though they are very well developed in parts of Britain, and they occur in North America. When well developed, they exhibit three main divisions: a lower set of sandstones, a middle group, generally calcareous, and an upper series of sandstones, constituting respectively the Lower, Middle, and Upper Permians.

In Russia, Germany, and Britain, the Permian rocks consist of the following members:—

1. The *Lower Permians*, consisting mainly of a great series of sandstones, of different colors, but usually red. The base of this series is often constituted by massive breccias with included fragments of the older rocks, upon which they may happen to repose; and similar breccias sometimes occur in the upper portion of the series as well. The thickness of this group varies a good deal, but may amount to 3000 or 4000 feet.

2. The *Middle Permians*, consisting, in their typical development, of laminated marls, or "marl-slate," surmounted by beds of magnesian limestone (the "Zechstein" of the German geologists). Sometimes the limestones are degenerate or wholly deficient, and the series may consist of sandy shales and gypsiferous clays. The magnesian limestone, however, of the Middle Permians is, as a rule, so well marked a feature that it was long spoken of as *the* Magnesian Limestone.

3. The *Upper Permians*, consisting of a series of sandstones

and shales, or of red or mottled marls, often gypsiferous, and sometimes including beds of limestone.

In North America, the Permian rocks appear to be confined to the region west of the Mississippi, being especially well developed in Kansas. Their exact limits have not as yet been made out, and their total thickness is not more than a few hundred feet. They consist of sandstones, conglomerates, limestones, marls, and beds of gypsum.

The following diagrammatic section shows the general sequence of the Permian deposits in the north of England, where the series is extensively developed (fig. 133):—

Generalized Section of the Permian Rocks in the North of England.

Fig. 133.

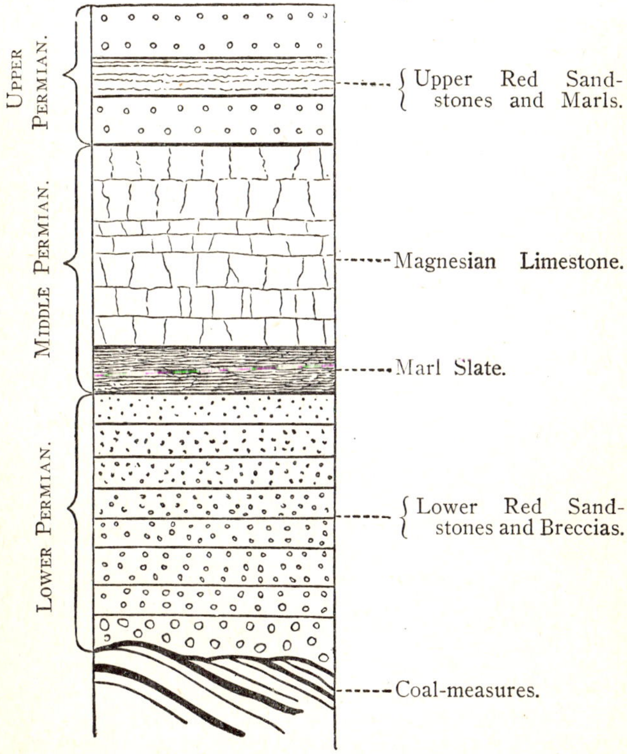

The record of the *life* of the Permian period is but a scanty one, owing doubtless to the special peculiarities of such of the deposits of this age with which we are as yet acquainted. Red rocks are, as a general rule, more or less completely unfossiliferous, and sediments of this nature are highly characteristic of the Permian. Similarly, magnesian limestones are rarely as highly charged with organic remains as is the case with normal calcareous deposits, especially when they have been subjected to concretionary action, as is observable to such a marked extent in the Permian limestones. Nevertheless, much interest is attached to the organic remains, as marking a kind of transition-period between the Palæozoic and Mesozoic epochs.

The *plants* of the Permian period, as a whole, have a distinctly Palæozoic aspect, and are far more nearly allied to those of the Coal-measures than they are to those of the earlier Secondary rocks; though the Permian *species* are mostly distinct from the Carboniferous, and there are some new genera. Thus, we find species of *Lepidodendron, Calamites, Equisetites, Asterophyllites, Annularia,* and other highly characteristic Carboniferous genera. On the other hand, the *Sigillarioids* of the Coal seem to have finally disappeared at the close of the

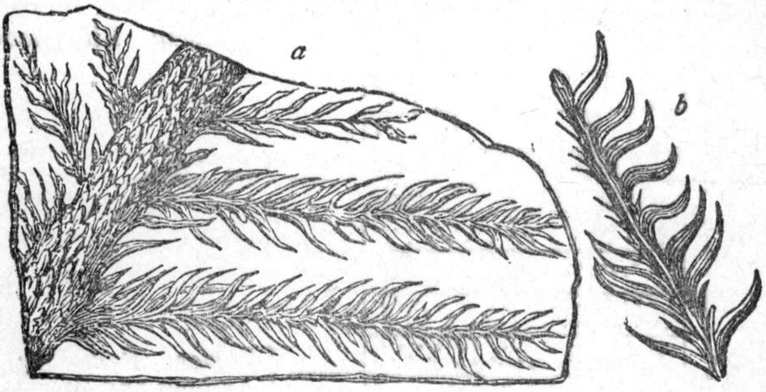

Fig. 134.—*Walchia piniformis*, from the Permian of Saxony. *a*, Branch; *b*, Twig. (After Gutbier.)

Carboniferous period. Ferns are abundant in the Permian rocks, and belong for the most part to the well-known Carboniferous genera *Alethopteris, Neuropteris, Sphenopteris,* and *Pecopteris.* There are also Tree-ferns referable to ancient genus *Psaronius.* The *Conifers* of the Permian period are numerous, and belong in part to Carboniferous genera. A characteristic genus, however, is *Walchia* (fig. 134), distinguished by its lax short leaves. This genus, though not exclusively Permian, is mainly so, the best-known species being the *W. piniformis.* Here, also, we meet with Conifers which produce true cones, and which differ, therefore, in an important degree from the Taxoid Conifers of the Coal-measures. Besides *Walchia,* a characteristic form of these is the *Ullmania selaginoides,* which occurs in the Magnesian Limestone of Durham, the Middle Permian of Westmorland, and the "Kupfer-schiefer" of Germany. The group of the *Cycads,* which we shall subsequently find to be so characteristic of the vegetation of the Secondary period, is, on the other hand, only doubtfully represented in the Permian deposits by the singular genus *Nœggerathia.*

The *Protozoans* of the Permian rocks are few in number, and for the most part imperfectly known. A few *Foraminifera* have been obtained from the Magnesian Limestone of England, and the same formation has yielded some ill-understood *Sponges.* It does not seem, however, altogether impossible that some of the singular "concretions" of this formation may ultimately prove to have an organic structure, though others would appear to be clearly of purely inorganic origin. From the Permian of Saxony, Professor Geinitz has described two species of *Spongillopsis,* which he believes to be most nearly allied to the existing fresh-water Sponges (*Spongilla*). This observation has an interest as bearing upon the mode of deposition and origin of the Permian sediments.

The *Cœlenterates* are represented in the Permian by but a few Corals. These belong partly to the *Tabulate* and partly to the *Rugose* division; but the latter great group, so abundantly represented in Silurian, Devonian, and Carboniferous seas, is now extraordinarily reduced in numbers, the British strata of this age yielding only species of the single genus *Polycœlia.* So far, therefore, as at present known, all the characteristic genera of the Rugose Corals of the Carboniferous had become extinct before the deposition of the limestones of the Middle Permian.

The *Echinoderms* are represented by a few *Crinoids,* and by a Sea-urchin belonging to the genus *Eocidaris.* The latter genus is nearly allied to the *Archæocidaris* of the Carboniferous, so that this Permian form belongs to a characteristically Palæozoic type.

A few *Annelides* (*Spirorbis, Vermilia,* &c.) have been described, but are of no special importance. Amongst the *Crustaceans,* however, we have to note the total absence of the great Palæozoic group of the *Trilobites;* whilst the little *Ostracoda* and *Phyllopods* still continue to be represented. We have also to note the first appearance here of the "Short-tailed" Decapods or Crabs (*Brachyura*), the highest of all the groups of *Crustacea,* in the person of *Hemitrochiscus paradoxus,* an extremely minute Crab from the Permian of Germany.

Amongst the *Mollusca,* the remains of *Polyzoa* may fairly be said to be amongst the most abundant of all the fossils of the Permian formation. The principal forms of these are the fronds of the Lace-corals (*Fenestella, Retepora,* and *Synocladia*), which are very abundant in the Magnesian Limestone of the north of England, and belong to various highly characteristic species (such as *Fenestella retiformis, Retepora Ehrenbergi,* and *Synocladia virgulacea*). The *Brachiopoda* are also represented in moderate numbers in the Permian. Along with species of the persistent genera *Discina Crania,* and *Lingula,* we still meet with representatives of the old groups *Spirifera, Athyris,* and *Streptorhynchus;* and the Carboniferous *Productæ* yet survive under well-marked and characteristic types, though in much-diminished numbers. The species of Brachiopods here figured (fig. 135) are characteristic of the Magnesian Limestone in Britain and of the corresponding strata on the Continent.

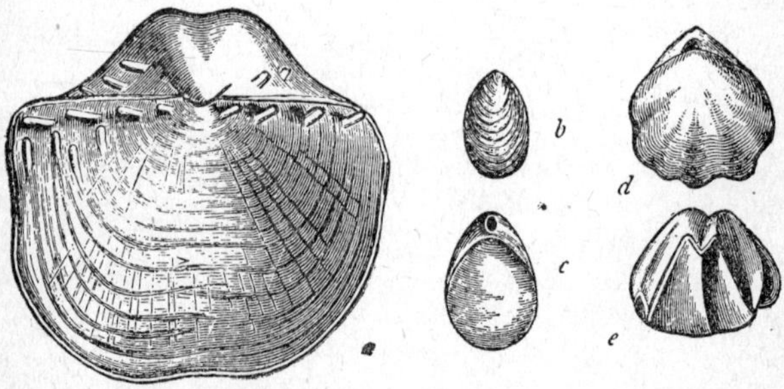

Fig. 135.—Brachiopods of the Permian formation. *a, Producta horrida; b, Lingula Credneri; c, Perebratula elongata; d* and *e, Camarophoria globulina.* (After King.)

Upon the whole, the most characteristic Permian *Brachiopods* belong to the genera *Producta, Strophalosia,* and *Camarophoria.*

The *Bivalves* (*Lamellibranchiata*) have a tolerably varied development in the Permian rocks; but nearly all the old types, except some of those which occur in the Carboniferous, have now disappeared. The principal Permian Bivalves belong to the groups of the Pearl Oysters (*Aviculidæ*) and the *Trigoniadæ,* represented by genera such as *Bakewellia* and *Schizodus;* the true Mussels (*Mytilidæ*), represented by species which have been referred to *Mytilus* itself; and the Arks (*Arcadæ*), represented by species of the genera *Arca* (fig. 136) and *Byssoarca.* The first and last of these three families have a very ancient origin; but the family of the *Trigoniadæ,* though feebly represented at the present day, is one which attains its maximum development in the Mesozoic period.

The *Univalves* (*Gasteropoda*) are rare, and do not demand special notice. It may be observed, however, that the Palæozoic genera *Euomphalus, Murchisonia, Loxonema,* and *Macrocheilus* are still in existence, together with the persistent genus *Pleurotomaria.* *Pteropods* of the old genera *Theca* and *Conularia* have been discovered; but the first of these characteristically Palæozoic types finally dies out here, and the second only survives but a short time longer. Lastly, a few *Cephalopods* have been

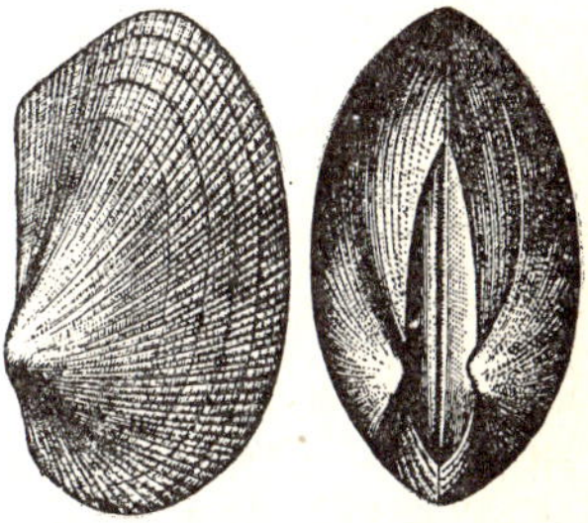

Fig. 136.—*Arca antiqua.* Permian.

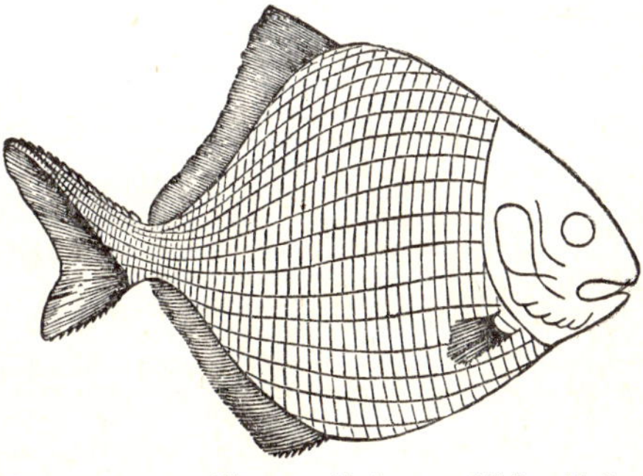

Fig. 137.—*Platysomus gibbosus*, a "heterocercal" Ganoid, from the Middle Permian of Russia.

found, still wholly referable to the Tetrabranchiate group, and belonging to the old genera *Orthoceras* and *Cyrtoceras* and the long-lived *Nautilus*.

Amongst *Vertebrates,* we meet in the Permian period not only with the remains of Fishes and Amphibians, but also, for the first time, with true Reptiles. The *Fishes* are mainly *Ganoids,* though there are also remains of a few Cestraciont Sharks. Not only are the *Ganoids* still the predominant group of Fishes, but all the known forms possess the unsymmetrical ("heterocercal") tail which is so characteristic of the Palæozoic Ganoids. Most of the remains of the Permian Fishes have been obtained from the "Marl-slate" of Durham and the corresponding "Kupfer-schiefer" of Germany, on the horizon of the Middle Permian; and the principal genera of the Ganoids are *Palæoniscus* and *Platysomus* (fig. 137).

The *Amphibians* of the Permian period belong principally to the order of the *Labyrinthodonts,* which commenced to be represented in the Carboniferous, and has a large development in the Trias. Under the name, however, of *Palæosiren Beinerti,* Professor Geinitz has described an Amphibian from the Lower Permian of Germany, which he believes to be most nearly allied to the existing "Mud-eel" (*Siren lacertina*) of North America, and therefore to be related to the Newts and Salamanders (*Urodela*).

Finally, we meet in the Permian deposits with the first undoubted remains of true Reptiles. These are distinguished, as a class, from the *Amphibians,* by the fact that they are air-breathers throughout the whole of their life, and therefore are at no time provided with gills; whilst they are exempt from that metamorphosis which all the *Amphibia* undergo in early life, consequent upon their transition from an aquatic to a more or less purely aerial mode of respiration. Their skeleton is well ossified; they usually have horny or bony plates, singly or in combination, developed in the skin; and their limbs (when present) are never either in the form of *fins* or *wings,* though sometimes capable of acting in either of these capacities, and liable to great modifications of form and structure. Though there can be no doubt whatever as to the occurrence of genuine Reptiles in deposits of unquestionable Permian age, there is still uncertainty as to the precise number of types which may have existed at this period. This uncertainty arises partly from the difficulty of deciding in all cases whether a given bone be truly Labyrinthodont or Reptilian, but more especially from the confusion which exists at present between the Permian and the overlying Triassic deposits. Thus there are various deposits in different regions which have yielded the remains of Reptiles, and which cannot in the meanwhile be definitely referred either to the Permian series or to the Trias by clear stratigraphical or palæonto-

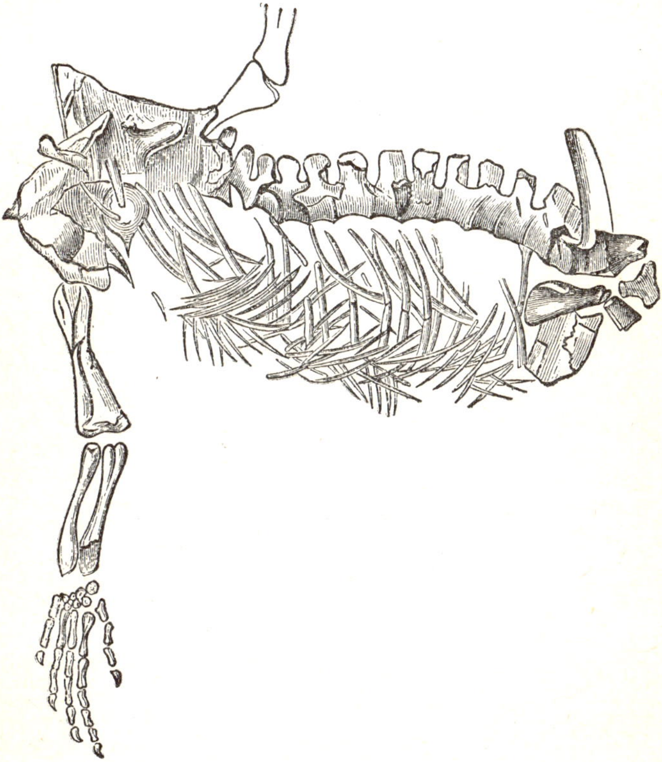

Fig. 138.—*Protorosaurus Speneri*, Middle Permian, Thuringia, reduced in size. (After Von Meyer.) [Copied from Dana.]

logical evidence. All that can be done in such cases is to be guided by the characters of the Reptiles themselves, and to judge by their affinities to remains from known Triassic or Permian rocks to which of these formations the beds containing them should be referred; but it is obvious that this method of procedure is seriously liable to lead to error. In accordance, however, with this, the only available mode of determination in some cases, the remains of *Thecodontosaurus* and *Palæosaurus* discovered in the dolomitic conglomerates near Bristol will be considered as Triassic, thus leaving *Protorosaurus* * as the principal and most important representative of the Permian Reptiles. † The type-species of the genus *Protorosaurus* is the *P. Speneri* (fig. 138) of the "Kupfer-schiefer" of Thuringia, but other allied species have been detected in the Middle Permian of Germany and the north of England. This Reptile attained a length of from three to four feet; and it has been generally referred to the group of the Lizards (*Lacertilia*), to which it is most nearly allied in its general structure, at the same time that it differs from all existing members of this group in the fact that its numerous conical and pointed teeth were implanted in distinct sockets in the jaws—this being a Crocodilian character. In other respects, however, *Protorosaurus* approximates closely to the living Monitors (*Varanidæ*); and the fact that the bodies of the vertebræ are slightly cupped or hollowed out at the ends would lead to the belief that the animal was aquatic in its habits. At the same time, the structure of the hind-limbs and their bony supports proves clearly that it must have also possessed the power of progression upon the land. Various other Reptilian bones have been described from the Permian formation, of which some are probably really referable to Labyrinthodonts, whilst others are regarded by Professor Owen as referable to the order of the "Theriodonts," in which the teeth are implanted in sockets, and resemble those of carnivorous quadrupeds in consisting of three groups in each jaw (namely, incisors, canines, and molars). Lastly, in red sandstones of Permian age in Dumfriesshire have been discovered the tracks of what would ap-

* Though commonly spelt as above, it is probable that the name of this Lizard was really intended to have been *Proterosaurus*—from the Greek *proteros*, first; and *saura*, lizard: and this spelling is followed by many writers.

† In an extremely able paper upon the subject (Quart. Journ. Geol. Soc., vol. xxvi.), Mr. Etheridge has shown that there are good physical grounds for regarding the dolomitic conglomerate of Bristol as of Triassic age, and as probably corresponding in time with the Muschelkalk of the Continent.

pear to have been *Chelonians* (Tortoises and Turtles); but it would not be safe to accept this conclusion as certain upon the evidence of footprints alone. The *Chelichnus Duncani,* however, described by Sir William Jardine in his magnificent work on the 'Ichnology of Annandale,' bears a great resemblance to the track of a Turtle.

No remains of Birds or Quadrupeds have hitherto been detected in deposits of Permian age.

Literature.

The following works may be consulted by the student with regard to the Permian formation and its fossils:—

(1) "On the Geological Relations and Internal Structure of the Magnesian Limestone and the Lower Portions of the New Red Sandstone Series, &c."—'Trans. Geol. Soc.,' ser. 2, vol. iii. Sedgwick.

(2) 'The Geology of Russia in Europe.' Murchison, De Verneuil, and von Keyserling.

(3) 'Siluria.' Murchison.

(4) 'Das Permische System in Sachsen.' Geinitz and Gutbier.

(5) 'Die Versteinerungen des Deutschen Zechsteingebirges.' Geinitz.

(6) 'Die Animalischen Ueberreste der Dyas.' Geinitz.

(7) 'Monograph of the Permian Fossils of England' Palæontographical Society). King.

(8) 'Monograph of the Permian Brachiopoda of Britain' (Palæontographical Society). Davidson.

(9) "On the Permian Rocks of the North-West of England and their Extension into Scotland"—'Quart. Journ. Geol. Soc.,' vol. xx. Murchison and Harkness.

(10) 'Catalogue of the Fossils of the Permian System of the Counties of Northumberland and Durham.' Howse.

(11) 'Petrefacta Germaniæ.' Goldfuss.

(12) 'Beiträge zur Petrefaktenkunde.' Münster.

(13) 'Ein Beitrag zur Palæontologie des Deutschen Zechsteingebirges.' Von Schauroth.

(14) 'Saurier aus dem Kupfer-schiefer der Zechstein-formation.' Von Meyer.

(15) 'Manual of Palæontology.' Owen.

(16) 'Recherches sur les Poissons Fossiles.' Agassiz.

(17) 'Ichnology of Annandale.' Sir William Jardine.
(18) 'Die Fossile Flora der Permischen Formation.' Gœppert
(19) 'Genera et Species Plantarum Fossilium.' Unger.
(20) "On the Red Rocks of England of older Date than the Trias"—'Quart. Journ. Geol. Soc.,' vol. xxvii. Ramsay.

CHAPTER XV.

THE TRIASSIC PERIOD.

We come now to the consideration of the great *Mesozoic,* or Secondary series of formations, consisting, in ascending order, of the Triassic, Jurassic, and Cretaceous systems. The Triassic group forms the base of the Mesozoic series, and corresponds with the higher portion of the New Red Sandstone of the older geologists. Like the Permian rocks, and as implied by its name, the *Trias* admits of a subdivision into three groups—a Lower, Middle, and Upper Trias. Of these subdivisions the middle one is wanting in Britain; and all have received German names, being more largely and typically developed in Germany than in any other country. Thus, the Lower Trias is known as the *Bunter Sandstein;* the Middle Trias is called the *Muschelkalk;* and the Upper Trias is known as the *Keuper.*

I. The lowest division of the Trias is known as the *Bunter Sandstein* (the *Grès bigarré* of the French), from the generally variegated colors of the beds which compose it (German, *bunt,* variegated). The Bunter Sandstein of the continent of Europe consists of red and white sandstones, with red clays, and thin limestones, the whole attaining a thickness of about 1500 feet. The term "marl" is very generally employed to designate the clays of the Lower and Upper Trias; but the term is inappropriate, as they may contain no lime, and are therefore not always genuine marls. In Britain the Bunter Sandstein consists of red and mottled sandstones, with unconsolidated conglomerates, or "pebble-beds," the whole having a thickness of 1000 to 2000 feet. The Bunter Sandstein, as a rule, is very barren of fossils.

II. The Middle Trias is not developed in Britain, but it is largely developed in Germany, where it constitutes what is known as the *Muschelkalk* (Germ. *Muschel,* mussel; *kalk,* limestone), from the abundance of fossil shells which it contains. The Muschelkalk (the *Calcaire coquillier* of the French) consists of compact grey or yellowish limestones, sometimes dolomitic, and including occasional beds of gypsum and rock-salt.

III. The Upper Trias, or *Keuper* (the *Marnes irisées* of the French), as it is generally called, occurs in England; but is not so well developed as it is in Germany. In Britain, the Keuper is 1000 feet or more in thickness, and consists of white and brown sandstones with red marls, the whole topped by red clays with rock-salt and gypsum.

The Keuper in Britain is extremely unfossiliferous; but it passes upwards with perfect conformity into a very remarkable group of beds, at one time classed with the Lias, and now known under the names of the Penarth beds (from Penarth, in Glamorganshire), the Rhætic beds (from the Rhætic Alps), or the *Avicula contorta* beds (from the occurrence in them of great numbers of this peculiar Bivalve). These singular beds have been variously regarded as the highest beds of the Trias, or the lowest beds of the Lias, or as an intermediate group. The phenomena observed on the Continent, however, render it best to consider them as Triassic, as they certainly agree with the so-called Upper St. Cassian or Kössen beds which form the top of the Trias in the Austrian Alps.

The Penarth beds occur in Glamorganshire, Gloucestershire, Warwickshire, Staffordshire, and the north of Ireland; and they generally consist of a small thickness of grey marls, white limestones, and black shales, surmounted conformably by the lowest beds of the Lias. The most characteristic fossils which they contain are the three Bivalves *Cardium Rhæticum, Avicula contorta,* and *Pecten Valoniensis;* but they have yielded many other fossils, amongst which the most important are the remains of Fishes and small Mammals (*Microlestes*).

In the Austrian Alps the Trias terminates upwards in an extraordinary series of fossiliferous beds, replete with marine fossils. Sir Charles Lyell gives the following table of these remarkable deposits:—

Strata below the Lias in the Austrian Alps, in descending order.

1. Kössen beds, (Synonyms, Upper St Cassian beds of Escher and Merian.)	Grey and black limestone, with calcareous marls having a thickness of about 50 feet. Among the fossils, Brachiopoda very numerous; some few species common to the genuine Lias; many peculiar. *Avicula contorta, Pecten Valoniensis, Cardium Rhæticum, Avicula inæquivalvis, Spirifer Münsteri,* Dev. Strata containing the above fossils alternate with the Dachstein beds, lying next below.	
2. Dachstein beds.	White or greyish limestone, often in beds three or four feet thick. Total thickness of the formation above 2000 feet. Upper part fossiliferous, with some strata composed of corals (*Lithodendron.*) Lower portion without fossils. Among the characteristic shells are *Hemicardium Wulfeni, Megalodon triqueter,* and other large bivalves.	
3. Hallstadt beds (or St Cassian).	Red, pink, or white marbles, from 800 to 1000 feet in thickness, containing more than 800 species of marine fossils, for the most part mollusca. Many species of *Orthoceras.* True *Ammonites,* besides *Ceratites* and *Goniatites, Belemnites* (rare), *Porcellia, Pleurotomaria, Trochus, Monotis Salinaria,* &c	
4. *A.* Guttenstein beds. *B.* Werfen beds, base of Upper Trias? Lower Trias of some geologists.	*A.* Black and grey limestone 150 feet thick, alternating with the underlying Werfen beds. *B.* Red and green shale and sandstone, with salt and gypsum.	Among the fossils are *Ceratites cassianus, Myacites fassaensis, Naticella costata, &c.*

In the United States, rocks of Triassic age occur in several areas between the Appalachians and the Atlantic seaboard; but they show no such triple division as in Germany, and their exact place in the system is uncertain. The rocks of these areas consist of red sandstones, sometimes shaly or conglomeratic, occasionally with beds of impure limestone. Other more extensive areas where Triassic rocks appear at the surface, are found west of the Mississippi, on the slopes of the Rocky Mountains, where the beds consist of sandstones and gypsiferous marls. The American Trias is chiefly remarkable for having yielded the remains of small Marsupial (*Dromatherium*), and numerous footprints, which have generally been referred to Birds (*Brontozoum*), along with the tracks of undoubted Reptiles (*Otozoum, Anisopus,* &c.)

The subjoined section (fig. 139) expresses, in a diagrammatic manner, the general sequence of the Triassic rocks when fully developed, as, for example, in the Bavarian Alps:—

GENERALIZED SECTION OF THE TRIASSIC ROCKS OF CENTRAL EUROPE.

Fig. 139.

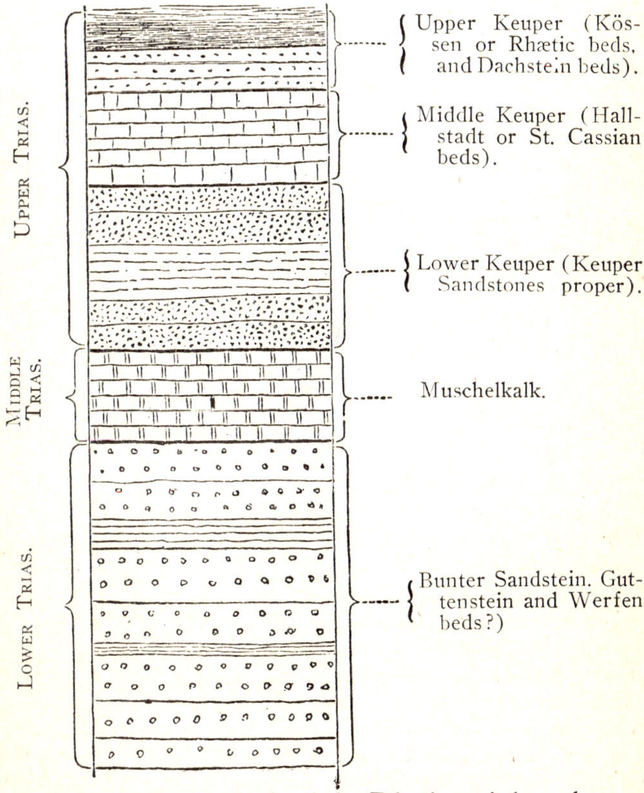

With regard to the *life* of the Triassic period, we have to notice a difference as concerns the different members of the group similar to that which has been already mentioned in connection with the Permian formation. The arenaceous deposits of the series, namely, resemble those of the Permian, not only in being commonly red or variegated in their color, but also in their conspicuous paucity of organic remains. They for the most part are either wholly unfossiliferous, or they contain the remains of plants or the bones of reptiles, such as may easily have been drifted from some neighboring shore. The few fossils which may be considered as properly belonging to these deposits are chiefly Crustaceans (*Estheria*) or Fishes, which may well have lived in the waters of estuaries or vast inland seas. We may therefore conclude, with considerable probability, that the barren sandy and marly accumulations of the Bunter Sandstein and Lower Keuper were not laid down in an open sea, but are probably brackish-water deposits, formed in estuaries or land-locked bodies of salt water. This at any rate would appear to be the case as regards these members of the series as developed in Britain and in their typical areas on the continent of Europe; and the origin of most of the North American Trias would appear to be much the same. Whether this view be correct or not, it is certain that the beds in question were laid down in *shallow* water, and in the immediate vicinity of *land*, as shown by the numerous drifted plants which they contain and the common occurrence in them of the footprints of air-breathing animals (Birds, Reptiles, and Amphibians). On the other hand, the middle and the highest members of the Trias are largely calcareous, and are replete with the remains of undoubted marine animals. There cannot, therefore, be the smallest doubt but that the Muschelkalk and the Rhætic or Kössen beds were slowly accumulated in an open sea, of at least a moderate depth; and they have preserved for us a very considerable selection from the marine fauna of the Triassic period.

The *plants* of the Trias are, on the whole, as distinctively Mesozoic in their aspect as those of the Permian are Palæozoic. In spite, therefore, of the great difficulty which is experienced in effecting a satisfactory stratigraphical separation between the Permian and the Trias, we have in this fact a proof that the two formations were divided by an interval of time sufficient to allow of enormous changes in the terrestrial vegetation of the world. The *Lepidodendroids, Asterophyllites,* and *Annulariæ,* of the Coal and Permian formations, have now apparently wholly disappeared; and the Triassic flora consists mainly of Ferns, Cycads, and Conifers, of which only the two last need special notice. The *Cycads* (fig. 140) are true exogenus plants, which in general form and habit of growth pre-

Fig. 140.—*Zamia spiralis*, a living Cycad. Australia.

sent considerable resemblance to young Palms, but which in reality are most nearly related to the Pines and Firs (*Coniferæ*). The trunk is unbranched, often much shortened, and bears a crown of feathery pinnate fronds. The leaves are usually "circinate"—they unroll in expanding, like the fronds of ferns. The seeds are not protected by a seed-vessel, but are borne upon the edge of altered leaves, or are carried on the scales of a cone. All the living species of Cycads are natives of warm countries, such as South America, the West Indies, Japan, Australia, Southern Asia, and South Africa. The remains of Cycads, as we have seen, are not known to occur in the Coal formation, or only to a very limited extent towards its close; nor are they known with certainty as occurring in Permian deposits. In the Triassic period, however, the remains of Cycads belonging to such genera as *Pterophyllum* (fig. 141, *b*), *Zamites*, and *Podozamites* (fig. 141 *c*), are sufficiently abundant to constitute quite a marked feature in the vegetation; and they continue to be abundantly represented throughout the whole Mesozoic series. The name "Age of Cycads," as applied to the Secondary epoch, is therefore, from a botanical point of view, an extremely appropriate one. The *Conifers* of the Trias are not uncommon, the principal form being *Voltzia* (fig. 141, *a*), which possesses some peculiar characters, but would appear to be mostly nearly related to the recent Cypresses.

As regards the *Invertebrate animals* of the Trias, our knowledge is still principally derived from the calcareous beds which constitute the center of the system (the Muschelkalk) on the continent of Europe, and from the St. Cassian and Rhætic beds still higher in the series; whilst some of the

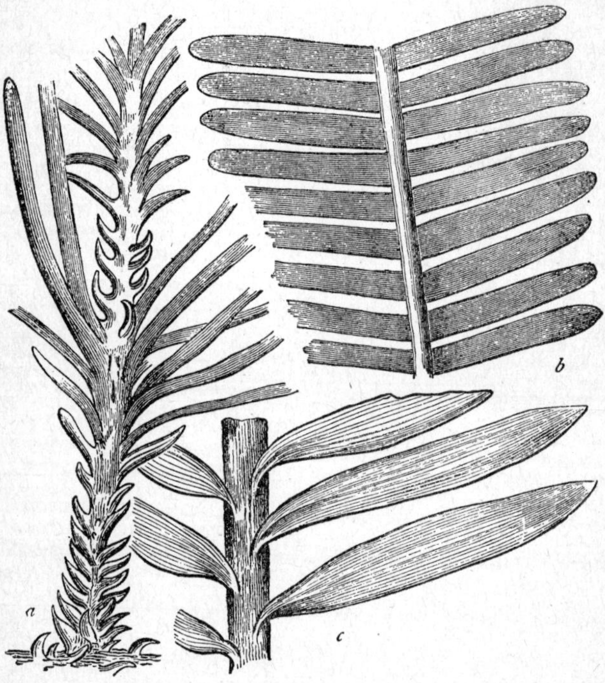

Fig. 141.—Triassic Conifers and Cycads. *a*, *Voltzia* (*Schizoneura*) *heterophylla*, portion of a branch, Europe and America; *b*, Part of the frond of *Pterophyllum Jægeri*, Europe; *c*, Part of the frond of *Podozamites lanceolatus*, America.

Triassic strata of California and Nevada have likewise yielded numerous remains of marine Invertebrates. The *Protozoans* are represented by *Foraminifera* and *Sponges,* and the *Cœlenterates* by a small number of *Corals;* but these require no special notice. It may be mentioned, however, that the great

Palæozoic group of the *Rugose* corals has no known representative here, its place being taken by corals of Secondary type (such as *Montlivaltia, Synastræa*, &c.)

The *Echinoderms* are represented principally by *Crinoids*, the remains of which are extremely abundant in some of the limestones. The best-known species is the famous "Lily-Encrinite" (*Encrinus liliiformis*, fig. 142), which is characteristic of the Muschelkalk. In this beautiful species, the flower-like head is supported upon a rounded stem, the joints of which are elaborately articulated with one another; and the fringed arms are composed each of a double series of alternating calcareous pieces. The Palæozoic Urchins, with their supernumerary rows of plates, the Cystideans, and the Pentremites have finally disappeared; but both Star-fishes and Brittle-stars continue to be represented. One of the latter—namely, the *Aspidura loricata* of Goldfuss (fig. 143)—is highly characteristic of the Muschelkalk.

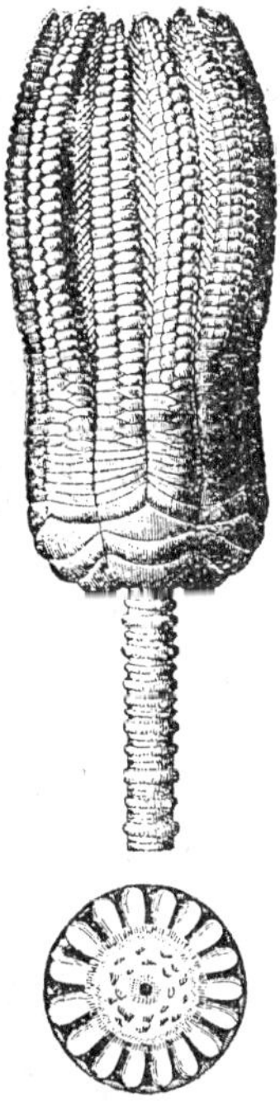

Fig. 142.—Head and upper part of the column of *Encrinus liliiformis*. The lower figure shows the articulating surface of one of the joints of the column. Muschelkalk, Germany.

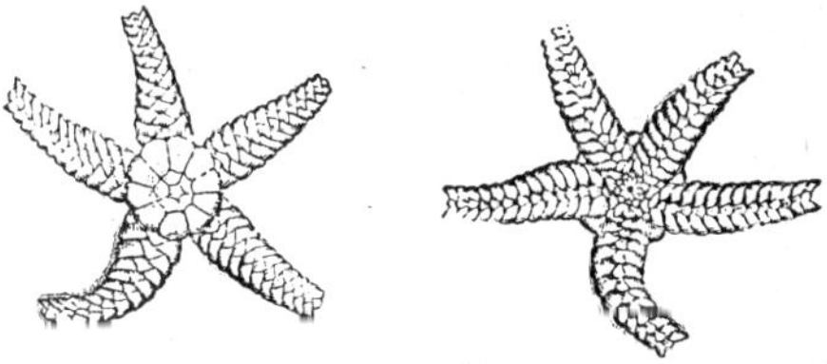

Fig. 143.—*Aspidura loricata*, a Triassic Ophiuroid. Muschelkalk, Germany.

The remains of *Articulate Animals* are not very abundant in the Trias, if we except the bivalved cases of the little Water-fleas (*Ostracoda*), which are occasionally very plentiful. There are also many species of the horny, concentrically-striated valves of the *Estheriæ* (see fig. 122, *b*), which might easily be taken for small Bivalve Molluscs. The "Long-tailed" Decapods, of the type of the Lobster, are not without examples, but they become much more numerous in the succeeding Juriassic period. Remains of insects have also been discovered.

Amongst the *Mollusca* we have to note the disappearance, amongst the lower groups, of many characteristic Palæozoic types. Amongst the *Polyzoans,* the characteristic "Lace-corals," *Fenestella, Retepora,** *Synocladia, Polypora,* &c., have

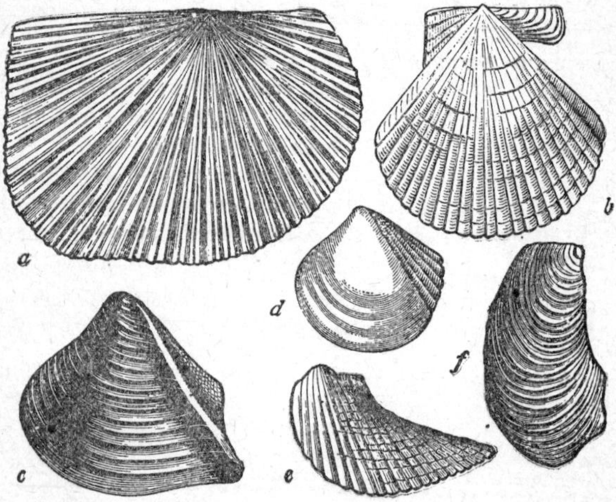

Fig. 144.—Triassic Lamellibranchs, *a, Daonella (Halobia) Lommelli ; b, Pecten Valoniensis; c, Myophoria lineata; d, Cardium Rhæticum; e, Avicula contorta; f, Avicula socialis.*

become apparently extinct. The same is true of many of the ancient types of *Brachiopods,* and conspicuously so of the great family of the *Productidæ,* which played such an important part in the seas of the Carboniferous and Permian periods.

Bivalves (*Lamellibranchiata*) and *Univalves* (*Gasteropoda*) are well represented in the marine beds of the Trias, and some of the former are particularly characteristic either of the

* The genus *Retepora* is really a recent one, represented by living forms; and the so-called *Reteporæ* of the Palæozoic rocks should properly receive another name (*Phyllopora*), as being of a different nature. The name *Retepora* has been here retained for these old forms simply in accordance with general usage.

formation as a whole or of minor subdivisions of it. A few of these characteristic species are figured in the accompanying illustration (fig. 144). Bivalve shells of the genera *Daonella* (fig. 144, *a*) and *Halobia* (*Monotis*) are very abundant, and are found in the Triassic strata of almost all regions. These groups belong to the family of the Pearl-oysters (*Aviculidæ*), and are singular from the striking resemblance borne by some of their included forms to the *Strophomenæ* amongst the Lamp-shells, though, of course, no real relation exists between the two. The little Pearl-oyster, *Avicula socialis* (fig. 144, *f*), is found throughout the greater part of the Triassic series, and is especially abundant in the Muschelkalk. The genus *Myophoria* (fig. 144, *c*), belonging to the *Trigoniadæ*, and related therefore to the Permian *Schizodus*, is characteristically Triassic, many species of the genus being known in deposits of this age. Lastly, the so-called "Rhætic" or "Kössen" beds are characterized by the occurrence in them of the Scallop, *Pecten Valoniensis* (fig. 144, *b*); the small Cockle, *Cardium Rhæticum* (fig. 144, *d*); and the curiously-twisted Pearl-oyster, *Avicula contorta* (fig. 144, *e*)—this last Bivalve being so abundant that the strata in question are often spoken of as the "Avicula contorta beds."

Passing over the groups of the *Heteropods* and *Pteropods*, we have to notice the *Cephalopoda*, which are represented in the Trias not only by the chambered shells of *Tetrabranchiates*, but also, for the first time, by the internal skeletons of *Dibranchiate* forms. The Trias, therefore, masks the first recognized appearance of true Cuttle-fishes. All the known examples of these belong to the great Mesozoic group of the *Belemnitidæ*; and as this family is much more largely developed in the succeeding Jurassic period, the consideration of its characters will be deferred till that formation is treated of. Amongst the chambered *Cephalopods* we find quite a number of the Palæozoic *Orthoceratites*, some of them of considerable size, along with the ancient *Cyrtoceras* and *Goniatites*; and these old types, singularly enough, occur in the higher portion of the Trias (St. Cassian beds), but have, for some unexplained reason, not yet been recognized in the lower and equally fossiliferous formation of the Muschelkalk. Along with these we meet for the first time with true *Ammonites,* which fill such an extensive place in the Jurassic seas, and which will be spoken of hereafter. The form, however, which is most characteristic of the Trias is *Ceratites* (fig. 145). In this genus the shell is curved into a flat spiral, the volutions of which are in contact; and it further agrees with both *Goniatites* and *Ammonites* in the fact that the septa or partitions between the air-chambers are not simple and plain (as in the *Nautilus* and its allies), but are folded and bent as they approach the outer wall of the shell. In the *Goniatite* these foldings of the septa are of a simply lobed or angulated nature, and in the *Ammonite* they are extremely complex; whilst in the *Ceratite* there is an intermediate state of things, the special feature of which is, that those foldings which are turned towards the mouth of the shell are merely rounded, whereas those which are turned away from the mouth are characteristically toothed. The genus *Ceratites,* though principally Triassic, has recently been recognized in strata of Carboniferous age in India.

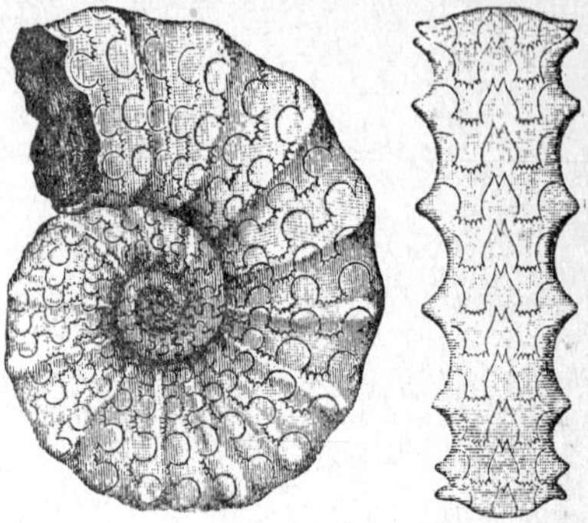

Fig. 145.—*Ceratites nodosus*, viewed from the side and from behind. Muschelkalk.

From the foregoing it will be gathered that one of the most important points in connection with the Triassic *Mollusca* is the remarkable intermixture of Palæozoic and Mesozoic types which they exhibit. It is to be remembered, also, that this intermixture has hitherto been recognized, not in the Middle Triassic limestones of the Muschelkalk, in which—as the oldest Triassic beds with marine fossils—we should naturally expect to find it, but in the St. Cassian beds, the age of which is considerably later than that of the Muschelkalk. The intermingling of old and new types of Shell-fish in the Upper Trias is well brought out in the annexed table, given by Sir

Charles Lyell in his 'Student's Elements of Geology' (some of the less important forms in the table being omitted here):—

GENERA OF FOSSIL MOLLUSCA IN THE ST. CASSIAN AND HALLSTADT BEDS.

Common to Older Rocks.	Characteristic of Triassic Rocks.	Common to Newer Rocks.
Orthoceras.	Ceratites.	Ammonites.
Bactrites.	Cochloceras.	Chemnitzia.
Macrocheilus.	Rhabdoceras.	Cerithium.
Loxonema.	Aulacoceras.	Monodonta.
Holopella.	Naticella.	Sphœra.
Murchisonia.	Platystoma.	Cardita.
Porcellia.	Halobia.	Myoconcha.
Athyris.	Hörnesia.	Hinnites.
Retzia.	Koninckia.	Monotis.
Cyrtina.	Scoliostoma.	Plicatula.
Euomphalus.	Myophoria.	Pachyrisma.
	(The last two are principally but not exclusively Triassic.)	Thecidium.

Thus, to emphasize the more important points alone, the Trias has yielded, amongst the Gasteropods, the characteristically Palæozoic *Loxonema, Holopella, Murchisonia, Euomphalus,* and *Porcellia,* along with typically Triassic forms like *Platystoma* and *Scoliostoma,* and the great modern groups *Chemnitzia* and *Cerithium.* Amongst the Bivalves we find the Palæozoic *Megalodon* side by side with the Triassic *Halobia* and *Myophoria,* these being associated with the *Carditæ, Hinnites, Plicatulæ,* and *Trigoniæ* of later deposits. The Brachiopods exhibit the Palæozoic *Athyris, Retzia,* and *Cyrtina,* with the Triassic *Koninckia* and the modern *Thecidium.* Finally, it is here that the ancient genera *Orthoceras, Cyrtoceras,* and *Goniatites* make their last appearance upon the scene of life, the place of the last of these being taken by the more complex and almost exclusively Triassic *Ceratites;* whilst the still more complex genus *Ammonites* first appears here in force, and is never again wanting till we reach the close of the Mesozoic period. The first representatives of the great Secondary family of the *Belemnites* are also recorded from this horizon.

Amongst the *Vertebrate Animals* of the Trias, the *Fishes* are represented by numerous forms belonging to the *Ganoids* and the *Placoids.* The Ganoids of the period are still all provided with unsymmetrical ("heterocercal") tails, and belong principally to such genera as *Palæoniscus* and *Catopterus.* The remains of Placoids are in the form of teeth and spines, the two principal genera being the two important Secondary groups *Acrodus* and *Hybodus.* Very nearly at the summit of the Trias in England, in the Rhætic series, is a singular stratum, which is well known as the "bone-bed," from the number of fish-remains which it contains. More interesting, however, than the above, are the curious palate-teeth of the Trias, upon which Agassiz founded the genus *Ceratodus.* The teeth of Ceratodus (fig. 146) are singular flattened plates, composed of spongy bones beneath, covered superficially with a layer of enamel. Each plate is approximately triangular, one margin (which we know to be the outer one) being prolonged into prongs or conical prominences, whilst the surface is more or less regularly undulated. Until recently, though the master-mind of Agassiz recognized that these singular bodies were undoubtedly the teeth of fishes, we were entirely ignorant as to their precise relation to the animal, or

Fig. 146.—*a*, Dental plate of *Ceratodus serratus*, Keuper ; *b*, Dental plate of *Ceratodus altus*, Keuper. (After Agassiz.)

as to the exact affinities of the fish thus armed. Lately, however, there has been discovered in the rivers of Queensland (Australia) a living species of *Ceratodus* (*C. Fosteri,* fig. 147), with teeth precisely similar to those of its Triassic predecessor; and we thus have become acquainted with the use of these structures and the manner in which they were implanted in the mouth. The palate carries two of these plates, with their longer straight sides turned towards each other, their sharply-sinuated sides turned outwards, and their short straight sides or bases directed backwards. Two similar plates in the lower jaw correspond to the upper, their undulated surfaces fitting exactly to those of the opposite teeth. There are also two sharp-edged front teeth, which are placed in the front of the mouth in the upper jaw; but these have not been recognized in the fossil specimens. The living *Ceratodus* feeds on vegetable matters, which are taken up or torn off from plants by the sharp front teeth, and then partially crushed between the undulated surfaces of the back teeth (Günther); and there need be little doubt but that the Triassic *Ceratodi* followed a similar mode of existence. From the study of the living *Ceratodus,* it is certain that the genus belongs to the same group as the existing Mud-fishes (*Dipnoi*); and we therefore learn that this, the highest, group of the entire class of Fishes existed in Triassic times under forms little or not at all different from species now alive; whilst it has become probable that the order can be traced back into the Devonian period.

The *Amphibians* of the Trias all belong to the old order of

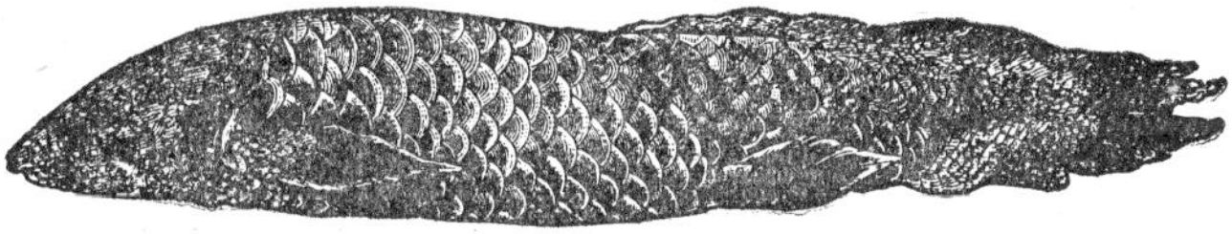

Fig. 147.—*Ceratodus Fosteri,* the Australian Mud-fish, reduced in size.

the *Labyrinthodonts,* and some of them are remarkable for their gigantic dimensions. They were first known by their footprints, which were found to occur plentifully in the Triassic sandstones of Britain and the continent of Europe, and which consisted of a double series of alternately-placed pairs of hand-shaped impressions, the hinder print of each pair being much larger than the one in front (fig. 148). So like were these impressions to the shape of the human hand, that the at that time unknown animal which produced them was at once christened *Cheirotherium,* or "Hand-beast." Further discoveries, however, soon showed that the footprints of *Cheirotherium* were really produced by species of Amphibians which, like the existing Frogs, possessed hind-feet of a much larger size than

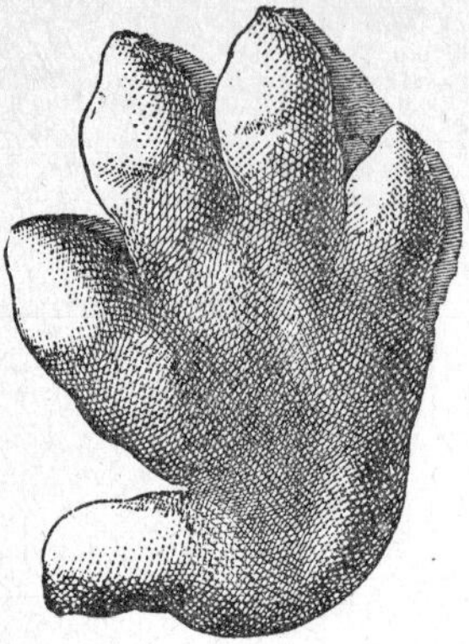

the fore-feet and to which the name of *Labyrinthodonts* was applied in consequence of the complex microscopic structure of the teeth (fig. 149). In the essential details of their structure, the Triassic Labyrinthodonts did not differ materially from their predecessors in the Coal-measures and Permian rocks. They possessed the same frog-like skulls (fig. 150), with a lizard-like body, a long tail, and comparatively feeble limbs. The hind-limbs were stronger and longer than the fore-limbs, and the lower

Fig. 148.—Footprints of a Labyrinthodont (*Cheirotherium*), from the Triassic Sandstones of Hessberg, near Hildburghausen, Germany, reduced one-eighth. The lower figure shows a slab, with several prints, and traversed by reticulated sun-cracks ; the upper figure shows the impression of one of the hind-feet, one-half of the natural size. (After Sickler.)

surface of the body was protected by an armor of bony plates. Some of the Triassic Labyrinthodonts must have attained dimensions utterly unapproached amongst existing Amphibians, the skull of *Labyrinthodon Jægeri* (fig. 150) being upwards of three feet in length and two feet in breadth. Restorations of some of these extraordinary creatures have been attempted in the guise of colossal Frogs; but they must in reality have more closely resembled huge Newts.

Remains of *Reptiles* are very abundant in Triassic deposits, and belong to very varied types. The most marked feature, in fact, connected with the Vertebrate fauna of the Trias, and of the Secondary rocks in general, is the great abundance of Reptilian life. Hence the Secondary period is often spoken of as the "Age of Reptiles." Many of the Triassic reptiles depart widely in their structure from any with which we are acquainted as existing on the earth at the present day, and it is only possible here to briefly note some of the more important of these ancient forms. Amongst the group of the Lizards (*Lacertilia*), represented by *Protorosaurus* in the older Permian strata, three types more or less certainly referable to this order may be mentioned. One of these is a small reptile which

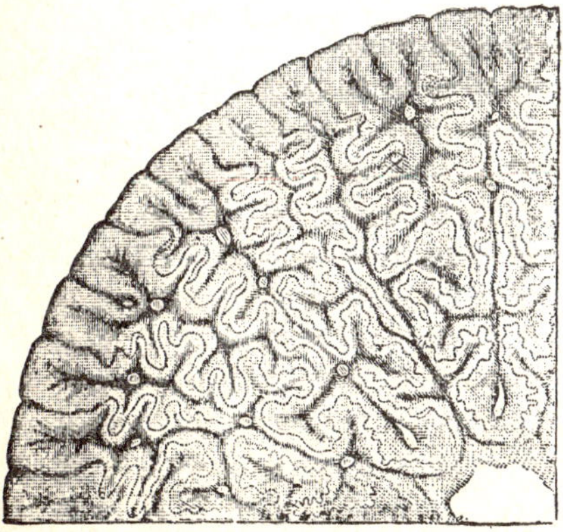

Fig. 149.—Section of the tooth of *Labyrinthodon* (*Mastodonsaurus*) *Jægeri*, showing the microscopic structure. Greatly enlarged. Trias.

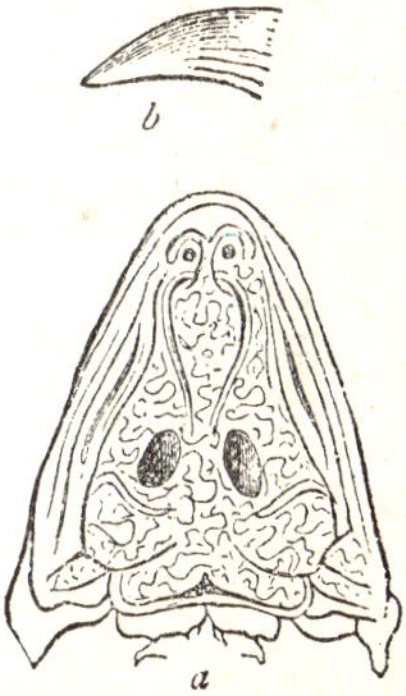

Fig. 150.—*a*, Skull of *Labyrinthodon Jægeri*, much reduced in size; *b*, Tooth of the same, Trias, Würtemberg.

was found many years ago in sandstones near Elgin, in Scotland, and which excited special interest at the time in consequence of the fact that the strata in question were believed to belong to the Old Red Sandstone formation. It is, however, now certain that the Elgin sandstones which contain *Telerpeton Elginense,* as this reptile is termed, are really to be regarded as of Triassic age. By Professor Huxley, *Telerpeton* is regarded as a Lizard, which cannot be considered as "in any sense a less perfectly-organized creature than the Gecko, whose swift and noiseless run over walls and ceilings surprises the traveler in climates warmer than our own." The "Elgin Sandstones" have also yielded another Lizard, which was originally described by Professor Huxley under the name of *Hyperodapedon,* the remains of the same genus having been subsequently discovered in Triassic strata in India and South Africa. The Lizards of this group must therefore have at one time enjoyed a very wide distribution over the globe; and the living *Sphenodon* of New Zealand is believed by Professor Huxley to be the nearest living ally of this family. The *Hyperodapedon* of the Elgin Sandstones was about six feet in length, with limbs adapted for terrestrial progression, but with the bodies of the vertebræ slightly biconcave, and having two rows of palatal teeth, which become worn down to the bone in old age. Lastly, the curious *Rhynchosaurus* of the Trias is also referred, by the eminent comparative anatomist above mentioned, to the order of the Lizards. In this singular reptile (fig. 151) the skull is somewhat bird-like, and the jaws appear to have been destitute of teeth, and to have been encased in a horny sheath like the beak of a Turtle or a Bird. It is possible, however, that the palate was furnished with teeth.

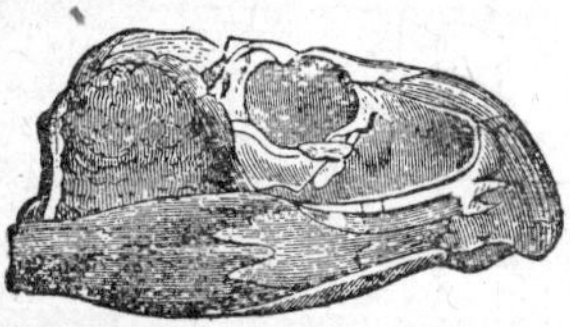

Fig. 151.—Skull of *Rhynchosaurus articeps.* Trias. (After Owen.)

The group of the Crocodiles and Alligators (*Crocodilia*), distinguished by the fact that the teeth are implanted in distinct sockets and the skin more or less extensively provided with bony plates, is represented in the Triassic rocks by the *Stagonolepis* of the Elgin Sandstones. The so-called "Thecodont" reptiles (such as *Belodon, Thecodontosaurus,* and *Palæosaurus,* fig 152, *c, d, e*) are also nearly related to the Crocodiles, though it is doubtful if they should be absolutely referred to this group. In these reptiles, the teeth are implanted in distinct sockets in the jaws, their crowns being more or less compressed and pointed, "with trenchant and finely serrate margins" (Owen). The bodies of the vertebræ are hollowed out at both ends, but the limbs appear to be adapted for progression on the land. The genus *Belodon* (fig. 152, *c*) is known to occur in the Keuper of Germany and in America; and *Palæosaurus* (fig. 153, *e*) has also been found in the Trias of the same region. Teeth of the latter, however, are found, along with remains of *Thecodontosaurus* (fig. 153, *d*), in a singular magnesian conglomerate near Bristol, which was originally believed to be of Permian age, but which appears to be undoubtedly Triassic.

The Trias has also yielded the remains of the great marine reptiles which are often spoken of collectively as the "Enaliosaurians" or "Sea-lizards," and which will be more particularly spoken of in treating of the Jurassic period, of which they are more especially characteristic. In all these reptiles the limbs are flattened out, the digits being enclosed in a continuous skin, thus forming powerful swimming-paddles, resembling the "flippers" of the Whales and Dolphins both in their general structure and in function. The tail is also long, and adapted to act as a swimming-organ; and there can be no doubt but that these extraordinary and often colossal reptiles frequented the sea, and only occasionally came to the land. The Triassic Enaliosaurs belong to a group of which the later genus *Plesiosaurus* is the type (the *Sauropterygia*). One of the best known of the Triassic genera is *Nothosaurus* (fig. 152, *a*), in which the neck was long and bird-like, the jaws being im-

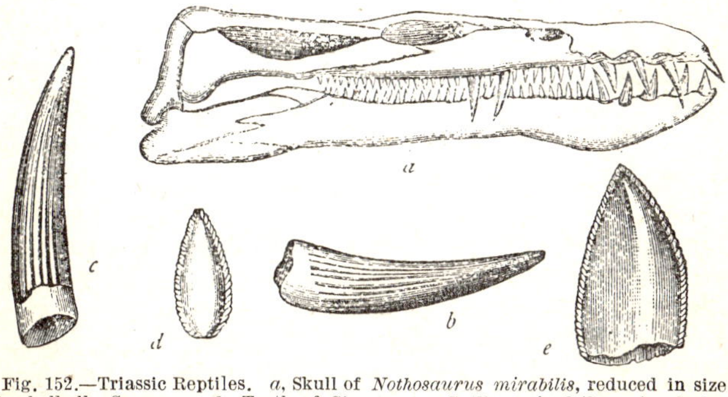

Fig. 152.—Triassic Reptiles. *a*, Skull of *Nothosaurus mirabilis*, reduced in size—Muschelkalk, Germany; *b*, Tooth of *Simosaurus Gaillardoti*, of the natural size—Muschelkalk, Germany; *c*, Tooth of *Belodon Carolinensis*—Trias, America; *d*, Tooth of *Thecodontosaurus antiquus*, slightly enlarged—Britain; *e*, Tooth of *Palæosaurus platyodon*, of the natural size—Britain.

mensely elongated, and carrying numerous powerful conical teeth implanted in distinct sockets. The teeth in *Simosaurus*

(152, *b*) are of a similar nature; but the orbits are of enormous size, indicating eyes of corresponding dimensions, and perhaps pointing to the nocturnal habits of the animal. In the singular *Placodus*, again, the teeth are in distinct sockets, but resemble those of many fishes in being rounded and obtuse (fig. 153), forming broad crushing plates adapted for the comminution of shell-fish. There is a row of these teeth all round the upper jaw proper, and a double series on the palate, but the lower jaw has only a single row of teeth. *Placodus* is found in the Muschelkalk, and the characters of its dental apparatus indicate that it was much more peaceful in its habits than its associates the Nothosaur and Simosaur.

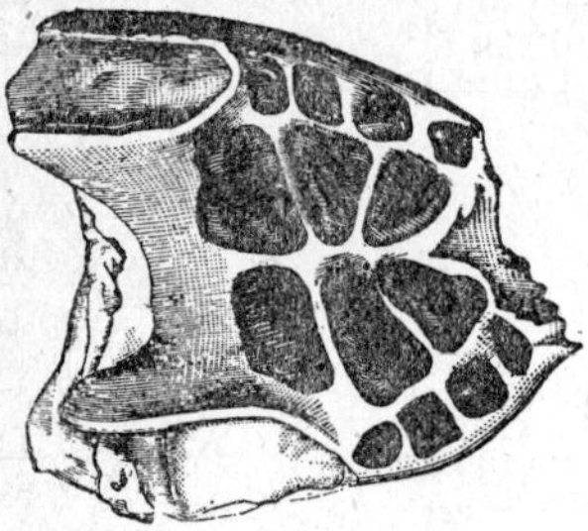

Fig. 153.—Under surface of the upper jaw and palate of *Placodus gigas*. Muschelkalk, Germany.

The Triassic rocks of South Africa and India have yielded the remains of some extraordinary Reptiles, which have been placed by Professor Owen in a separate order under the name of *Anomodontia*. The two principal genera of this group are *Dicynodon* and *Oudenodon*, both of which appear to have been large Reptiles, with well-developed limbs, organized for progression upon the dry land. In *Oudenodon* (fig. 154, B) the jaws seem to have been wholly destitute of teeth, and must have been encased in a horny sheath, similar to that with which we are familiar in the beak of a Turtle. In *Dicynodon* (fig. 154, A), on the other hand, the front of the upper jaw and the whole of the lower jaw were destitute of teeth, and the front of the mouth must have constituted a kind of beak; but the upper jaw possessed on each side a single huge conical tusk, which is directed downwards, and must have continued to grow during the life of the animal.

It may be mentioned that the above-mentioned Triassic sandstones of South Africa have recently yielded to the researches of Professor Owen a new and unexpected type of Reptile, which exhibits some of the structural peculiarities which we have been accustomed to regard as characteristic of the Carnivorous quadrupeds. The Reptile in question has been named *Cynodraco*, and it is looked upon by its distinguished discoverer as the type of a new order, to which he has given the name of *Theriodontia*. The teeth of this singular form agree with those of the Carnivorous quadrupeds in consisting of three distinct groups—namely, front teeth or *incisors*, eye teeth or *canines*, and back teeth or *molars*. The canines also are long and pointed, very much compressed, and having their lateral margins finely serrated, thus presenting a singular resemblance to the teeth of the extinct "Sabre-toothed Tiger" (*Machairodus*). The bone of the upper arm (humerus) further shows some remarkable resemblances to the same bone in the Carnivorous Mammals. As has been previously noticed, Professor Owen is of opinion that some of the Reptilian remains

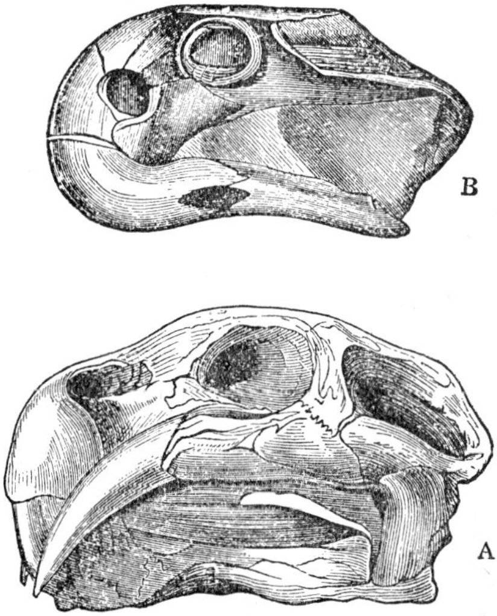

Fig. 154.—Triassic Anomodont Reptiles. A, Skull of *Dicynodon lacerticeps*, showing one of the great maxillary tusks ; B, Skull of *Oudenodon Bainii*, showing the toothless, beak-like jaws. From the Trias of South Africa. (After Owen.)

of the Permian deposits will also be found to belong to this group of the "Theriodonts."

Lastly, we find in the Triassic rocks the remains of Reptiles belonging to the great Mesozoic order of the *Deinosauria*. This order attains its maximum at a later period, and will be spoken of when the Jurassic and Cretaceous deposits come to be considered. The chief interest of the Triassic Reptiles of this group arises from the fact that they are known by their footprints as well as by their bones; and a question has arisen whether the supposed footprints of *birds* which occur in the Trias have not really been produced by Deinosaurs. This leads us, therefore, to speak at the same time as to the evidence which we have of the existence of the class of Birds during the Triassic period. No actual bones of any bird have as yet been detected in any Triassic deposit; but we have tolerably clear evidence of their existence at this time in the form of *footprints*. The impressions in question are found in considerable numbers in certain red Sandstones of the age of the Trias in the valley of the Connecticut River, in the United States. They vary much in size, and have evidently been produced by many different animals walking over long stretches of estuarine mud and sand exposed at low water. The footprints now under consideration form a double series of *single* prints, and therefore, beyond all question, are the tracks of a *biped*—that is, of an animal which walked upon

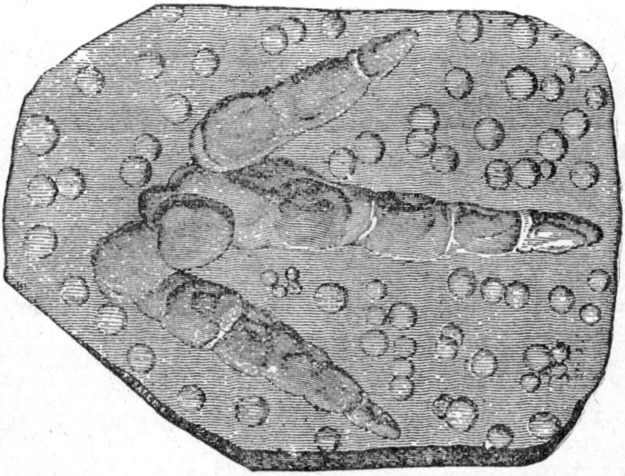

Fig. 155.—Supposed footprint of a Bird, from the Triassic Sandstones of the Connecticut River. The slab shows also numerous "rain-prints."

two legs. No living animals, save Man and the Birds, walk habitually on two legs; and there is, therefore, a *prima facie* presumption that the authors of these prints were Birds. Moreover, each impression consists of the marks of three toes turned forward (fig. 155), and therefore are precisely such as might be produced by Wading or Cursorial Birds. Further, the impressions of the toes show exactly the same numerical progression in the number of the joints as is observable in living Birds—that is to say, the innermost of the three toes consists of three joints, the middle one of four, and the outer one of five joints. Taking this evidence collectively, it would have seemed, until lately, quite certain that these tracks could only have been formed by Birds. It has, however, been shown that the Deinosaurian Reptiles possess, in some cases at any rate, some singularly bird-like characters, amongst which is the fact that the animal possessed the power of walking, temporarily at least, on its hind-legs, which were much longer and stronger than the fore-limbs, and which were sometimes furnished with no more than three toes. As the bones and teeth of Deinosaurs have been found in the Triassic deposits of North America, it may be regarded as certain that *some* of the bipedal tracks originally ascribed to Birds must have really been produced by these Reptiles. It seems at the same time almost a certainty that others of the three-toed impressions of the Connecticut sandstones were in truth produced by Birds, since it is doubtful if the bipedal mode of progression was more than an occasional thing amongst the Deinosaurs, and the greater number of the many known tracks exhibit no impressions of fore-feet. Upon the whole, therefore, we may, with much probability, conclude that the great class of Birds (*Aves*) was in existence in the Triassic period. If this be so, not only must there have been quite a number of different forms, but some of them must have been of very large size. Thus the largest footprints hitherto discovered in the Connecticut sandstones are 22 inches long and 12 inches wide, with a proportionate length of stride. These measurements indicate a foot four times as large as that of the African Ostrich; and the animal which produced them—whether a Bird or a Deinosaur—must have been of colossal dimensions.

Fig. 156.—Lower jaw of *Dromatherium sylvestre*, Trias, North Carolina. (After Emmons.)

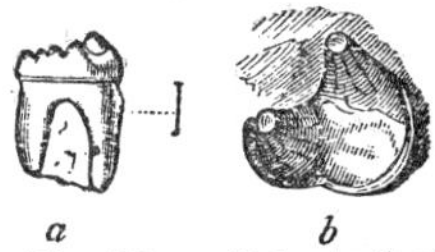

Fig. 157.—*a*, Molar tooth of *Microlestes antiquus*, magnified; *b*, Crown of the same, magnified still further. Trias, Germany.

Finally, the Trias completes the tale of the great classes of the Vertebrate sub-kingdom by presenting us with remains of the first known of the true Quadrupeds or *Mammalia.* These are at present only known by their teeth, or, in one instance, by one of the halves of the lower jaw; and these indicate minute Quadrupeds, which present greater affinities with the

Fig. 158.—The Banded Ant-eater (*Myrmecobius fasciatus*) of Australia.

little Banded Ant-eater (*Myrmecobius fasciatus,* fig. 158) of Australia than with any other living form. If this conjecture be correct, these ancient Mammals belonged to the order of the Marsupials or Pouched Quadrupeds (*Marsupialia*), which are now exclusively confined to the Australian province, South America, and the southern portion of North America. In the Old World, the only known Triassic Mammals belong to the genus *Microlestes,* and to the probably identical *Hypsiprymnopsis* of Professor Boyd Dawkins. The teeth of *Microlestes* (fig. 157) were originally discovered by Plieninger in 1847 in the "bone-bed" which is characteristic of the summit of the Rhætic series both in Britain and on the continent of Europe; and the known remains indicate two species. In Britain, teeth of *Microlestes* have been discovered by Mr. Charles Moore in deposits of Upper Triassic age, filling a fissure in the Carboniferous limestone near Frome, in Somersetshire; and a molar tooth of *Hypsiprymnopsis* was found by Professor Boyd Dawkins in Rhætic marls below the "bone-bed," at Watchet, also in Somersetshire. In North America, lastly, there has been found in strata of Triassic age one of the branches of the lower jaw of a small Mammal, which has been described under the name of *Dromatherium sylvestre* (fig. 156). The fossil exhibits ten small molars placed side by side, one canine, and three incisors, separated by small intervals, and it indicates a small insectivorous animal, probably most nearly related to the existing *Myrmecobius*.

LITERATURE.

The following list comprises a few of the more important sources of information as to the Triassic strata and their fossil contents:—

(1) 'Geology of Oxford and the Valley of the Thames.' Phillips.

(2) 'Memoirs of the Geological Survey of Great Britain and Ireland.'

(3) 'Report on the Geology of Londonderry,' &c. Portlock.

(4) "On the Zone of Avicula contorta," &c.—'Quart. Journ. Geol. Soc.,' vol. xvi., 1860. Dr. Thomas Wright.

(5) "On the Zones of the Lower Lias and the Avicula contorta Zone"—'Quart. Journ. Geol. Soc.,' vol. xvii., 1861. Charles Moore.

(6) "On Abnormal Conditions of Secondary Deposits," &c. —'Quart. Journ. Geol. Soc.,' vol. xxiii., 1876-77. Charles Moore.

(7) 'Geognostische Beschreibung des Bayerischen Alpengebirges.' Gümbel.

(8) 'Lethæa Rossica.' Pander.

(9) 'Lethæa Geognostica.' Bronn.

(10) 'Petrefacta Germaniæ.' Goldfuss.

(11) 'Petrefaktenkunde.' Quenstedt.

(12) 'Monograph of the Fossil Estheriæ' (Palæontographical Society). Rupert Jones.

(13) "Fossil Remains of Three Distinct Saurian Animals, recently discovered in the Magnesian Conglomerate near Bristol"—'Trans. Geol. Soc.,' ser. 2, vol. v., 1840. Riley and Stutchbury.

(14) 'Die Saurier des Muschelkalkes.' Von Meyer.

(15) 'Beiträge zur Palæontologie Württembergs.' Von Meyer and Plieninger.

(16) 'Manual of Palæontology.' Owen.

(17) 'Odontography.' Owen.

(18) 'Report of Fossil Reptiles' (British Association, 1841). Owen.

(19) "On Dicynodon"—'Trans. Geol. Soc.,' vol. iii., 1845. Owen.
(20) 'Descriptive Catalogue of Fossil Reptilia and Fishes in the Museum of the Royal College of Surgeons, England.' Owen.
(21) "On Species of Labyrinthodon from Warwickshire"—'Trans. Geol. Soc.,' ser. 2, vol. vi. Owen.
(22) "On a Carnivorous Reptile" (Cynodraco major), &c.—'Quart. Journ. Geol. Soc.,' vol. xxxii., 1876. Owen.
(23) "On Evidences of Theriodonts in Permian Deposits," &c.—'Quart. Journ. Geol Soc.,' vol. xxxii., 1876. Owen.
(24) "On the Stagonolepis Robertsoni," &c.—'Quart. Journ. Geol. Soc.,' vol. xv., 1859. Huxley.
(25) "On a New Specimen of Telerpeton Elginense"—'Quart. Journ. Geol. Soc.,' vol. xxiii., 1866. Huxley.
(26) "On Hyperodapedon"—'Quart. Journ. Geol. Soc.,' vol. xxv., 1869. Huxley.
(27) "On the Affinities between the Deinosaurian Reptiles and Birds"—'Quart. Journ. Geol. Soc.,' vol xxvi., 1870. Huxley.
(28) "On the Classification of the Deinosauria." &c.—'Quart. Journ. Geol. Soc.,' vol. xxvi., 1870. Huxley.
(29) "Palæontologica Indica"—'Memoirs of the Geol. Survey of India.'
(30) "On the Geological Position and Geographical Distribution of the Dolomitic Conglomerate of the Bristol Area"—'Quart. Journ. Geol. Soc.,' vol. xxvi., 1870. R. Etheridge, sen.
(31) "Remains of Labyrinthodonta from the Keuper Sandstone of Warwick"—'Quart. Journ. Geol. Soc.,' vol. xxx., 1874. Miall.
(32) 'Manual of Geology.' Dana.
(33) 'Synopsis of Extinct Batrachia and Reptilia of North America.' Cope.
(34) 'Fossil Footmarks.' Hitchcock.
(35) 'Ichnology of New England.' Hitchcock.
(36) 'Traité de Paléontologie Végétale.' Schimper.
(37) 'Histoire des Végétaux Fossiles.' Brongniart.
(38) 'Monographie der Fossilen Coniferen.' Gœppert.

CHAPTER XVI.

THE JURASSIC PERIOD.

Resting upon the Trias, with perfect conformity, and with an almost undeterminable junction, we have the great series of deposits which are known as the *Oolitic Rocks,* from the common occurrence in them of oolitic limestones, or as the *Jurassic Rocks,* from their being largely developed in the mountain-range of the Jura, on the western borders of Switzerland. Sediments of this series occupy extensive areas in Great Britain, on the continent of Europe, and in India. In North America, limestones and marls of this age have been detected in "the Black Hills, the Laramie range, and other eastern ridges of the Rocky Mountains; also over the Pacific slope, in the Uintah, Wahsatch, and Humboldt Mountains, and in the Sierra Nevada" (Dana); but in these regions their extent is still unknown, and their precise subdivisions have not been determined. Strata belonging to the Jurassic period are also known to occur in South America, in Australia, and in the Arctic zone. When fully developed, the Jurassic series is capable of subdivision into a number of minor groups, of which some are clearly distinguished by their mineral characters, whilst others are separated with equal certainty by the differences of the fossils that they contain. It will be sufficient for our present purpose, without entering into the more minute subdivisions of the series, to give here a very brief and general account of the main sub-groups of the Jurassic rocks, as developed in Britain—the arrangement of the Jura-formation of the continent of Europe agreeing in the main with that of England.

I. The Lias.—The base of the Jurassic series of Britain is formed by the great calcareo-argillaceous deposit of the "Lias," which usually rests conformably and almost inseparably upon the Rhætic beds (the so-called "White Lias"), and passes up, generally conformably, into the calcareous sandstones of the Inferior Oolite. The Lias is divisible into the three principal groups of the Lower, Middle, and Upper Lias, as below, and these in turn contain many well-marked "zones;" so that the Lias has some claims to be considered as an independent formation, equivalent to all the remaining Oolitic rocks. The *Lower Lias* (*Terrain Sinemurien* of D'Orbigny) sometimes attains a thickness of as much as 600 feet, and consists of a great series of bluish or greyish laminated clays, alternating with thin bands of blue or grey limestone—the whole, when seen in quarries or cliffs from a little distance, assuming a characteristically striped and banded appearance. By means of particular species of *Ammonites,* taken along with other fossils which are confined to particular zones, the Lower Lias may be subdivided into several well-marked horizons.

The *Middle Lias,* or *Marlstone Series* (*Terrain Liasien* of D'Orbigny), may reach a thickness of 200 feet, and consists of sands, arenaceous marls, and argillaceous limestones, sometimes with ferruginous beds. The *Upper Lias* (*Terrain Toarcien* of D'Orbigny) attains a thickness of 300 feet, and consists principally of shales below, passing upwards into arenaceous strata.

II. THE LOWER OOLITES.—Above the Lias comes a complex series of partly arenaceous and argillaceous, but principally calcareous strata, of which the following are the more important groups: *a,* the *Inferior Oolite* (*Terrain Bajociên* of D'Orbigny), consisting of more than 200 feet of oolitic limestones, sometimes more or less sandy; *b,* The *Fuller's Earth,* a series of shales, clays, and marls, about 120 feet in thickness; *c,* The *Great Oolite* or *Bath Oolite* (*Terrain Bathonien* of D'Orbigny), consisting principally of oolitic limestones, and attaining a thickness of about 130 feet. The well-known "Stonesfield Slates" belong to this horizon; and the locally developed "Bradford Clay," "Cornbrash," and "Forest-marble" may be regarded as constituting the summit of this group.

III. THE MIDDLE OOLITES.—The central portion of the Jurassic series of Britain is formed by great argillaceous deposit, capped by calcareous strata, as follows: *a,* The *Oxford Clay* (*Terrain Callovien* and *Terrain Oxfordien* of D'Orbigny), consisting of dark-colored laminated clays, sometimes reaching a thickness of 700 feet, and in places having its lower portion developed into a hard calcareous sandstone ("Kelloway Rock"); *b,* The *Coral-Rag* (*Terrain Corallien* of D'Orbigny, "Nerinean Limestone" of the Jura, "Diceras Limestone" of the Alps), consisting, when typically developed, of a central mass of oolitic limestone, underlaid and surmounted by calcareous grits.

IV. THE UPPER OOLITES.—*a,* The base of the Upper Oolites of Britain is constituted by a great thickness (600 feet or more) of laminated, sometimes carbonaceous or bituminous clays, which are known as the *Kimmeridge Clay* (*Terrain Kimméridgien* of D'Orbigny); *b,* The *Portland Beds* (*Terrain Portlandien* of D'Orbigny) succeed the Kimmeridge clay, and consist inferiorly of sandy beds surmounted by oolitic limestones ("Portland Stone"), the whole series attaining a thickness of 150 feet or more, and containing marine fossils; *c,* The *Purbeck Beds* are apparently peculiar to Great Britain, where they form the summit of the entire Oolitic series, attaining a total thickness of from 150 to 200 feet. The Purbeck beds consist of arenaceous, argillaceous, and calcareous strata, which can be shown by their fossils to consist of a most remarkable alternation of fresh-water, brackish-water, and purely marine sediments, together with old land-surfaces, or vegetable soils, which contain the upright stems of trees, and are locally known as "Dirt-beds."

One of the most important of the Jurassic deposits of the continent of Europe, which is believed to be on the horizon of the Coral-rag or of the lower part of the Upper Oolites, is the "*Solenhofen Slate*" of Bavaria, an exceedingly fine-grained limestone, which is largely used in lithography, and is celebrated for the number and beauty of its organic remains, and especially for those of Vertebrate animals.

The subjoined sketch-section (fig. 159) exhibits in a diagrammatic form the general succession of the Jurassic rocks of Britain.

Regarded as a whole, the Jurassic formation is essentially marine; and though remains of drifted plants, and of insects and other air-breathing animals, are not uncommon, the fossils of the formation are in the main marine. In the Purbeck series of Britain, anticipatory of the great river-deposit of the Wealden, there are fresh-water, brackish-water, and even terrestrial strata, indicating that the floor of the oolitic ocean was undergoing upheaval, and that the marine conditions which had formerly prevailed were nearly at an end. In places also, as in Yorkshire and Sutherlandshire, are found actual beds of coal: but the great bulk of the formation is an indubitable sea-deposit; and its limestones, oolitic as they commonly are, nevertheless are composed largely of the comminuted skeletons of marine animals. Owing to the enormous number and variety of the organic remains which have been yielded by the richly fossiliferous strata of the Oolitic series, it will not be possible here to do more than to give an outline-sketch of the principal forms of life which characterize the Jurassic period as a whole. It is to be remembered, however, that every minor group of the Jurassic formation has its own peculiar fossils, and that by the labors of such eminent observers as Quenstedt, Oppel, D'Orbigny, Wright, De la Beche, Tate, and others, the entire series of Jurassic sediments admits of a more complete and more elaborate subdivision into zones characterized by special life-forms than has as yet been found practicable in the case of any other rock-series.

GENERALIZED SECTION OF THE JURASSIC ROCKS OF ENGLAND.

Fig. 159.

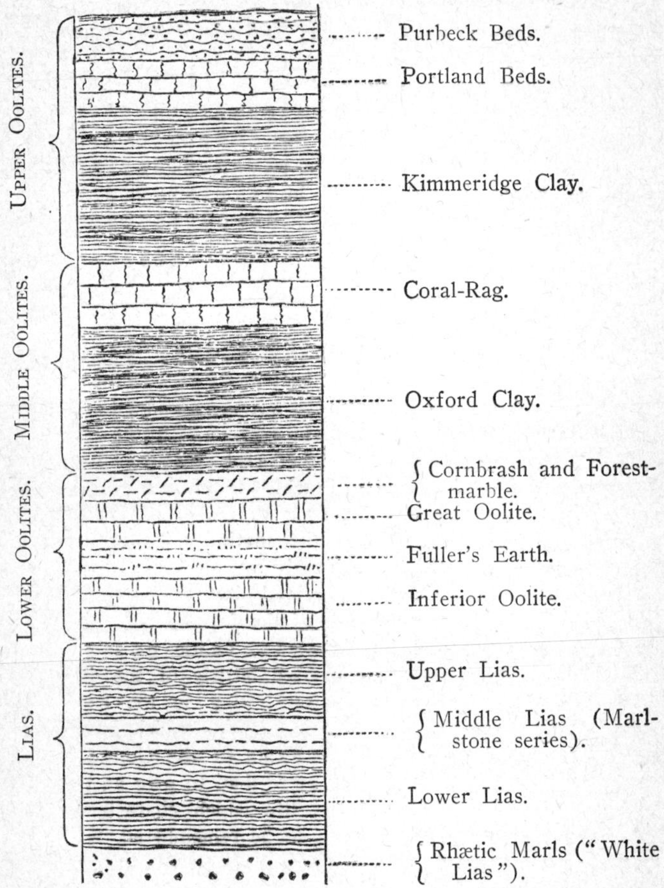

The *plants* of the Jurassic period consist principally of Ferns, Cycads, and Conifers—agreeing in this respect, therefore, with those of the preceding Triassic formation. The *Ferns* are very abundant, and belong partly to old and partly to new genera. The *Cycads* are also very abundant, and, on the whole, constitute the most marked feature of the Jurassic vegetation, many genera of this group being known (*Pterophyllum, Otozamites, Zamites, Crossozamia, Williamsonia, Bucklandia,* &c.) The so-called "dirt-bed" of the Purbeck series consists of an ancient soil, in which stand erect the trunks of Conifers and the silicified stools of Cycads of the genus *Mantellia* (fig. 160). The *Coniferæ* of the Jurassic are represented by various forms more or less nearly allied to the existing *Araucariæ;* and these are known not only by their stems or branches, but also in some cases by their cones. We meet, also, with the remains of undoubted Endogenous plants, the most important of which are the fruits of forms allied to the existing Screw-pines (*Pandaneæ*), such as *Podocarya* and *Kaidacarpum.* So far, however, no remains of Palms have been found; nor are we acquainted with any Jurassic plants which could be certainly referred to the great "Angiospermous" group of the Exogens, including the majority of our ordinary plants and trees.

Amongst animals, the *Protozoans* are well represented in the Jurassic deposits by numerous *Foraminifers* and *Sponges;* as are the *Cœlenterates* by numerous *Corals.* Remains

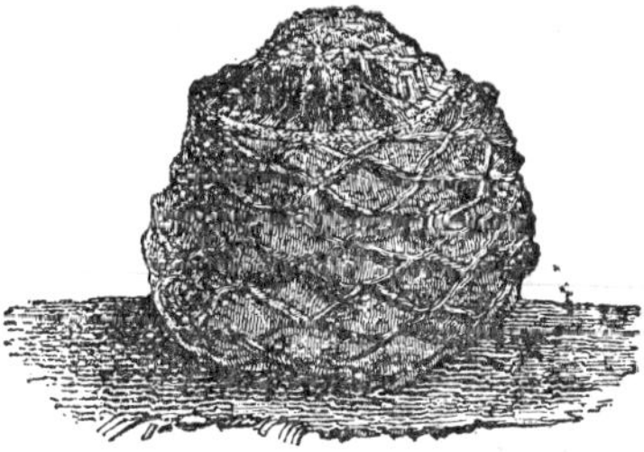

Fig. 160.—*Mantellia* (*Cycadeoidea*) *megalophylla*, **a Cycad from the Purbeck "dirt-bed." Upper Oolites, England.**

of these last-mentioned organisms are extremely abundant in some of the limestones of the formation, such as the "Coral-rag" and the Great Oolite; and the former of these may fairly be considered as an ancient "reef." The *Rugose Corals* have not hitherto been detected in the Jurassic rocks; and the "*Tabulate Corals,*" so-called, are represented only by examples of the modern genus *Millepora.* With this exception, all the Jurassic Corals belonging to the great group which predominates in recent seas (*Zoantharia sclerodermata*); and the majority belong to the important reef-building family of the "Star-corals" (*Astræidæ*). The form here figured (*Thecosmilia annularis,* fig. 161) is one of the characteristic species of the Coral-rag.

The *Echinoderms* are very numerous and abundant fossils in the Jurassic series, and are represented by Sea-lilies, Sea-urchins, Star-fishes, and Brittle-stars. The *Crinoids* are still common, and some of the limestones of the series are largely composed of the *débris* of these organisms. Most of the Jurassic forms resemble those with which we are already familiar, in having the body permanently attached to some foreign object by means of a longer or shorter jointed stalk

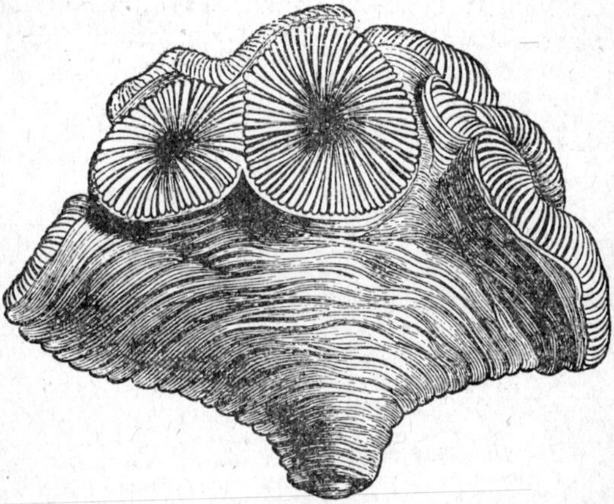

Fig. 161.—*Thecosmilia annularis.* Coral-rag, England.

or "column." One of the most characteristic Jurassic genera of these "stalked" Crinoids (though not exclusively confined to this period) is *Pentacrinus* (fig. 162). In this genus, the column is five-sided, with whorls of "side-arms;" and the arms are long, slender, and branched. The genus is represented at the present day by the beautiful "Medusa-head Pentacrinite" (*Pentacrinus caput-medusæ*). Another characteristic Oolitic genus is *Apiocrinus,* comprising the so-called "Pear Encrinites." In this group the column is long and rounded, with a dilated base, and having its uppermost joints expanded so as to form, with the cup itself, a pear-shaped mass, from the summit of which spring the comparatively short arms. Besides the "stalked" Crinoids, the Jurassic rocks have yielded the remains of the higher group of the "free" Crinoids, such as

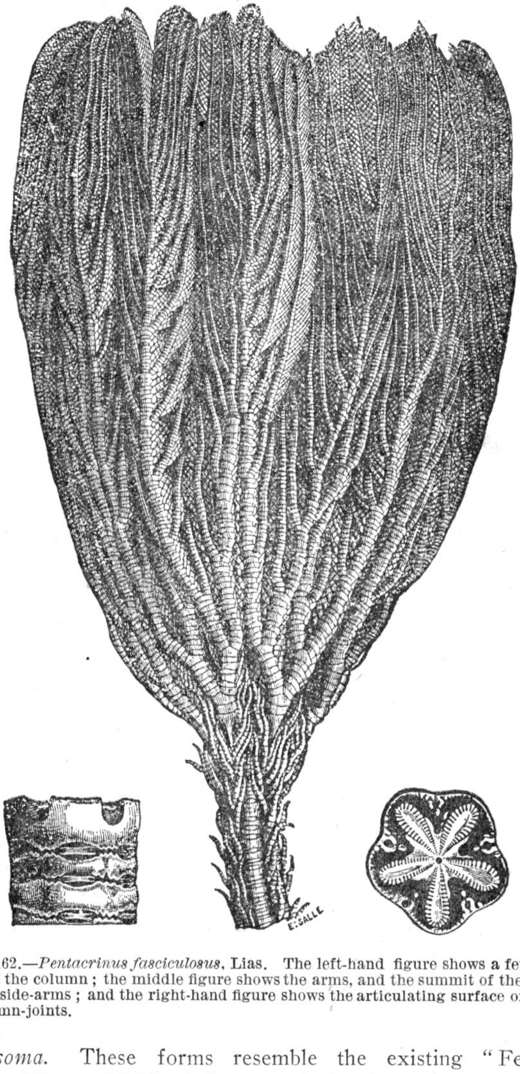

Fig. 162.—*Pentacrinus fasciculosus*, Lias. The left-hand figure shows a few of the joints of the column; the middle figure shows the arms, and the summit of the column with its side-arms; and the right-hand figure shows the articulating surface of one of the column-joints.

Saccosoma. These forms resemble the existing "Featherstars" (*Comatula*) in being attached when young to some foreign body by means of a jointed stem, from which they detach themselves when fully grown to lead an independent existence. In this later stage of their life, therefore, they closely resemble the Brittle-stars in appearance. True Star-fishes (*Asteroids*) and Brittle-stars (*Ophiuroids*) are abundant in the Jurassic rocks, and the Sea-urchins (*Echinoids*) are so numerous and so well preserved as to constitute quite a marked feature of some beds of the series. All the Oolitic urchins agree with the modern *Echinoids* in having the shell composed of no more than twenty rows of plates. Many different genera are known, and a characteristic species of the Middle Oolites (*Hemicidaris crenularis,* fig. 163) is here figured.

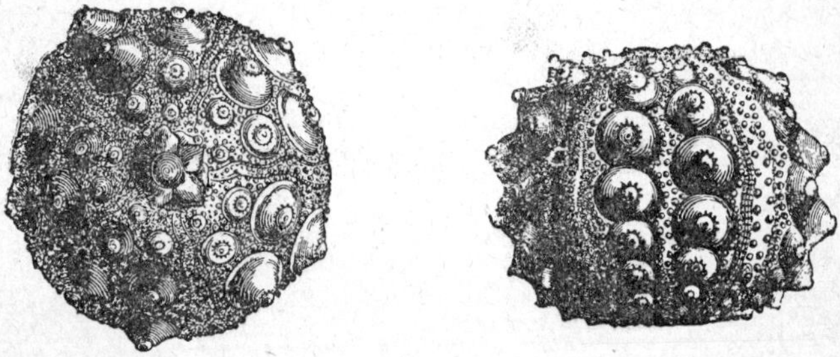

Fig. 163.—*Hemicidaris crenularis*, showing the great tubercles on which the spines were supported. Middle Oolites.

Passing over the *Annelides,* which, though not uncommon, are of little special interest, we come to the *Articulates,* which also require little notice. Amongst the *Crustaceans,* whilst the little Water-fleas (*Ostracoda*) are still abundant, the most marked feature is the predominance which is now assumed by the *Decapods*—the highest of the known groups of the class. True Crabs (*Brachyura*) are by no means unknown; but the principal Oolitic Decapods belonged to the "Long-tailed" group (*Macrura*), of which the existing Lobsters, Prawns, and Shrimps are members. The fine-grained lithographic slates of Solenhofen are especially famous as a depot for the remains of these Crustaceans, and a characteristic species from this locality (*Eryon arctiformis,* fig. 164) is here represented. Amongst the air-breathing *Articulates,* we meet in the Oolitic rocks with the remains of Spiders (*Arachnida*), Centipedes (*Myriapoda*), and numerous true Insects (*Insecta*). In connection with the last-mentioned of these groups, it is of interest to note the occurrence of the oldest known fossil Butterfly —the *Palæontina Oolitica* of the Stonesfield slate—the relationships of which appear to be with some of the living Butterflies of Tropical America.

Coming to the *Mollusca*, the *Polyzoans*, numerous and

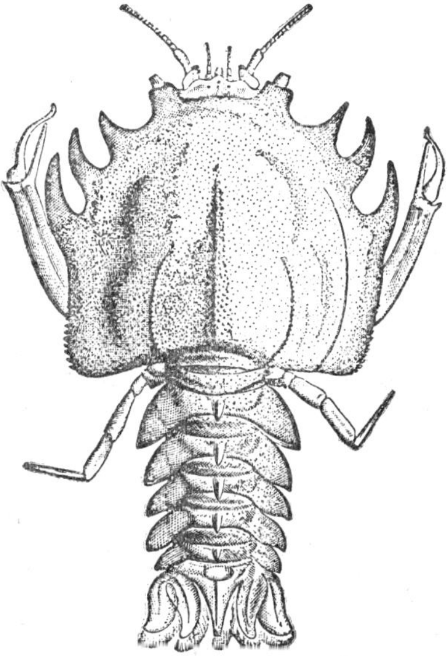

Fig. 164.—*Eryon arctiformis*, a "Long-tailed Decapod," from the Middle Oolites (Solenhofen Slate.)

beautiful as they are, must be at once dismissed; but the *Brachiopods* deserve a moment's attention. The Jurassic Lamp-shells (fig. 165) do not fill by any means such a predominant place in the marine fauna of the period, as in many Palæozoic deposits, but they are still individually numerous. The two ancient genera *Leptæna* (fig. 165, *a*) and *Spirifera* (fig. 165, *b*), dating the one from the Lower and the other from the Upper Silurian, appear here for the last time upon the scene, but they have not hitherto been recognized in deposits later than the Lias. The great majority of the Jurassic *Brachiopods,* however, belong to the genera *Terebratula* (fig. 165 *c, e, f*) and *Rhynchonella* (fig. 165, *d*), both of which are represented by living forms at the present day. The *Terebratulæ*, in particular, are very abundant, and the species are often confined to special horizons in the series.

Remains of *Bivalves* (*Lamellibranchiata*) are very numerous in the Jurassic deposits, and in many cases highly characteristic. In the marine beds of the Oolites, which constitute by far the greater portion of the whole formation, the Bivalves are of course marine, and belong to such genera as *Trigonia, Lima, Pholadomya, Cardinia, Avicula, Hippopodium,* &c.; but in the Purbeck beds, at the summit of the series, we find bands of Oysters alternating with strata containing fresh-water or brackish-water Bivalves, such as *Cyrenæ* and *Corbulæ*. The predominant Bivalves of the Jurassic, however, are the *Oysters,*

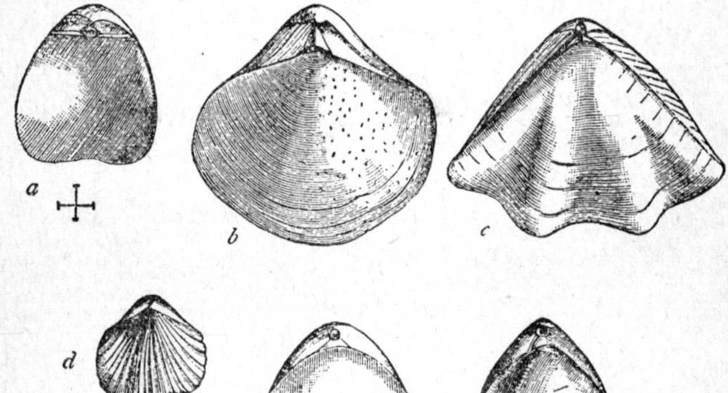

Fig. 165.—Jurassic Brachiopods. *a, Leptæna Liassica*, enlarged, the small cross below the figure indicating the true size of the shell—Lias, *b, Spirifera rostrata*, Lias; *c, Terebratula quadrifida*, Lias; *d, d', Rhynchonella varians*, Fuller's Earth and Kelloway Rock; *e. Terebratula sphæroidalis*, Inferior Oolite; *f, Terebratula digona*, Bradford Clay, Forest-marble, and Great Oolite. (After Davidson).

which occur under many forms, and often in vast numbers, particular species being commonly restricted to particular horizons. Thus of the true Oysters, *Ostrea distorta* is characteristic of the Purbeck series, where it forms a bed twelve feet in thickness, known locally as the "Cinder-bed;" *Ostrea expansa* abounds in the Portland beds; *Ostrea deltoidea* is characteristic of the Kimmeridge clay; *Ostrea gregaria* predominates in the Coral-rag; *Ostrea acuminata* characterizes the small group of the Fuller's Earth; whilst the plaited *Ostrea*

Marshii (fig. 166) is a common shell in the Lower and Middle Oolites. Besides the more typical Oysters, the Oolitic rocks abound in examples of the singularly unsymmetrical forms belonging to the genera *Exogyra* and *Gryphæa* (fig. 167). In the former of these are included Oysters with the beaks "reversed"—that is to say, turned towards the hinder part of the shell; whilst in the latter are Oysters in which the lower valve of the shell is much the largest, and has a large incurved beak, whilst the upper valve is small and concave. One of the most characteristic *Exogyræ* is the *E. Virgula* of the Oxford Clay, and of the same horizon on the Continent; and the *Gryphæa incurva* (fig. 167) is equally abundant in, and characteristic of, the formation of the Lias. Lastly, we may

Fig. 166.—*Ostrea Marshii.* Middle and Lower Oolites.

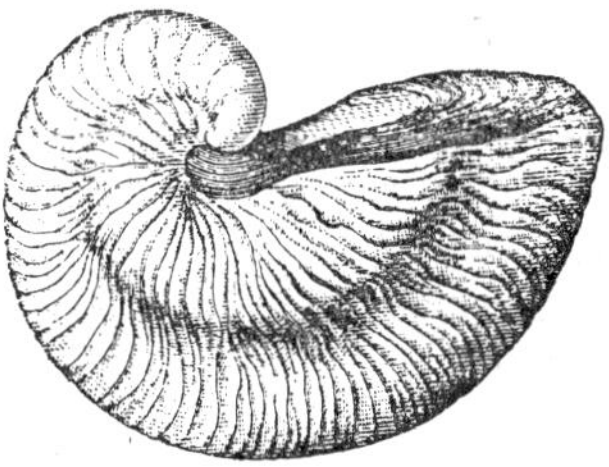

Fig. 167.—*Gryphæa incurva.* Lias.

notice the extraordinary shells belonging to the genus *Diceras* (fig. 168), which are exclusively confined to the Middle Oolites. In this formation in the Alps they occur in such abundance as to give rise to the name of "Calcaire à Dicerates," applied to beds of the same age as the Coral-rag of Britain. The genus *Diceras* belongs to the same family as the "Thorny Clams" (*Chama*) of the present day—the shell being composed of nearly equally-sized valves, the beaks of which are extremely prominent and twisted into a spiral. The shell was attached to some foreign body by the beak of one of its valves.

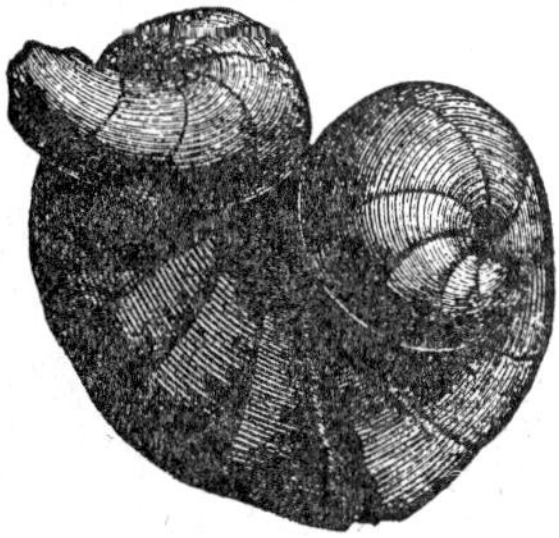

Fig. 168.—*Diceras arietina.* Middle Oolite.

Amongst the Jurassic Univalves (*Gasteropoda*) there are many examples of the ancient and long-lived *Pleurotomaria;* but on the whole the Univalves begin to have a modern aspect. The round-mouthed ("holostomatous"), vegetable-eating Sea-snails, such as the Limpets (*Patellidæ*), the Nerites (*Nerita*), the *Turritellæ, Chemnitziæ,* &c., still hold a predominant place. The two most noticeable genera of this group are *Cerithium* and *Nerinæa*—the former of these attaining great importance in the Tertiary and Recent seas, whilst the latter (fig. 169) is highly characteristic of the Jurassic series, though not exclusively confined to it. One of the limestones of the Jura, believed to be of the age of the Coral-rag (Middle Oolite) of Britain, abounds to such an extent in the turreted shells of *Nerinæa* as to have gained the name of "Calcaire à Nérinées." In addition to forms such as the preceding, we now for the first time meet, in any force, with the Carnivorous Univalves, in which the mouth of the shell is notched or produced into a canal, giving rise to the technical name of "siphonostomatous," applied to the shell. Some of the carnivorous forms belong to extinct types, such as the *Purpuroidea* of the Great Oolite; but others are referable to well-known existing genera. Thus we meet here with species of the familiar groups of the Whelks (*Buccinum*), the Spindle-shells (*Fusus*), the Spider-shells (*Pteroceras*), *Murex, Rostellaria,* and others which are not at present known to occur in any earlier formation.

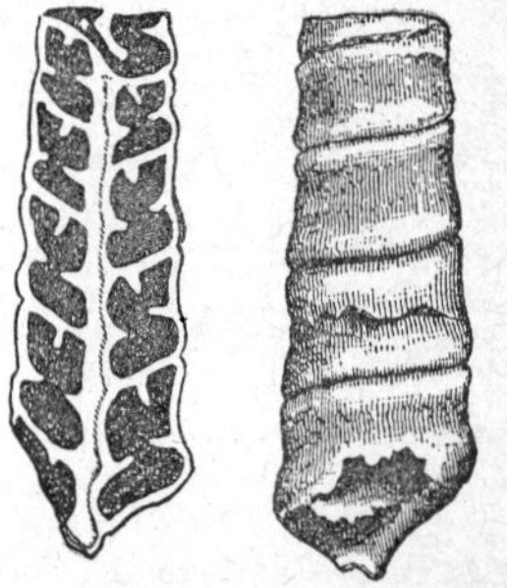

Fig. 169. — *Nerinæa Goodhallii*, one-fourth of the natural size. The left-hand figure shows the appearance presented by the shell when vertically divided. Coral-rag, England.

Amongst the Wing-shells (*Pteropoda*), it is sufficient to mark the final appearance in the Lias of the ancient genus *Conularia.*

Lastly, the order of the *Cephalopoda,* in both its Tetrabranchiate and Dibranchiate sections, undergoes a vast development in the Jurassic period. The old and comparatively simple genus *Nautilus* is still well represented, one species being very similar to the living Pearly Nautilus (*N. pompilius*); but the *Orthocerata* and *Goniatites* of the Trias have finally disappeared; and the great majority of the Tetrabranchiate forms are referable to the comprehensive genus *Ammonites,* with its many sub-genera and its hundreds of recorded species. The shell in *Ammonites* is in the form of a flat spiral, all the coils of which are in contact (figs. 170 and 171). The inner-

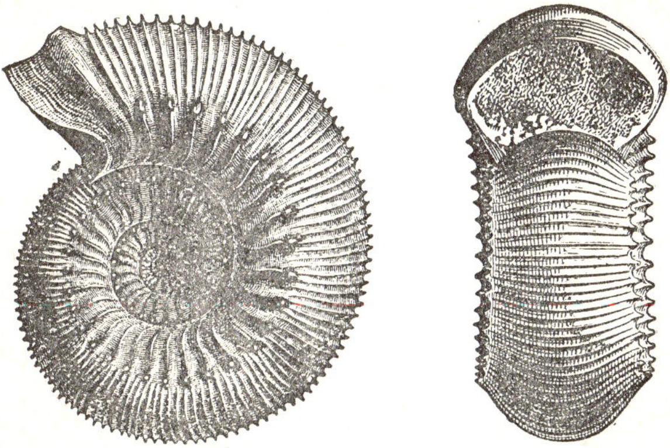

Fig. 170.—*Ammonites Humphresianus*. Inferior Oolite.

most whorls of the shell are more or less concealed; and the body-chamber is elongated and narrow, rather than expanded towards the mouth. The tube or siphuncle which runs through the air-chambers is placed on the dorsal or *convex* side of the

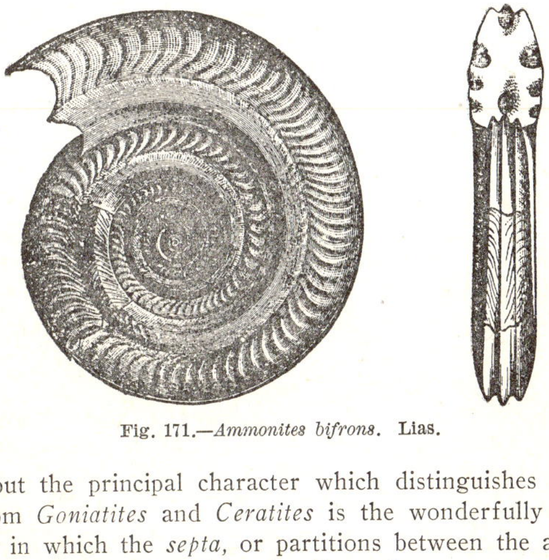

Fig. 171.—*Ammonites bifrons*. Lias.

shell; but the principal character which distinguishes *Ammonites* from *Goniatites* and *Ceratites* is the wonderfully complex manner in which the *septa*, or partitions between the air-chambers, are folded and undulated. To such an extent does this take place, that the edges of the septa, when exposed by the removal of the shell-substance, present in an exaggerated manner the appearance exhibited by an elaborately-dressed shirt-frill when viewed edgewise. The species of *Ammonites* range from the Carboniferous to the Chalk; but they have not been found in deposits older than the Secondary, in any region except India; and they are therefore to be regarded as essentially Mesozoic fossils. Within these limits, each formation is characterized by particular species, the number of individuals being often very great, and the size which is sometimes attained being nothing short of gigantic. In the Lias, particular species of *Ammonites* may succeed one another regularly, each having a more or less definite horizon, which it does not transgress. It is thus possible to distinguish a certain number of zones, each characterized by a particular Ammonite, together with other associated fossils. Some of these zones are very persistent and extend over very wide areas, thus affording valuable aid to the geologist in his determination of rocks. It is to be remembered, however, that there are other species which are not thus restricted in their vertical range, even in the same formations in which definite zones occur.

The Cuttle-fishes or *Dibranchiate Cephalopods* constitute a feature in the life of the Jurassic period little less conspicuous and striking than that afforded by the multitudinous and varied chambered shells of the *Ammonitidæ*. The remains by which these animals are recognized are necessarily less perfect, as a rule, than those of the latter, as no external shell is present (except in rare and more modern groups), and the internal skeleton is not necessarily calcareous. Nevertheless, we have an ample record of the Cuttle-fishes of the Jurassic period, in the shape of the fossilized jaws or beak, the ink-bag, and, most commonly of all, the horny or calcareous structure which is embedded in the soft tissues, and is variously known as the "pen" or "bone." The beaks of Cuttle-fishes, though not abundant, are sufficiently plentiful to have earned for themselves the general title of "Rhyncholites;" and in their form and function they resemble the horny, parrot-like beak of the existing Cephalopods. The ink-bag or leathery sac in which the Cuttle-fishes store up the black pigment with which they obscure the water when attacked, owes its preservation to the fact that the coloring-matter which it contains is finely-divided carbon, and therefore nearly indestructible except by heat. Many of these ink-bags have been found in the Lias; and the coloring-matter is sometimes so well preserved that it has been, as an experiment, employed in painting as a fossil "sepia." The "pens" of the Cuttle-fishes are not commonly preserved, owing to their horny consistence, but they are not unknown. The form here figured (*Beloteuthis subcostata,* fig. 172) belonged to an old type essentially similar to our modern Calamaries, the skeleton of which consists of a horny shaft and two lateral wings, somewhat like a feather in general shape. When, on the other hand, the internal skeleton is calcareous, then it is very easily preserved in a fossil condition; and the abundance of remains of this nature in the Secondary rocks, combined with their apparent total absence in Palæozoic strata, is a strong presumption in favor of the view that the order of the Cuttle-fishes did not come into existence till the commencement of the Mesozoic period. The great majority of the skeletons of this kind which are found in the Jurassic rocks belong to the great extinct family of the "Belemnites" (*Belemnitidæ*), which, so far as known, is entirely confined to rocks of Secondary age. From its pointed, generally cylindro-conical form, the skeleton of the *Belemnite* is popularly known as a "thunderbolt" (fig. 173, C). In its perfect condition—in which it is, however, rarely obtainable—the skeleton consists of a chambered conical shell (the "phragmacone"), the partitions between the chambers of which are pierced by a marginal tube or "Siphuncle." This conical shell—curiously similar in its structure to the *external* shell of the Nautilus—is extended forwards into a horny "pen," and is sunk in a corresponding conical pit (fig. 173, B), excavated in the substance of a nearly cylindrical fibrous body or "guard," which projects backwards for a longer or shorter distance, and is the part most usually found in a fossil condition. Many different kinds of *Belemnites* are known, and their guards literally swarm in many parts of the Jurassic series, whilst some specimens attain very considerable dimensions. Not only is the internal skeleton known, but specimens of *Belemnites* and the nearly allied *Belemnoteuthis* have been found in some of the fine-grained sediments of the Jurassic formation, from which much has been learnt even as to the anatomy of

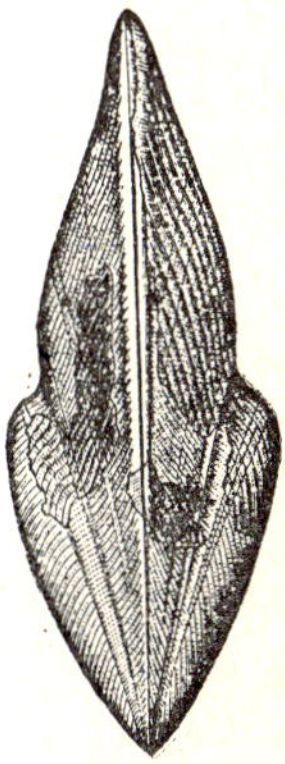

Fig. 172. — *Beloteuthis subcostata.* Jurassic (Lias).

the soft parts of the animal. Thus we know that the Belemnites were in many respects comparable with the existing

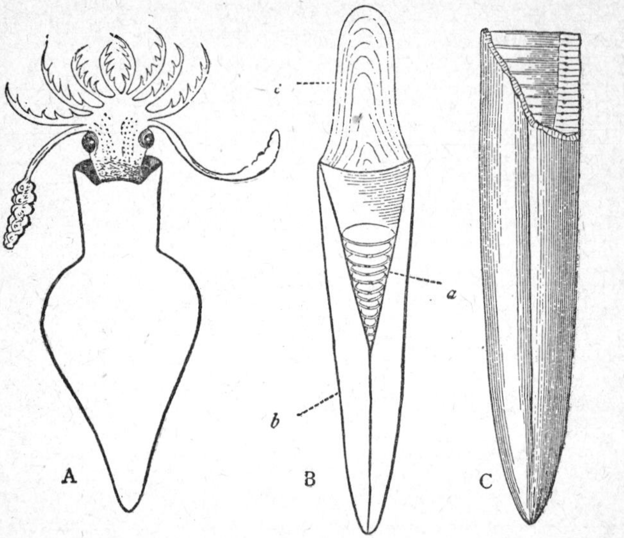

Fig. 173.—A, Restoration of the animal of the Belemnite ; B, Diagram showing the complete skeleton of a Belemnite, consisting of the chambered phragmacone (*a*), the guard (*b*), and the horny pen (*c*); C, Specimen of *Belemnites canaliculatus*, from the Inferior Oolite. (After Phillips.)

Calamaries or Squids, the body being furnished with lateral fins, and the head carrying a circle of ten "arms," two of

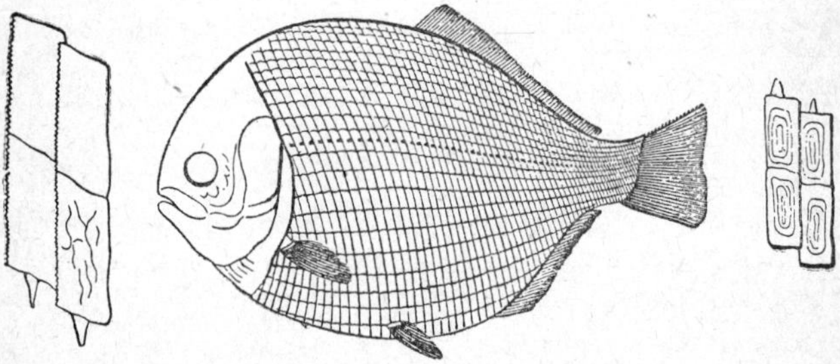

Fig. 174.—*Tetragonolepis* (restored), and scales of the same. Lias.

which were longer than the others (fig. 173, A). The suckers on the arms were provided, further, with horny hooks; there was a large ink-sac; and the mouth was armed with horny mandibles resembling in shape the beak of a parrot.

Coming next to the *Vertebrates,* we find that the Jurassic *Fishes* are still represented by *Ganoids* and *Placoids.* The Ganoids, however, unlike the old forms, now for the most part possess nearly or quite symmetrical ("homocercal") tails. A characteristic genus is *Tetragonolepis* (fig. 174), with its deep, compressed body, its rhomboidal, closely-fitting scales, and its single long dorsal fin. Amongst the *Placoids* the teeth of true Sharks (*Notidanus*) occur for the first time; but by far the greater number of remains referable to this group are still the fin-spines and teeth of "Cestracionts," resembling the living Port-Jackson Shark. Some of these teeth are pointed (*Hybodus*); but others are rounded, and are adapted for crushing shell-fish. Of these latter, the commonest are the teeth of *Acrodus* (fig. 175), of which the hinder ones are of an elongated form, with a rounded surface, covered with fine transverse striæ proceeding from a central longitudinal line. From their general form and striation, and their dark color, these teeth are commonly called "fossil leeches" by the quarrymen.

Fig. 175.—Tooth of *Acrodus nobilis*. Lias.

The Amphibian group of the *Labyrinthodonts,* which was so extensively developed in the Trias, appears to have become extinct, no representative of the order having hitherto been detected in rocks of Jurassic age.

Much more important than the Fishes of the Jurassic series are the *Reptiles,* which are both very numerous, and belong to a great variety of types, some of these being very extraordinary in their anatomical structure. The predominant group is that of the "Enaliosaurs" or "Sea-lizards," divided into two great orders, represented respectively by the *Ichthyosaurus* and the *Plesiosaurus.*

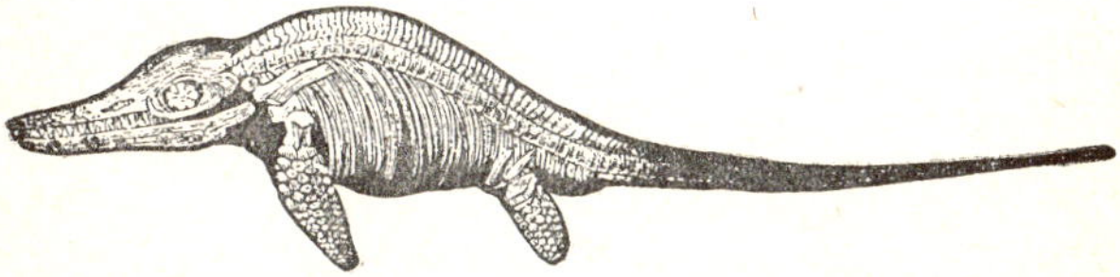

Fig. 176.—*Ichthyosaurus communis*. Lias.

The *Ichthyosauri* or "Fish-Lizards" are exclusively Mesozoic in their distribution, ranging from the Lias to the Chalk, but abounding especially in the former. They were huge Reptiles, of a fish-like form, with a hardly conspicuous neck (fig. 176), and probably possessing a simply smooth or wrinkled skin, since no traces of scales or bony integumentary plates have ever been discovered. The tail was long, and was probably furnished at its extremity with a powerful expansion of the skin, constituting a tail-fin similar to that possessed by the Whales. The limbs are also like those of Whales in the essentials of their structure, and in their being adapted to act as swimming-paddles. Unlike the Whales, however, the Ichthyosaurs possessed the hind-limbs as well as the fore-limbs, both pairs having the bones flattened out and the fingers completely enclosed in the skin, the arm and leg being at the same time greatly shortened. The limbs are thus converted into efficient "flippers," adapting the animal for an active existence in the sea. The different joints of the backbone (vertebræ) also show the same adaptation to an aquatic mode of life, being hollowed out at both ends, like the biconcave vertebræ of Fishes. The spinal column in this way was endowed with the flexibility necessary for an animal intended to pass the greater part of its time in water. Though the *Ichthyosaurs* are undoubtedly marine animals, there is, however, reason to believe that they occasionally came on shore, as they possess a strong bony arch, supporting the fore-limbs, such as would permit of partial, if laborious, terrestrial progression. The head is of enormous size, with greatly prolonged jaws, holding numerous powerful conical teeth lodged in a common groove. The nature of the dental apparatus is such as to leave no doubt as to the rapacious and predatory habits of the Ichthyosaurs—an inference which is further borne out by the examination of their petrified droppings, which are known to geologists as "coprolites," and which contain numerous fragments of the bones and scales of the Ganoid fishes which inhabited the same seas. The orbits are of huge size; and as the eyeball was protected, like that of birds, by a ring of bony plates in its outer coat, we even know that the pupils of the eyes were of correspondingly large dimensions. As these bony plates have the function of protecting the eye from injury under sudden changes of pressure in the surrounding medium, it has been inferred, with great probability, that the Ichthyosaurs were in the habit of diving to considerable depths in the sea. Some of the larger specimens of *Ichthyosaurus* which have been discovered in the Lias indicate an animal of from 20 to nearly 40 feet in length; and many species are known to have existed, whilst fragmentary remains of their skeletons are very abundant in some localities. We may therefore safely conclude that these colossal Reptiles were amongst the most formidable of the many tyrants of the Jurassic seas.

The *Plesiosaurus* (fig. 177) is another famous Oolitic Reptile, and, like the preceding, must have lived mainly or exclusively in the sea. It agrees with the Ichthyosaur in some important features of its organization, especially in the fact that both pairs of limbs are converted into "flippers" or swimming-paddles, whilst the skin seems to have been equally destitute of any scaly or bony investiture. Unlike the *Ichthy-*

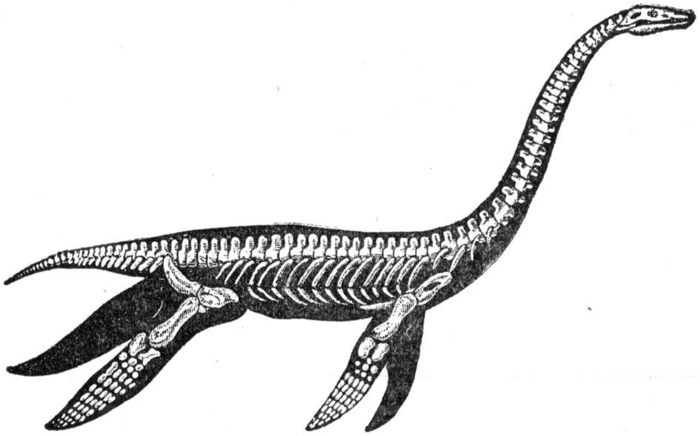

Fig. 177.—*Plesiosaurus dolichodeirus*, restored. Lias.

osaur, however, the *Plesiosaur* had the paddles placed far back, the tail being extremely short, and the neck greatly lengthened out, and composed of from twenty to forty vertebræ. The bodies of the vertebræ, also, are not deeply biconcave, but are flat, or only slightly cupped. The head is of relatively small size, with smaller orbits than those of the *Ichthyosaur,* and with a snout less elongated. The jaws, however, were armed with numerous conical teeth, inserted in distinct sockets. As regards the habits of the *Plesiosaur,* Dr. Conybeare arrives at the following conclusions: "That it was aquatic is evident from the form of its paddles; that it was marine is almost equally so from the remains with which it is universally associated;

that it may have occasionally visited the shore, the resemblance of its extremities to those of the Turtles may lead us to conjecture: its movements, however, must have been very awkward on land; and its long neck must have impeded its progress through the water, presenting a strong contrast to the organization which so admirably fits the *Ichthyosaurus* to cut through the waves." As its respiratory organs were such that it must of necessity have required to obtain air frequently, we may conclude "that it swam upon or near the surface, arching back its long neck like a swan, and occasionally darting it down at the fish which happened to float within its reach. It may perhaps have lurked in shoal water along the coast, concealed amongst the sea-weed; and raising its nostrils to a level with the surface from a considerable depth, may have found a secure retreat from the assaults of powerful enemies; while the length and flexibility of its neck may have compensated for the want of strength in its jaws, and its incapacity for swift motion through the water."

About twenty species of *Plesiosaurus* are known, ranging from the Lias to the Chalk, and specimens have been found indicating a length of from eighteen to twenty feet. The nearly related "*Pliosaurs,*" however, with their huge heads and short necks, must have occasionally reached a length of at least forty feet—the skull in some species being eight, and the paddles six or seven feet long, whilst the teeth are a foot in length.

Another extraordinary group of Jurassic Reptiles is that of the "Winged Lizards" or *Pterosauria.* These are often spoken of collectively as "Pterodactyles," from *Pterodactylus,* the type-genus of the group. As now restricted, however, the genus *Pterodactylus* is more Cretaceous than Jurassic, and it is associated in the Oolitic rocks with the closely allied genera *Dimorphodon* and *Rhamphorhynchus.* In all three of these genera we have the same general structural organization, involving a marvellous combination of characters, which we are in the habit of regarding as peculiar to Birds on the one hand, to Reptiles on another hand, and to the Flying Mammals or Bats in a third direction. The "Pterosaurs" are "Flying" Reptiles, in the true sense of the term, since they were indubitably possessed of the power of active locomotion in the air, after the manner of Birds. The so-called "Flying" Reptiles of the present day, such as the little *Draco volans* of the East Indies and Indian Archipelago, possess, on the other hand, no power of genuine flight, being merely able to sustain themselves in the air through the extensive leaps which they take from tree to tree, the wing-like expansions of the skin simply exercising the mechanical function of a parachute. The apparatus of flight in the "Pterosaurs" is of the most remarkable character, and most resembles the "wing" of a Bat, though very different in some important particulars. The "wing" of the Pterosaurs is like that of Bats, namely, in consisting of a thin leathery expansion of the skin which is attached to the sides of the body, and stretches between the fore and hind limbs, being mainly supported by an enormous elongation of certain of the digits of

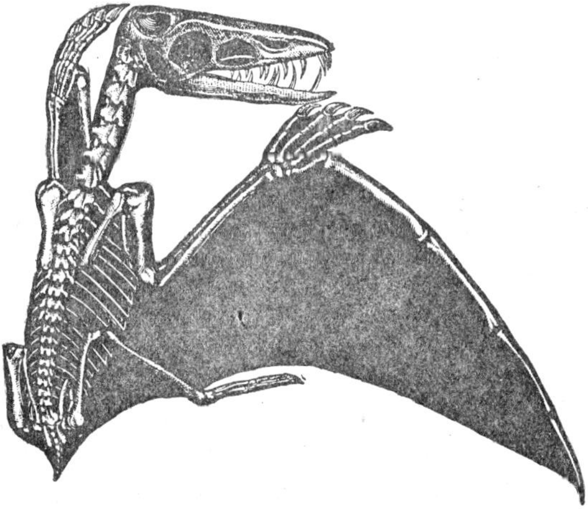

Fig. 178.—*Pterodactylus crassirostris.* From the Lithographic Slates of Solenhofen (Middle Oolite.) The figure is "restored," and it seems certain that the restoration is incorrect in the comparatively unimportant particular, that the hand should consist of no more than four fingers, three short and one long, instead of five, as represented.

the hand. In the Bats, it is the four outer fingers which are thus lengthened out; but in the Pterosaurs, the wing-membrane is borne by a single immensely-extended finger (fig. 178). No trace of the actual wing-membrane itself has, of course, been found fossilized; but we could determine that the "Pterodactyles" possessed the power of flight, quite apart from the extraordinary conformation of the hand. The proofs of this are to be found partly in the fact that the breast-bone was furnished with an elevated ridge or keel, serving for the attachment of the great muscles of flight, and still more in the fact that the bones were hollow and were filled with air—a peculiarity wholly confined amongst living animals to Birds only. The skull of the Pterosaurs is long, light, and singularly bird-like in appearance—a resemblance which is further increased by the comparative length of the neck and the size of the vertebræ of this region (fig. 178). The jaws, however, unlike those of any existing Bird, were, with one exception to be noticed hereafter, furnished with conical teeth sunk in distinct sockets; and there was always a longer or shorter tail composed of distinct vertebræ; whereas in all existing Birds the tail is abbreviated, and the terminal vertebræ are amalgamated to form a single bone, which generally supports the great feathers of the tail.

Modern naturalists have been pretty generally agreed that the *Pterosaurs* should be regarded as a peculiar group of the Reptiles; though they have been and are still regarded by high authorities, like Professor Seeley, as being really referable to the Birds, or as forming a class by themselves. The chief points which separate them from Birds, as a class, are the character of the apparatus of flight, the entirely different structure of the fore-limb, the absence of feathers, the composition of the tail out of distinct vertebræ, and the general presence of conical teeth sunk in distinct sockets in the jaws. The gap between the Pterosaurs and the Birds has, however, been greatly lessened of late by the discovery of fossil animals (*Ichthyornis* and *Hesperornis*) with the skeleton proper to Birds combined with the presence of teeth in the jaws, and by the still more recent discovery of other fossil animals (*Pteranodon*) with a Pterosaurian skeleton, but without teeth; whilst the undoubtedly feathered *Archæopteryx* possessed a long tail composed of separate vertebræ. Upon the whole, therefore, the relationships of the Pterosaurs cannot be regarded as absolutely settled. It seems certain, however, that they did not possess feathers—this implying that they were cold-blooded animals; and their affinities with Reptiles in this, as in other characters, are too strong to be overlooked.

The *Pterosaurs* are wholly Mesozoic, ranging from the Lias to the Chalk inclusive; and the fine-grained Lithographic Slate of Solenhofen has proved to be singularly rich in their remains. The genus *Pterodactylus* itself has the jaws toothed to the extremities with equal-sized conical teeth, and its species range from the Middle Oolites to the Cretaceous series, in connection with which they will be again noticed, together with the toothless genus *Pteranodon*. The genus *Dimorphodon* is Liassic, and is characterized by having the front teeth long and pointed, whilst the hinder teeth are small and lancet-shaped. Lastly, the singular genus *Rhamphorhynchus,* also from the Lower Oolites, is distinguished by the fact that there are teeth present in the hinder portions of both jaws; but the front portions are toothless, and may have constituted a horny beak. Like most of the other Jurassic Pterosaurs, *Rhamphorhynchus* (fig. 179) does not seem to have been much bigger than a pigeon, in this respect falling far below the giant "Dragons" of the Cretaceous period. It differed from its relatives, not

Fig. 179.—*Rhamphorhynchus Bucklandi,* restored. Bath Oolite, England. (After the late Professor Phillips.)

only in the armature of the mouth, but also in the fact that the tail was of considerable length. With regard to its habits and mode of life, Professor Phillips remarks that, "gifted with ample means of flight, able at least to perch on rocks and scuffle along the shore, perhaps competent to dive, though not so well as a Palmiped bird, many fishes must have yielded to the cruel beak and sharp teeth of Rhamphorhynchus. If we ask to which of the many families of Birds the analogy of structure and probable way of life would lead us to assimilate Rhamphorhynchus, the answer must point to the swimming races with long wings, clawed feet, hooked beak, and habits of violence and voracity; and for preference, the shortness of the legs, and other circumstances, may be held to claim for the Stonesfield fossil a more than fanciful similitude to the groups of Cormorants, and other marine divers, which constitute an effective part of the picturesque army of robbers of the sea."

Another extraordinary and interesting group of the Mesozoic Reptiles is constituted by the *Deinosauria,* comprising a series of mostly gigantic forms, which range from the Trias to the Chalk. All the "Deinosaurs" are possessed of the two pairs of limbs proper to Vertebrate animals, and these organs are in the main adapted for walking on the dry land. Thus, whilst the Mesozoic seas swarmed with the huge Ichthyosaurs and Plesiosaurs, and whilst the air was tenanted by the Dragon-like Pterosaurs, the land-surfaces of the Secondary period were peopled by numerous forms of Deinosaurs, some of them of even more gigantic dimensions than their marine brethren. The limbs of the *Deinosaurs* are, as just said, adapted for progression on the land; but in some cases, at any rate, the hind-limbs were much longer and stronger than the fore-limbs; and there seems to be no reason to doubt that many of these forms possessed the power of walking, temporarily or permanently, on their hind-legs, thus presenting a singular resemblance to Birds. Some very curious and striking points connected with the structure of the skeleton have also been shown to connect these strange Reptiles with the true Birds; and such high authorities as Professors Huxley and Cope are of opinion that the Deinosaurs are distinctly related to this class, being in some respects intermediate between the proper Reptiles and the great wingless Birds, like the Ostrich and Cassowary. On the other hand, Professor Owen has shown that the Deinosaurs possess some weighty points of relationship with the so-called

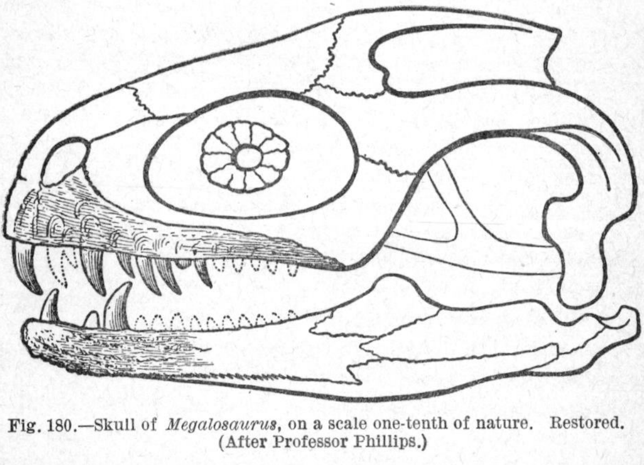

Fig. 180.—Skull of *Megalosaurus*, on a scale one-tenth of nature. Restored. (After Professor Phillips.)

"Pachydermatous" Quadrupeds, such as the Rhinoceros and Hippopotamus. The most important Jurassic genera of *Deinosauria* are *Megalosaurus* and *Cetiosaurus*, both of which extend their range into the Cretaceous period, in which flourished, as we shall see, some other well-known members of this order.

Megalosaurus attained gigantic dimensions, its thigh and shank bones measuring each about three feet in length, and its total length, including the tail, being estimated at from forty to fifty feet. As the head of the thigh-bone is set on nearly at right angles with the shaft, whilst all the long bones of the skeleton are hollowed out internally for the reception of the marrow, there can be no doubt as to the terrestrial habits of the animal. The skull (fig. 180) was of large size, four or five feet in length, and the jaws were armed with a series of powerful pointed teeth. The teeth are conical in shape, but are strongly compressed towards their summits, their lateral edges being finely serrated. In their form and their saw-like edges, they resemble the teeth of the "Sabre-toothed Tiger" *Machairodus*), and they render it certain that the Megalosaur was in the highest degree destructive and carnivorous in its habits. So far as is known, the skin was not furnished with any armor of scales or bony plates; and the fore-limbs are so disproportionately small as compared with the hind-limbs, that this huge Reptile—like the equally huge Iguanodon—may be conjectured to have commonly supported itself on its hind-legs only.

The *Cetiosaur* attained dimensions even greater than those of the Megalosaur, one of the largest thigh-bones measuring over five feet in length and a foot in diameter in the middle, and the total length of the animal being probably not less than fifty feet. It was originally regarded as a gigantic Crocodile, but it has been shown to be a true Deinosaur. Having obtained a magnificent series of remains of this reptile, Professor Phillips has been able to determine many very interesting points as to the anatomy and habits of this colossal animal, the total length of which he estimates as being probably not less than sixty or seventy feet. As to its mode of life, this accomplished writer remarks:—

"Probably when 'standing at ease' not less than ten feet in height, and of a bulk in proportion, this creature was unmatched in magnitude and physical strength by any of the largest inhabitants of the Mesozoic land or sea. Did it live in the sea, in fresh waters, or on the land? This question cannot be answered, as in the case of Ichthyosaurus, by appeal to the accompanying organic remains; for some of the bones lie in marine deposits, others in situations marked by estuarine conditions, and, out of the Oxfordshire district, in Sussex, in fluviatile accumulations. Was it fitted to live exclusively in water? Such an idea was at one time entertained, in consequence of the biconcave character of the caudal vertebræ, and it is often suggested by the mere magnitude of the creature, which would seem to have an easier life while floating in water, than when painfully lifting its huge bulk, and moving with slow steps along the ground. But neither of these arguments is valid. The ancient earth was trodden by larger quadrupeds than our elephant; and the biconcave character of vertebræ, which is not uniform along the column in Cetiosaurus, is perhaps as much a character of geological period as of a mechanical function of life. Good evidence of continual life in water is yielded in the case of Ichthyosaurus and other Enaliosaurs, by the articulating surfaces of their limb-bones, for these, all of them, to the last phalanx, have that slight and indefinite adjustment of the bones, with much intervening cartilage, which fits the leg to be both a flexible and forcible instrument of natation, much superior to the ordinary oar-blade of the boatman. On the contrary, in Cetiosaur, as well as in Megalosaur and Iguanodon, all the articulations are definite, and made so as to correspond to determinate movements in particular directions, and these are such as to be suited for walking. In particular, the femur, by its head projecting freely from the acetabulum, seems to claim a movement of free stepping more parallel to the line of the body, and more approaching to the vertical than the sprawling gait of the crocodile. The large claws concur in this indication of terrestrial habits. But, on the other hand, these characters are not contrary to the belief that the animal may have been amphibious; and the great vertical height of the anterior part of the tail seems to support this explanation, but it does not go further. . . . We have therefore a marsh-loving or river-side animal, dwelling amidst filicine, cycadaceous, and coniferous shrubs and trees full of insects and small mammalia. What was its usual diet? If *ex ungue leonem*, surely *ex dente cibum*. We have indeed but one tooth, and that small and incomplete. It resembles more the tooth of Iguanodon than that of any other reptile; for this reason it seems probable that the animal was nourished by similar vegetable food which abounded in the vicinity, and was not obliged to contend with Megalosaurus for a scanty supply of more stimulating diet."

All the groups of Jurassic Reptiles which we have hitherto been considering are wholly unrepresented at the present day, and do not even pass upwards into the Tertiary period. It may be mentioned, however, that the Oolitic deposits have also yielded the remains of Reptiles belonging to three of the existing orders of the class—namely, the Lizards (*Lacertilia*),

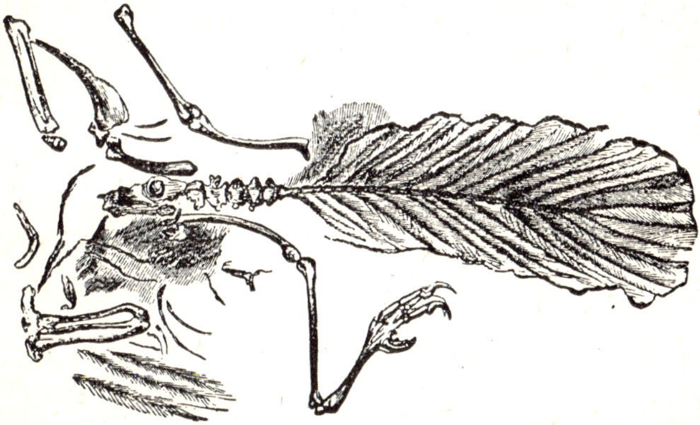

Fig. 181.—*Archæopteryx macrura*, showing tail and tail-feathers, with detached bones Reduced. From the Lithographic Slate of Solenhofen.

the Turtles (*Chelonia*), and the Crocodiles (*Crocodilia*). The Lizards occur both in the marine strata of the Middle Oolites and also in the fresh-water beds of the Purbeck series; and

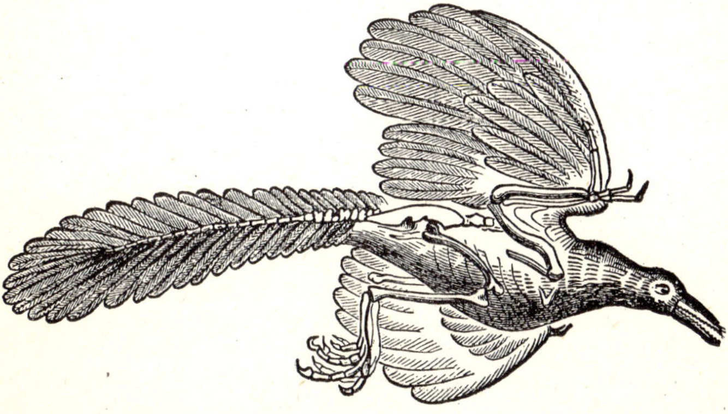

Fig. 182.—Restoration of *Archæopteryx macrura*. (After Owen).

they are of such a nature that their affinities with the typical Lacertilians of the present day cannot be disputed. The Chelonians, up to this point only known by the doubtful evidence of footprints in the Permian and Triassic sandstones, are here represented by unquestionable remains, indicating the existence of marine Turtles (the *Chelone planiceps* of the Portland Stone). No remains of Serpents (*Ophidians*) have as yet been detected in the Jurassic; but strata of this age have yielded the remains of numerous *Crocodilians,* which probably inhabited the sea. The most important member of this group is *Teleosaurus,* which attained a length of over thirty feet, and is in some respects allied to the living Gavials of India.

The great class of the Birds, as we have seen, is represented in rocks earlier than the Oolites simply by the not absolutely certain evidence of the three-toed footprints of the Connecticut Trias. In the Lithographic Slate of Solenhofen (Middle Oolite), there has been discovered, however, the at present unique skeleton of a Bird well known under the name of the *Archæopteryx macrura* (figs. 181, 182). The only known specimen—now in the British Museum—unfortunately does not exhibit the skull; but the fine-grained matrix has preserved a number of the other bones of the skeleton, along with the impressions of the tail and wing feathers. From these remains we know that *Archæopteryx* differed in some remarkable peculiarities of its structure from all existing members of the class of Birds. This extraordinary Bird (fig. 182) appears to have been about as big as a Rook—the tail being long and extremely slender, and composed of separate vertebræ, each of which supports a single pair of quill-feathers. In the flying Birds of the present day, as before mentioned, the terminal vertebræ of the tail are amalgamated to form a single bone ("ploughshare-bone"), which supports a cluster of tail-feathers; and the tail itself is short. In the embryos of existing Birds the tail is long, and is made up of separate vertebræ, and the same character is observed in many existing Reptiles. The tail of *Archæopteryx,* therefore, is to be regarded as the permanent retention of an embryonic type of structure, or as an approximation to the characters of the Reptiles. Another remarkable point in connection with *Archæopteryx,* in which it differs from all known Birds is, that the wing was furnished with two free claws. From the presence of feathers, *Archæopteryx* may be inferred to have been hot-blooded; and this character, taken along with the structure of the skeleton of the wing, may be held as sufficient to justify its being considered as belonging to a class of Birds. In the structure of the tail, however, it is singularly Reptilian; and there is reason to believe that its jaws were furnished with teeth sunk in distinct sockets, as is the case in no existing Bird. This conclusion, at any rate, is rendered highly probable by the recent discovery of "Toothed Birds" (*Odontornithes*) in the Cretaceous rocks of North America.

The *Mammals* of the Jurassic period are known to us by a number of small forms which occur in the "Stonesfield Slate" (Great Oolite) and in the Purbeck beds (Upper Oolite). The remains of these are almost exclusively separated halves of the lower jaw, and they indicate the existence

Fig. 183.—Lower jaw of *Amphitherium* (*Thylacotherium*) *Prevostii*, Stonesfield Slate (Great Oolite.)

during the Oolitic period in Europe of a number of small "Pouched animals" (*Marsupials*). In the horizon of the Stonesfield Slate four genera of these little Quadrupeds have been described—viz., *Amphilestes, Amphitherium, Phascolotherium,* and *Stereognathus.* In *Amphitherium* (fig. 183), the molar teeth are furnished with small pointed eminences of "cusps;" and the animal was doubtless insectivorous. By

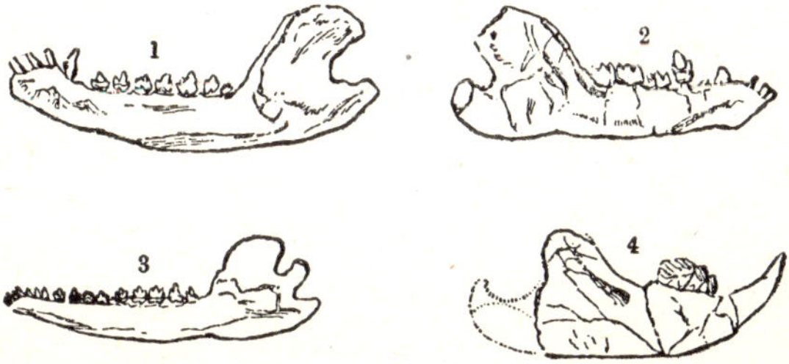

Fig. 184 —Oolitic Mammals.—1, Lower jaw and teeth of *Phascolotherium*, Stonesfield Slate ; 2, Lower jaw and teeth of *Amphitherium*, Stonesfield Slate ; 3, Lower jaw and teeth of *Triconodon*, Purbeck beds ; 4. Lower jaw and teeth of *Plagiaulax*, Purbeck beds. All the figures are of the natural size.

Professor Owen, the highest living authority on the subject, *Amphitherium* is believed to be a small Marsupial, most nearly allied to the living Banded Ant-eater (*Myrmecobius*) of Australia (fig. 158). *Amphilestes* and *Phascolotherium* (fig.

184) are also believed by the same distinguished anatomist and palæontologist to have been insect-eating Marsupials, and the latter is supposed to find its nearest living ally in the Opossums (*Didelphys*) of America. Lastly, the *Stereognathus* of the Stonesfield Slate is in a dubious position. It may have been a Marsupial; but, upon the whole, Professor Owen is inclined to believe that it must have been a hoofed and herbivorous Quadruped belonging to the series of the higher Mammals (*Placentalia*). In the Middle Purbeck beds, near to the close of the Oolitic period, we have also evidence of the existence of a number of small Mammals, all of which are probably Marsupials. Fourteen species are known, all of small size, the largest being no bigger than a Polecat or Hedgehog. The genera to which these little quadrupeds have been referred are *Plagiaulax, Spalacotherium, Triconodon,* and *Galestes.* The first of these (fig. 184, 4) is believed by Professor Owen to have been carnivorous in its habits; but other authorities maintain that it was most nearly allied to the living Kangaroo-rats (*Hypsiprymnus*) of Australia, and that it was essentially herbivorous. The remaining three genera appear to have been certainly insectivorous, and find their nearest living representatives in the Australian Phalangers and the American Opossums.

Finally, it is interesting to notice in how many respects the Jurassic fauna of Western Europe approached to that now inhabiting Australia. At the present day, Australia is almost wholly tenanted by Marsupials; upon its land-surface flourish *Araucariæ* and Cycadaceous plants, and in its seas swim the Port-Jackson Shark (*Cestracion Philippi*); whilst the Molluscan genus *Trigonia* is nowadays exclusively confined to the Australian coasts. In England, at the time of the deposition of the Jurassic rocks, we must have had a fauna and flora very closely resembling what we now see in Australia. The small Marsupials, *Amphitherium, Phascolotherium,* and others, prove that the Mammals were the same in order; cones of Araucarian pines, with tree-ferns and fronds of Cycads, occur throughout the Oolitic series; spine-bearing fishes, like the Port-Jackson Shark, are abundantly represented by genera such as *Acrodus* and *Strophodus;* and lastly, the genus *Trigonia,* now exclusively Australian, is represented in the Oolites by species which differ little from those now existing. Moreover, the discovery during recent years of the singular Mud-fish, the *Ceratodus Fosteri,* in the rivers of Queensland, has added another and a very striking point of resemblance to those already mentioned; since this genus of Fishes, though pre-eminently Triassic, nevertheless extended its range into the Jurassic. Upon the whole, therefore, there is reason to conclude that Australia has undergone since the close of the Jurassic period fewer changes and vicissitudes than any other known region of the globe; and that this wonderful continent has therefore succeeded in preserving a greater number of the characteristic life-features of the Oolites than any other country with which we are acquainted.

LITERATURE.

The following list comprises some of the more important sources of information as to the rocks and fossils of the Jurassic series:—

(1) 'Geology of Oxford and the Thames Valley.' Phillips.
(2) 'Geology of Yorkshire,' vol. ii. Phillips.
(3) 'Memoirs of the Geological Survey of Great Britain.'
(4) 'Geology of Cheltenham.' Murchison, 2d ed. Buckman.
(5) 'Introduction to the Monograph of the Oolitic Asteriadæ' (Palæontographical Society). Wright.
(6) "Zone of Avicula contorta and the Lower Lias of the South of England"—'Quart. Journ. Soc.,' vol. xvi., 1860. Wright.
(7) "Oolites of Northamptonshire"—'Quart. Journ. Geol. Soc.,' vols. xxvi. and xxix. Sharp.
(8) 'Manual of Geology.' Dana.
(9) 'Der Jura.' Quenstedt.
(10) 'Das Flötzgebirge Württembergs.' Quenstedt.
(11) 'Jura Formation.' Oppel.
(12) 'Paléontologie du Département de la Moselle.' Terquem.
(13) 'Cours élémentaire de Paléontologie.' D'Orbigny.
(14) 'Paléontologie Française.' D'Orbigny.
(15) 'Fossil Echinodermata of the Oolitic Formation.' (Palæontographical Society). Wright.
(16) 'Brachiopoda of the Oolitic Formation' (Palæontographical Society). Davidson.
(17) 'Mollusca of the Great Oolite' (Palæontographical Society). Morris and Lycett.
(18) 'Monograph of the Fossil Trigoniæ' (Palæontographical Society). Lycett.
(19) 'Corals of the Oolitic Formation' (Palæontographical Society). Edwards and Haime.
(20) 'Supplement to the Corals of the Oolitic Formation' (Palæontographical Society). Martin Duncan.
(21) 'Monograph of the Belemnitidæ' (Palæontographical Society). Phillips.
(22) 'Structure of the Belemnitidæ' (Mem. Geol. Survey). Huxley.

(23) 'Sur les Belemnites.' Blainville.
(24) 'Cephalopoden.' Quenstedt.
(25) 'Mineral Conchology.' Sowerby.
(26) 'Jurassic Cephalopoda' (Palæontologica Indica). Waagen.
(27) 'Manual of the Mollusca.' Woodward.
(28) 'Petrefaktenkunde.' Schlotheim.
(29) 'Bridgewater Treatise.' Buckland.
(30) 'Versteinerungen des Oolithengebirges.' Rœmer.
(31) 'Catalogue of British Fossils.' Morris.
(32) 'Catalogue of Fossils in the Museum of Practical Geology.' Etheridge.
(33) 'Beiträge zur Petrefaktenkunde.' Münster.
(34) 'Petrefacta Germaniæ.' Goldfuss.
(35) 'Lethæa Rossica.' Eichwald.
(36) 'Fossil Fishes' (Decades of the Geol. Survey). Sir Philip Egerton.
(37) 'Manual of Palæontology.' Owen.
(38) 'British Fossil Mammals and Birds.' Owen.
(39) 'Monographs of the Fossil Reptiles of the Oolitic Formation.' (Palæontographical Society). Owen.
(40) 'Fossil Mammals of the Mesozoic Formations' Palæontographical Society). Owen.
(41) 'Catalogue of Ornithosauria.' Seeley.
(42) "Classification of the Deinosauria"—'Quart. Journ. Geol. Soc.,' vol. xxvi., 1870. Huxley.

CHAPTER XVII.

THE CRETACEOUS PERIOD.

The next series of rocks in ascending order is the great and important series of the *Cretaceous Rocks,* so called from the general occurrence in the system of chalk (Lat. *creta,* chalk). As developed in Britain and Europe generally, the following leading subdivisions may be recognized in the Cretaceous series:—

1. Wealden, 2. Lower Greensand or Neocomian,	Lower Cretaceous.
3. Gault, 4. Upper Greensand, 5. Chalk, 6. Maestricht beds,	Upper Cretaceous.

I. *Wealden.*—The *Wealden* formation, though of considerable importance, is a local group, and is confined to the south-east of England, France, and some other parts of Europe. Its name is derived from the *Weald,* a district comprising parts of Surrey, Sussex, and Kent, where it is largely developed. Its lower portion, for a thickness of from 500 to 1000 feet, is arenaceous, and is known as the Hastings Sands. Its Upper portion, for a thickness of 150 to nearly 300 feet, is chiefly argillaceous, consisting of clays with sandy layers, and occasionally courses of limestone. The geological importance of the Wealden formation is very great, as it is undoubtedly the delta of an ancient river, being composed almost wholly of fresh-water beds, with a few brackish-water and even marine strata, intercalated in the lower portion. Its geographical extent, though uncertain, owing to the enormous denudation to which it has been subjected, is nevertheless great, since it extends from Dorsetshire to France, and occurs also in North Germany. Still, even if it were continuous between all these points, it would not be larger than the delta of such a modern river as the Ganges. The river which produced the Wealden series must have flowed from an ancient continent occupying what is now the Atlantic Ocean; and the time occupied in the formation of the Wealden must have been very great, though we have, of course, no data by which we can accurately calculate its duration.

The fossils of the Wealden series are, naturally, mostly the remains of such animals as we know at the present day as inhabiting rivers. We have, namely, fresh-water Mussels (*Unio*), River-snails (*Paludina*), and other fresh-water shells, with numerous little bivalved Crustaceans, and some fishes.

II. *Lower Greensand* (*Néocomien* of D'Orbigny).—The Wealden beds pass upward, often by insensible gradations, into the *Lower Greensand.* The name Lower Greensand is not an appropriate one, for green sands only occur sparingly and occasionally, and are found in other formations. For this reason it has been proposed to substitute for Lower Greensand the name *Neocomian,* derived from the town of Neufchâtel—anciently called *Neocomum*—in Switzerland. If this name were adopted, as it ought to be, the Wealden beds would be called the Lower Neocomian.

The Lower Greensand or Neocomian of Britain has a thickness of about 850 feet, and consists of alternations of sands, sandstones, and clays, with occasional calcareous bands. The general color of the series is dark brown, sometimes red; and the sands are occasionally green, from the presence of silicate of iron.

The fossils of the Lower Greensand are purely marine, and among the most characteristic are the shells of *Cephalopods.*

The most remarkable point, however, about the fossils of the Lower Cretaceous series, is their marked divergence from the fossils of the Upper Cretaceous rocks. Of 280 species of fossils in the Lower Cretaceous series, only 51, or about 18 per cent, pass on into the Upper Cretaceous. This break in the life of the two periods is accompanied by a decided physical break as well; for the Gault is often, if not always, unconformably superimposed on the Lower Greensand. At the same time, the Lower and Upper Cretaceous groups form a closely-connected and inseparable series, as shown by a comparison of their fossils with those of the underlying Jurassic rocks and the overlying Tertiary beds. Thus, in Britain no marine fossil is known to be common to the marine beds of the Upper Oolites and the Lower Greensand; and of more than 500 species of fossils in the Upper Cretaceous rocks, almost every one died out before the formation of the lowest Tertiary strata, the only survivors being one Brachiopod and a few *Foraminifera.*

III. *Gault* (*Aptien* of D'Orbigny).—The lowest member of the Upper Cretaceous series is a stiff, dark-grey, blue, or brown clay, often worked for brick-making, and known as the *Gault,* from a provincial English term. It occurs chiefly in the southeast of England, but can be traced through France to the flanks of the Alps and Bavaria. It never exceeds 100 feet in thickness; but it contains many fossils, usually in a state of beautiful preservation.

IV. *Upper Greensand* (*Albien* of D'Orbigny; *Unterquader* and *Lower Plänerkalk* of Germany).—The Gault is succeeded upward by the *Upper Greensand,* which varies in thickness from 3 up to 100 feet, and which derives its name from the occasional occurrence in it of green sands. These, however, are local and sometimes wanting, and the name "Upper Greensand" is to be regarded as a *name* and not a description. The group consists, in Britain, of sands and clays, sometimes with bands of calcareous grit or siliceous limestone, and occasionally containing concretions of phosphate of lime, which are largely worked for agricultural purposes.

V. *White Chalk.*—The top of the Upper Greensand becomes argillaceous, and passes up gradually into the base of the great formation known as the true *Chalk,* divided into the three subdivisions of the chalk-marl, white chalk without flints, and white chalk with flints. The first of these is simply argillaceous chalk, and passes up into a great mass of obscurely-stratified white chalk in which there are no flints (*Turonien* of D'Orbigny; *Mittelquader* of Germany). This in turn, passes up into a great mass of white chalk, in which the stratification is marked by nodules of black flint arranged in layers (*Sénonien* of D'Orbigny; *Oberquader* of Germany). The thickness of these three subdivisions taken together is sometimes over 1000 feet, and their geographical extent is very great. White Chalk, with its characteristic appearance, may be traced from the north of Ireland to the Crimea, a distance of about 1140 geographical miles; and, in an opposite direction, from the south of Sweden to Bordeaux, a distance of about 840 geographical miles.

VI. In Britain there occur no beds containing Chalk fossils, or in any way referable to the Cretaceous period, above the true White Chalk with flints. On the banks of the Maes, however, near Maestricht in Holland, there occurs a series of yellowish limestones, of about 100 feet in thickness, and undoubtedly superior to the White Chalk. These *Maestricht beds* (*Danien* of D'Orbigny) contain a remarkable series of fossils, the characters of which are partly Cretaceous and partly Tertiary. Thus, with the characteristic Chalk fossils, *Belemnites, Baculites,* Sea-urchins, &c., are numerous Univalves Molluscs, such as Cowries and Volutes, which are otherwise exclusively Tertiary or Recent.

Holding a similar position to the Maestricht beds, and showing a similar intermixture of Cretaceous forms with later types, are certain beds which occur in the island of Seeland, in Denmark, and which are known as the *Faxöe Limestone.*

Of a somewhat later date than the Maestricht beds is the *Pisolitic Limestone* of France, which rests unconformably on the White Chalk, and contains a large number of Tertiary fossils along with some characteristic Cretaceous types.

The subjoined sketch-section exhibits the general succession of the Cretaceous deposits in Britain:—

GENERALIZED SECTION OF THE CRETACEOUS SERIES OF BRITAIN.

Fig. 185.

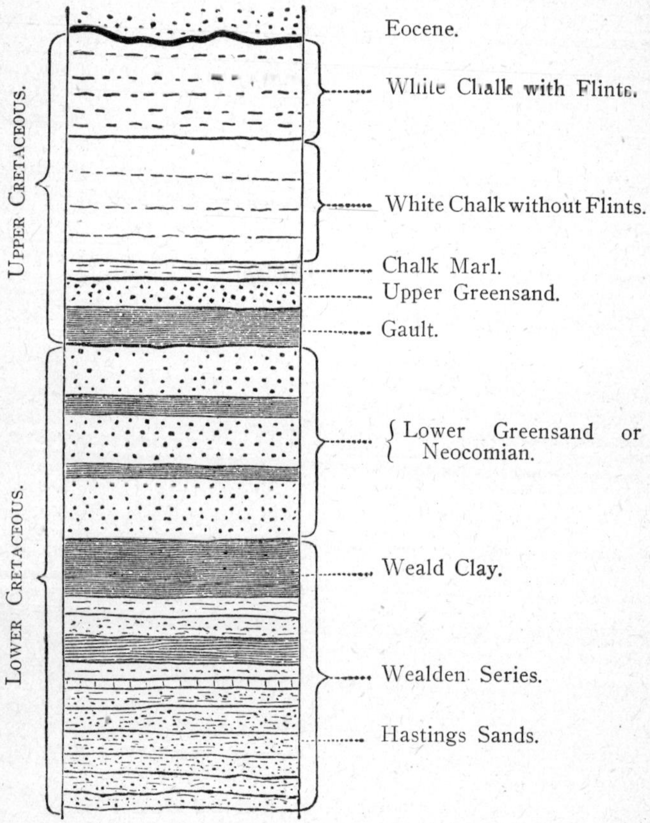

In North America, strata of Lower Cretaceous age are well represented in Missouri, Wyoming, Utah, and in some other areas; but the greater portion of the American deposits of this period are referable to the Upper Cretaceous. The rocks of this series are mostly sands, clays, and limestones—*Chalk* itself being unknown except in Western Arkansas. Amongst the sandy accumulations, one of the most important is the so-called "marl" of New Jersey, which is truly a "Greensand," and contains a large proportion of glauconite (silicate of iron and potash). It also contains a little phosphate of lime, and is largely worked for agricultural purposes. The greatest thickness attained by the Cretaceous rocks of North America is about 9000 feet, as in Wyoming, Utah, and Colorado. According to Dana, the Cretaceous rocks of the Rocky Mountain territories pass upwards "without interruption into a coal-bearing formation, several thousand feet thick, on which the following Tertiary strata lie unconformably." The lower portion of this "Lignitic formation" appears to be Cretaceous, and contains one or more beds of Coal; but the upper part of it perhaps belongs to the Lower Tertiary. In America, therefore, the lowest Tertiary strata appear to rest conformably upon the highest Cretaceous; whereas in Europe, the succession at this point is invariably an unconformable one. Owing, however, to the fact that the American "Lignitic formation" is a shallow-water formation, it can hardly be expected to yield much material whereby to bridge over the great palæontological gap between the White Chalk and Eocene in the Old World.

Owing to the fact that so large a portion of the Cretaceous formation has been deposited in the sea, much of it in deep water, the *plants* of the period have for the most part been found special members of the series, such as the Wealden beds, the Aix-la-Chapelle sands, and the Lignitic beds of North America. Even the purely marine strata, however, have yielded plant remains, some of these are peculiar and proper to the deep-sea deposits of the series. Thus the little calcareous discs termed "coccoliths," which are known to be of the nature of calcareous sea-weeds (*Algæ*) have been detected in the White Chalk; and the flints of the same formation commonly contain the spore-cases of the microscopic *Desmids* (the so-called Xanthidia), along with the siliceous cases of the equally diminutive *Diatoms*.

The plant-remains of the Lower Cretaceous greatly resemble those of the Jurassic period, consisting mainly of Ferns, Cycads, and Conifers. The Upper Cretaceous rocks, however, both in Europe and in North America, have yielded an abundant flora which resembles the existing vegetation of the globe in consisting mainly of Angiospermous Exogens and of Monocotyledons.* In Europe the plant-remains in question have been found chiefly in certain sands in the neighborhood of Aix-la-Chapelle, and they consist of numerous Ferns, Conifers (such as. *Cycadopteris*), Screw Pines (*Pandanus*), Oaks (*Quercus*), Walnut (*Juglans*), Fig. (*Ficus*), and many *Proteaceæ*, some of which are referred to existing genera (*Dryandra, Banksia, Grevillea,* &c.)

In North America, the Cretaceous strata of New Jersey, Alabama, Nebraska, Kansas, &c., have yielded the remains of numerous plants, many of which belong to existing genera. Amongst these may be mentioned Tulip-trees (*Liriodendron*), Sassafras (fig. 186), Oaks (*Quercus*), Beeches (*Fagus*), Plane-trees (*Platanus*), Alders (*Alnus*), Dog-wood (*Cornus*), Willows (*Salix*), Poplars (*Populus*), Cypresses (*Cupressus*), Bald Cypresses (*Taxodium*), Magnolias, &c. Besides these, however, there occur other forms which have now entirely disappeared from North America—as, for example, species of *Cinnamomum* and *Araucaria*.

It follows from the above, that the Lower and Upper Cretaceous rocks are, from a botanical point of view, sharply separated from one another. The Palæozoic period, as we have seen, is characterized by the prevalence of "Flowerless" plants (*Cryptogams*), its higher vegetation consisting almost exclusively of Conifers. The Mesozoic period, as a whole, is characterized by the prevalence of the Cryptogamic group of the Ferns, and the Gymnospermic groups of the Conifers and the Cycads. Up to the close of the Lower Cretaceous, no Angiospermous Exogens are certainly known to have existed, and Monocotyledonous plants or Endogens are very poorly represented. With the Upper Cretaceous, however, a new era of plant-life, of which our present is but the culmination, commenced, with a great and apparently sudden development of new forms. In place of the Ferns, Cycads, and Conifers of the earlier

* The "Flowering plants" are divided into the two great groups of the Endogens and Exogens. The *Endogens* (such as Grasses, Palms, Lilies, &c.) have no true bark, nor rings of growth, and the stem is said to be "endogenous;" the young plant also possesses but a single seed-leaf or "cotyledon." Hence these plants are often simply called "*Monocotyledons.*" The *Exogens*, on the other hand, have a true bark; and the stem increases by annual additions to the outside, so that rings of growth are produced. The young plant has two seed-leaves or "cotyledons," and these plants are therefore called "*Dicotyledons.*" Amongst the Exogens, the Pines (*Conifers*) and the Cycads have seeds which are unprotected by a seed-vessel, and they are therefore called "*Gymnosperms.*" All the other Exogens, including the ordinary trees, shrubs, and flowering plants, have the seeds enclosed in a seed-vessel, and are therefore called "*Angiosperms.*" The derivation of these terms will be found in the Glossary at the end of the volume.

Mesozoic deposits, we have now an astonishingly large number of true Angiospermous Exogens, many of them belonging to existing types; and along with these are various Monocotyledonous plants, including the first examples of the great and important group of the Palms. It is thus a matter of interest to reflect that plants closely related to those now inhabiting the earth, were in existence at a time when the ocean was tenanted by Ammonites and Belemnites, and when land and sea and air were peopled by the extraordinary extinct Reptiles of the Mesozoic period.

As regards animal life, the *Protozoans* of the Cretaceous period are exceedingly numerous, and are represented by *Fora-*

Fig. 186.—Cretaceous Angiosperms. *a. Sassafras Cretaceum; b, Liriodendron Meekii; c, Leguminosites Marcouanus; d, Salix Meekii.* **(After Dana.)**

minifera and *Sponges.* As we have already seen, the White Chalk itself is a deep-sea deposit, almost entirely composed of the microscopic shells of *Foraminifers,* along with Sponge-spicules, and organic *débris* of different kinds (see p. 23, fig. 7). The green grains which are abundant in several minor subdivisions of the Cretaceous, are also in many instances really casts in glauconite of the chambered shells of these minute organisms. A great many species of *Foraminifera* have been recognized in the Chalk; but the three principal genera are

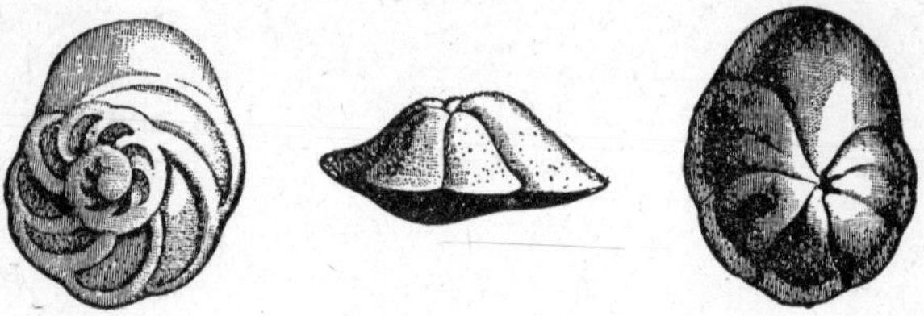

Fig. 187.—*Rotalia Boueana.*

Globigerina, Rotalia (fig. 187), and *Textularia*—groups which are likewise characteristic of the "ooze" of the Atlantic and Pacific Oceans at great depths. The flints of the Chalk also commonly contain the shells of *Foraminifera.* The Upper Greensand has yielded in considerable numbers the huge *Foraminifera* described by Dr. Carpenter under the name of *Parkeria,* the spherical shells of which are composed of sand-grains agglutinated together, and sometimes attain a diameter of two and a quarter inches. The Cretaceous *Sponges* are extremely numerous, and occur under a great number of varieties of shape and structure; but the two most characteristic genera are *Siphonia* and *Ventriculites,* both of which are exclusively confined to strata of this age. The *Siphoniæ* (fig. 188) consist of a pear-shaped, sometimes lobed head, supported by a longer or shorter stem, which breaks up at its base into a number of root-like processes of attachment. The water gained access to the interior of the Sponge by a number of minute openings covering the surface, and ultimately escaped by a single, large, chimney-shaped aperture at the summit. In some respects these sponges present a singular resemblance to the beautiful "Vitreous Sponges" (*Holtenia* or *Pheronema*) of the deep Atlantic; and, like these, they were probably denizens of a deep sea. The *Ventriculites* of the Chalk (fig. 189) is, however, a genus still more closely allied to the wonderful flint Sponges, which have been shown, by the researches of the Porcupine, Lightning, and Challenger expeditions, to live half buried in the calcareous ooze of the abysses of our great oceans. Many forms of this genus are known, having "usually the form of graceful vases, tubes, or funnels, variously ridged or grooved, or otherwise ornamented on the surface, frequently expanded above into a cup-like lip, and continued below into a bundle of fibrous roots. The minute structure of these bodies shows an extremely delicate tracery of fine tubes, sometimes empty, sometimes filled with loose calcareous matter dyed with peroxide of iron."—(Sir Wyville Thomson.) Many of the Chalk sponges, originally calcareous, have been converted into flint subsequently; but the Ventriculites are really composed of this substance, and are therefore genuine "Siliceous Sponges," like the existing Venus's Flower-basket (*Euplectella*). Like the latter, the skeleton was doubtless originally composed, in the young state, of disconnected six-rayed spicules, which ultimately become fixed together to constitute a continuous frame-work. The sea-water, as in the recent forms, must have been admitted to the interior of the Sponge by numerous apertures on its exterior, subsequently escaping by a single large opening at its summit.

Fig. 188.—*Siphonia ficus.*
Upper Greensand, Europe.

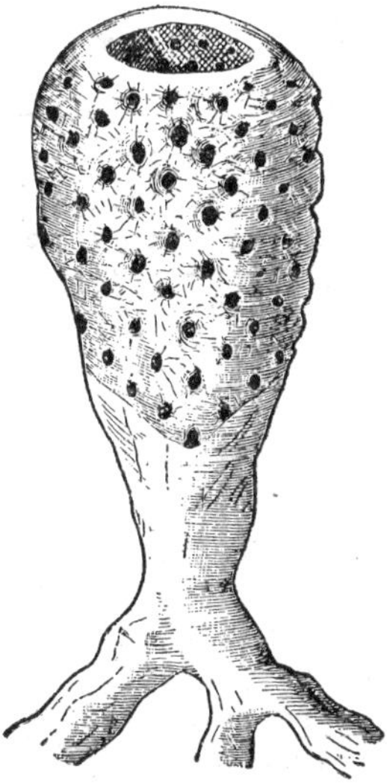

Fig. 189.—*Ventriculites simplex.*
White Chalk, Britain.

Amongst the *Cœlenterates,* the "Hydroid Zoophytes" are represented by a species of the encrusting genus *Hydractinia,* the horny polypary of which is so commonly found at the present day adhering to the exterior of shells. The occurrence of this genus is of interest, because it is the first known instance in the entire geological series of the occurrence of an unques-

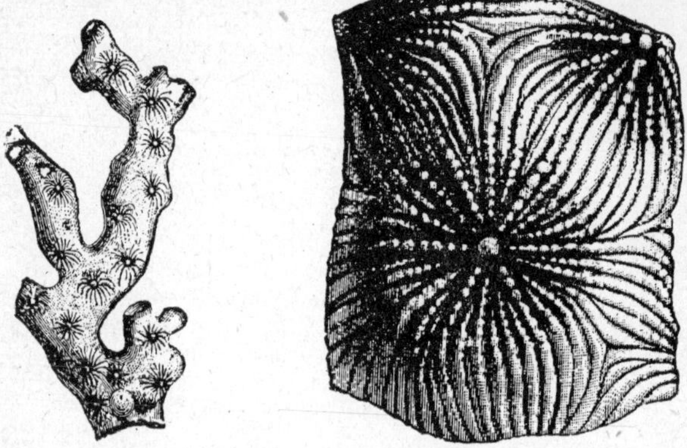

Fig. 190.—*Synhelia Sharpeana.* Chalk, England.

tionable Hydroid of a modern type, though many of the existing forms of these animals possess structures which are perfectly fitted for preservation in the fossil condition. The corals of the Cretaceous series are not very numerous, and for the most part are referable to types such as *Trochocyathus, Stephanophyllia, Parasmilia, Synhelia* (fig. 190), &c., which belong to the same great group of corals as the majority of existing forms. We have also a few "Tabulate Corals" (*Polytremacis*), hardly, if at all, generically separable from very ancient forms (*Heliolites*); and the Lower Greensand has yielded the remains of the little *Holocystis elegans,* long believed to be the last of the great Palæozoic group of the *Rugosa.*

As regards the *Echinoderms,* the group of the *Crinoids* now exhibits a marked decrease in the number and variety of its types. The "stalked" forms are represented by *Pentacrinus* and *Bourgueticrinus,* and the free forms by Feather-stars like our existing *Comatulæ;* whilst a link between the stalked and free groups is constituted by the curious "Tortoise Encrinite (*Marsupites*). By far the most abundant Cretaceous Echinoderms, however, are Sea-urchins (*Echinoids*); though several Star-fishes are known as well. The remains of Sea-urchins are so abundant in various parts of the Cretaceous series, especially in the White Chalk, and are often so beautifully preserved, that they constitute one of the most marked features of the fauna of the period. From the many genera of Sea-urchins which occur in strata of this age, it is difficult to select characteristic types; but the genera *Galerites* (fig. 191), *Discoidea* (fig. 192), *Micraster, Ananchytes, Diadema, Salenia,* and *Cidaris,* may be mentioned as being all important Cretaceous groups.

Coming to the *Annulose Animals* of the Cretaceous period, there is little special to remark. The *Crustaceans* belong for the most part to the highly-organized groups of the Lobsters

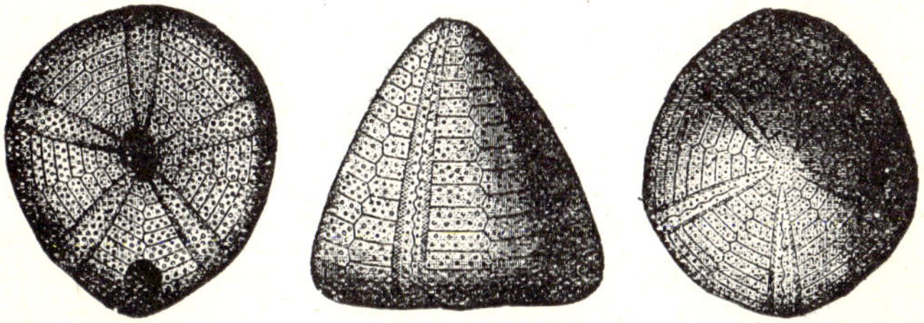

Fig. 191.—*Galerites albogalerus*, viewed from below, from the side, and from above. White Chalk.

and the Crabs (the Macrurous and Brachyurous Decapods); but there are also numerous little *Ostracodes,* especially in the fresh-water strata of the Wealden. It should further be noted

Fig. 192.—*Discoidea cylindrica;* under, side, and upper aspect. Upper Greensand.

that there occurs here a great development of the singular *Crustaceous* family of the Barnacles (*Lepadidæ*), whilst the allied family of the equally singular Acorn-shells (*Balanidæ*) is feebly represented as well.

Passing on to the *Mollusca,* the class of the Sea-mats and Sea-mosses (*Polyzoa*) is immensely developed in the Cretaceous period, nearly two hundred species being known to occur in the Chalk. Most of the Cretaceous forms belong to the family of the *Escharidæ,* the genera *Eschara* and *Escharina* (fig. 193) being particularly well represented. Most of the Cretaceous *Polyzoans* are of small size, but some attain considerable dimensions, and many simulate Corals in their general form and appearance.

The Lamp-shells (*Brachiopods*) have now reached a further stage of the progressive decline, which they have been undergoing ever since the close of the Palæozoic period. Though individually not rare, especially in certain minor subdivisions of the series, the number of generic types has now become distinctly diminished, the principal forms belonging to the genera *Terebratula, Terebratella* (fig. 194), *Terebratulina, Rhynchonella,* and *Crania* (fig. 195). In the last mentioned of these, the shell is attached to foreign bodies by the substance of one of the valves (the ventral), whilst the other or free valve is more or less limpet-shaped. All the above-mentioned genera are in existence at the present day; and one *species*—namely, *Terebratulina striata*—appears to be undistinguishable from one now living—the *Terebratulina caput-serpentis.*

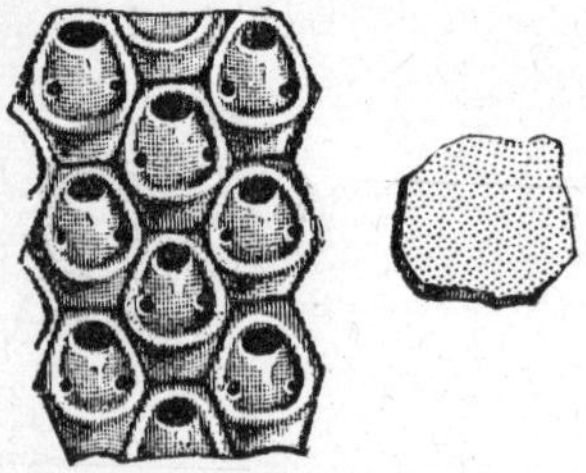

193.—A small fragment of *Escharina Oceani*, of the natural size ; and a portion of the same enlarged. Upper Greensand.

Whilst the Lamp-shells are slowly declining, the Bivalves (*Lamellibranchs*) are greatly developed, and are amongst the

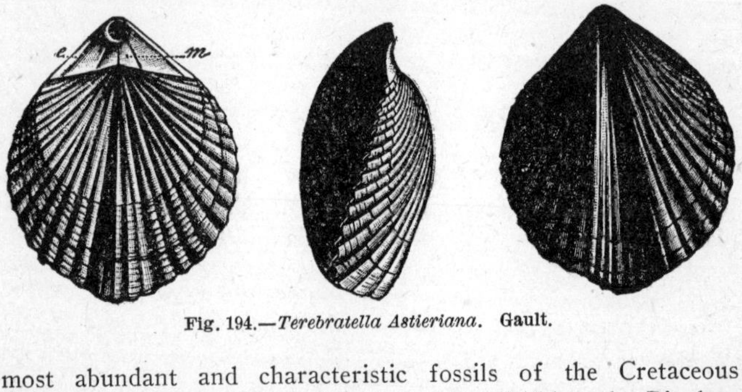

Fig. 194.—*Terebratella Astieriana*. Gault.

most abundant and characteristic fossils of the Cretaceous period. In the great river-deposit of the Wealden, the Bivalves are forms proper to fresh water, belonging to the existing River-mussels (*Unio*), *Cyrena* and *Cyclas;* but most of the Cretaceous Lamellibranchs are marine. Some of the most

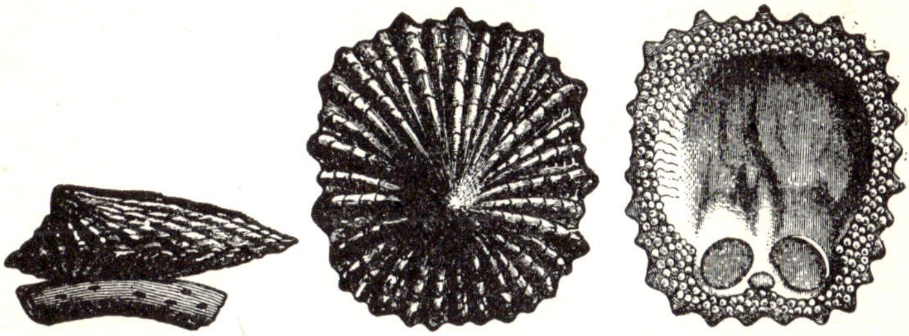

Fig. 195.—*Crania Ignabergensis.* The left-hand figure shows the perfect shell, attached by its ventral valve to a foreign body; the middle figure shows the exterior of the limpet-shaped dorsal valve; and the right-hand figure represents the interior of the attached valve. White Chalk.

abundant and characteristic of these belong to the great family of the Oysters (*Ostreidæ*). Amongst these are the genera *Gryphæa* and *Exogyra,* both of which we have seen to occur

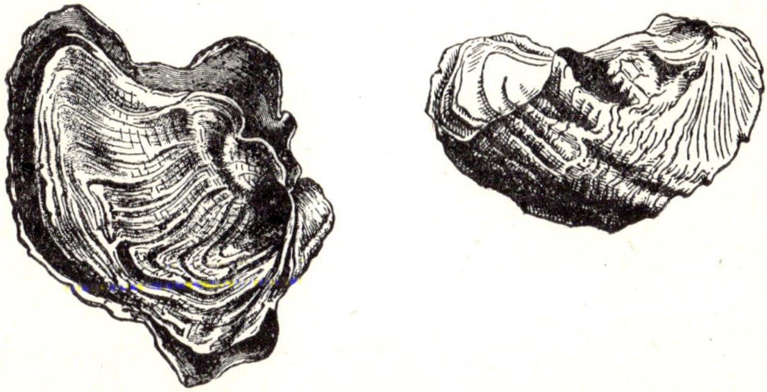

Fig. 196.—*Ostrea Couloni.* Lower Greensand.

abundantly in the Jurassic; and there are also numerous true Oysters (*Ostrea,* fig. 196) and Thorny Oysters (*Spondylus,* fig. 197). The genus *Trigonia,* so characteristic of the Mesozoic deposits in general, is likewise well represented in the Cretaceous strata. No single genus of Bivalves is, however, so highly characteristic of the Cretaceous period as *Inoceramus,* a group belonging to the family of the Pearl-mussels (*Aviculidæ*). The shells of this genus (fig. 198) have the valves unequal in size, the larger valve often being much twisted, and both valves being marked with radiating ribs or concentric furrows. The hinge-line is long and straight, with numerous pits for the attachment of the ligament which serves to open the shell. Some of the *Inocerami* attain a length of two or three feet, and fragments of the shell are often found perforated by boring

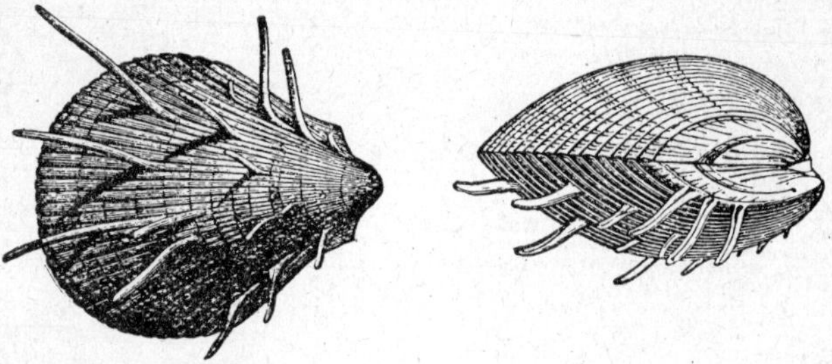

Fig. 197.—*Spondylus spinosus*. White Chalk.

Sponges. Another extraordinary family of Bivalves, which is exclusively confined to the Cretaceous rocks, is that of the *Hippuritidæ*. All the members of this group (fig. 199) were

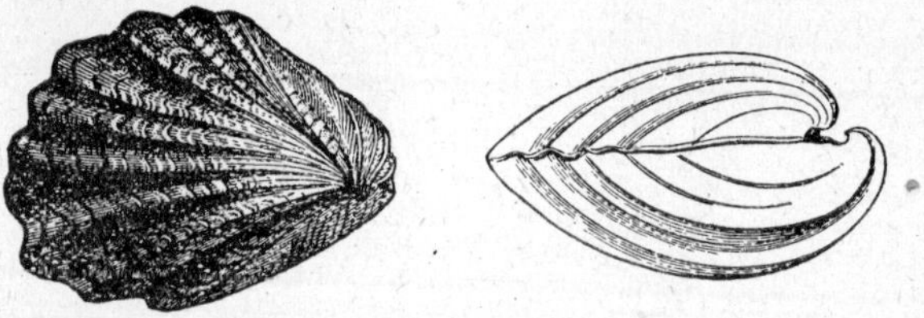

Fig. 198.—*Inoceramus sulcatus*. Gault.

attached to foreign objects, and lived associated in beds, like Oysters. The two valves of the shell are always altogether unlike in sculpturing, appearance, shape, and size; and the cast of the interior of the shell is often extremely unlike the form of the outer surface. The type-genus of the family is *Hippurites* itself (fig. 199), in which the shell is in the shape of a straight or slightly-twisted horn, sometimes a foot or more in length, constituted by the attached lower valve, and closed above by a small lid-like free upper valve. About a hundred species of the family of the *Hippuritidæ* are known, all of these being Cretaceous, and occurring in Britain (one species only), in Southern Europe, the West Indies, North America, Algeria, and Egypt. Species of this family occur in such numbers in certain compact marbles in the south of Europe, of the age of the Upper Cretaceous (Lower Chalk), as to have given orgin to the name of "Hippurite Limestones," applied to these strata.

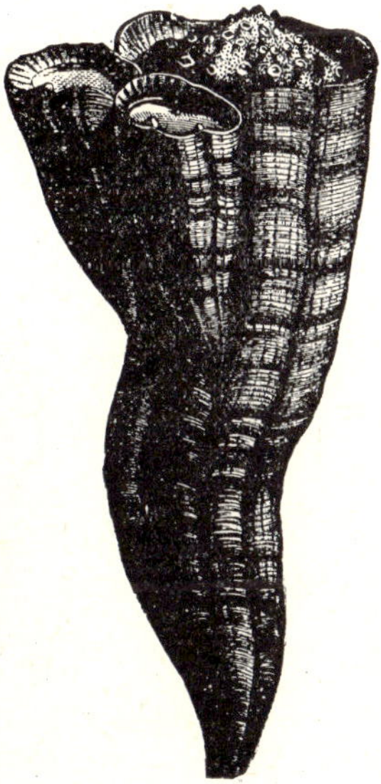

Fig. 199.—*Hippurites Toucasiana.* A large individual, with two smaller ones attached to it. Upper Cretaceous. South of Europe.

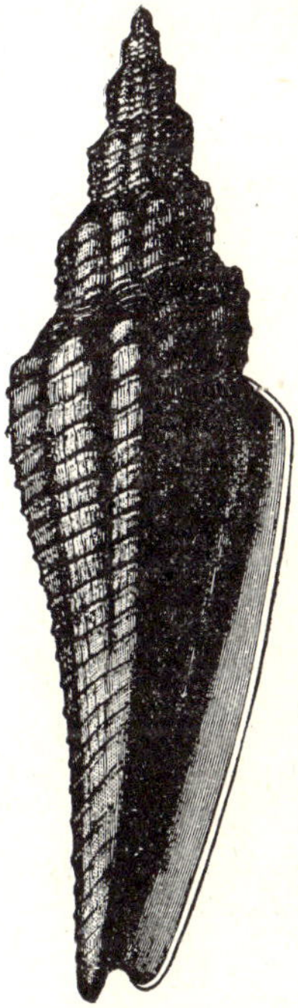

Fig. 200.—*Voluta elongata.* White Chalk.

The Univalves (*Gasteropods*) of the Cretaceous period are not very numerous, nor particularly remarkable. Along with species of the persistent genus *Pleurotomaria* and the Mesozoic *Nerinæa,* we meet with examples of such modern types as *Turritella* and *Natica,* the Staircase-shells (*Solarium*), the Wentle-traps (*Scalaria*), the Carrier-shells (*Phorus*), &c. Towards the close of the Cretaceous period, and especially in such transitional strata as the Maestricht beds, the Faxöe

Limestone, and the Pisolitic Limestone of France, we meet with a number of carnivorous ("siphonostomatous") Univalves, in which the mouth of the shell is notched or produced into a canal. Amongst these it is interesting to recognize examples of such existing genera as the Volutes (*Voluta,* fig. 200), the Cowries (*Cypræa*), the Mitre-shells (*Mitra*), the Wing-shells (*Strombus*), the Scorpion-shells (*Pleroceras*), &c.

Upon the whole, the most characteristic of all the Cretaceous Molluscs are the *Cephalopods,* represented by the remains of both *Tetrabranchiate* and *Dibranchiate* forms. Amongst the former, the long-lived genus *Nautilus* (fig. 201) again reappears, with its involute shell, its capacious body-chamber, its simple septa between the air-chambers, and its nearly or quite central siphuncle. The majority of the chambered *Cephalopods* of the Cretaceous belong, however, to the complex and beautiful family of the *Ammonitidæ,* with their elaborately folded and lobed septa and dorsally-placed siphuncle. This

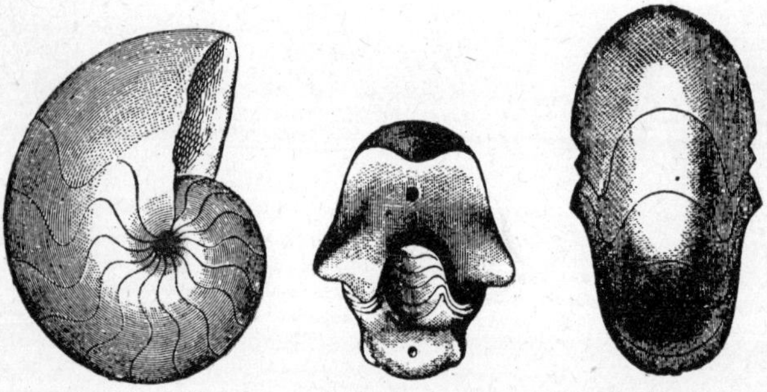

Fig. 201.—Different views of *Nautilus Danicus.* Faxöe Limestone (Upper Cretaceous), Denmark.

family disappears wholly at the close of the Cretaceous period; but it approaching extinction, so far from being signalized by any slow decrease and diminution in the number of specific or generic types, seems to have been attended by the development of whole series of new forms. The genus *Ammonites* itself, dating from the Carboniferous, has certainly passed its prime, but it is still represented by many species, and some of these attained enormous dimensions (two or three feet in diameter). The genus *Ancyloceras* (fig. 202), though likewise of more ancient origin (Jurassic), is nevertheless very characteristic of the Cretaceous. In this genus the first portion of the shell is in the form of a flat spiral, the coils of which are not in contact; and its last portion is produced at a tangent, becoming ultimately bent back in the form of a crosier. Be-

Fig. 202.—*Ancyloceras Matheronianus*. Gault.

sides these pre-existent types, the Cretaceous rocks have yielded a great number of entirely new forms of the *Ammonitidæ,* which are not known in any deposits of earlier or later date. Amongst the more important of these may be mentioned *Crioceras, Turrilites, Scaphites, Hamites, Ptychoceras,* and *Baculites.* In the genus *Crioceras* (fig. 204, *d*), the shell consists of an open spiral, the volutions of which are not in contact, thus resembling a partially-unrolled *Ammonite* or the inner portion of an *Ancyloceras.* In *Turrilites* (fig. 203), the shell is precisely like that of the *Ammonite* in its structure; but instead of forming a flat spiral, it is coiled into an elevated turreted shell, the whorls of which are in contact with one another. In the genus *Scaphites* (fig. 204, *e*), the shell resembles that of *Ancyloceras* in consisting of a series of volutions coiled into a flat spiral, the last being detached from the others, produced, and ultimately bent back in the form of a crosier; but the whorls of the enrolled part of the shell are in contact, instead of being separate as in the latter. In the genus *Hamites* (fig. 204, *f*), the shell is an extremely elongated cone, which is bent upon itself more than once, in a hook-like manner, all the volution being separate. The genus *Ptychoceras* (fig. 204, *a*) is very like *Hamites,* except that the shell is only bent once; and the two portions thus bent are in contact with one another. Lastly, in the genus *Baculites* (fig. 204, *b* and *c*) the shell is simply a straight elongated cone, not bent in any way, but possessing the folded septa which characterize the whole Ammonite family. The *Baculite* is the simplest of all the forms of the *Ammonitidæ;* and all the other forms, however complex, may be regarded as being simply produced by the bending or folding of such a conical septate shell in different ways. The *Baculite,* therefore, corresponds, in the series

Fig, 203.—*Turrilites catenatus.* The lower figure represents the entire shell; the upper figure represents the base of the shell seen from below. Gault.

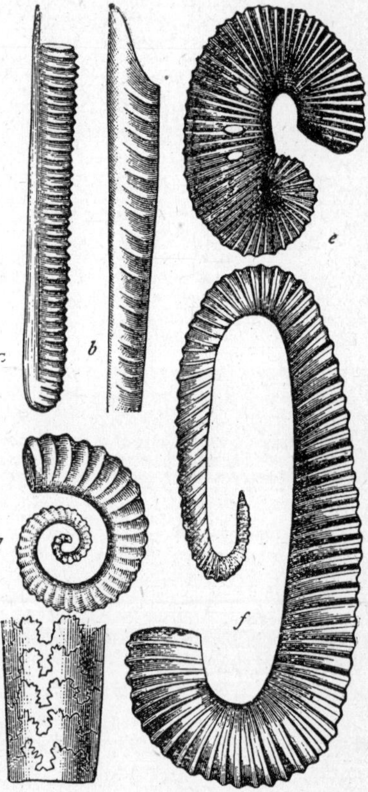

Fig. 204.—*a, Ptychoceras Emericianum,* reduced —Lower Greensand; *b, Baculites anceps,* reduced —Chalk; *c,* Portion of the same, showing the folded edges of the septa; *d, Crioceras cristatum,* reduced —Gault; *e, Scaphites æqualis,* natural size—Chalk; *f, Hamites rotundus,* restored.—Gault.

of the *Ammonitidæ,* to the *Orthoceras* in the series of the *Nautilidæ.* All the above-mentioned genera are characteristically, or exclusively, Cretaceous, and they are accompanied by a number of other allied forms, which cannot be noticed here. Not a single one of these genera, further, has hitherto been detected in any strata higher than the Cretaceous. We may therefore consider that these wonderful, varied, and elaborate forms of *Ammonitidæ* constitute one of the most conspicuous features in the life of the Chalk period.

The *Dibranchiate Cephalopods* are represented partly by the beak-like jaws of unknown species of Cuttle-fishes and partly by the internal skeletons of Belemnites. Amongst the latter, the genus *Belemnites* itself holds its place in the lower part of the Cretaceous series; but it disappears in the upper portion of the series, and its place is taken by the nearly-allied genus *Belemnitella* (fig. 205), distinguished by the possession of a straight fissure in the upper end of the guard. This also disappears at the close of the Cretaceous period; and no member of the great Mesozoic family of the *Belemnitidæ* has hitherto been discovered in any Tertiary deposit, or is known to exist at the present day.

Fig. 205.—Guard of *Belemnitella mucronata*—White Chalk.

Passing on next to the *Vertebrate Animals* of the Cretaceous period, we find the *Fishes* represented as before by the Ganoids and the Placoids, to which, however, we can now add the first known examples of the great group of the *Bony Fishes* or *Teleosteans*, comprising the great majority of existing forms. The *Ganoid* Fishes of the Cretaceous (*Lepidotus*, *Pycnodus*, &c.) present no features of special interest. Little, also, need be said about the *Placoid* fishes of this period. As in the Jurassic deposits, the remains of these consist partly of the teeth of genuine Sharks (*Lamna*, *Odontaspis*, &c.), and partly of the teeth and defensive spines of Cestracionts, such as the living Port-Jackson Shark. The pointed and sharp-edged teeth of true Sharks are very abundant in some beds, such as the Upper Greensand, and are beautifully preserved. The teeth of some forms (*Carcharias*, &c.) attain occasionally a length of three or four inches, and indicate the existence in the Cretaceous seas of huge predaceous fishes, probably larger than any existing Sharks. The remains of *Cestracionts* consist partly of the flattened teeth of genera such as *Acrodus* and *Ptychodus* (the latter confined to rocks of this age), and partly of the pointed teeth of *Hybodus*, a genus which dates from the Trias. In this genus the teeth (fig. 206) consist of a principal central cone, flanked by minor lateral cones; and the fin-spines (fig. 207) are longitudinally grooved, and carry a series of small spines on their hinder or concave margin. Lastly, the great modern order of the Bony Fishes or *Teleosteans*

Fig. 206.—Tooth of *Hybodus*.

Fig. 207.—Fin-spine of *Hybodus*. Lower Greensand.

makes its first appearance in the Upper Cretaceous rocks, where it is represented by forms belonging to no less than three existing groups—namely, the Salmon family (*Salmonidæ*), the Herring family (*Clupeidæ*), and the Perch family (*Percidæ*). All these fishes have thin, horny, overlapping scales, symmetrical ("homocercal") tails, and bony skeletons.

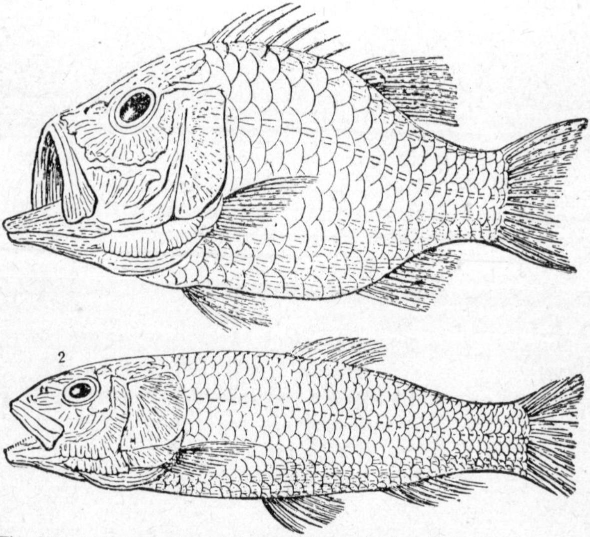

Fig. 208.—1, *Beryx Lewesiensis*, a Percoid fish from the Chalk; 2, *Osmeroides Mantelli*, a Salmonoid fish from the Chalk.

The genus *Beryx* (fig. 208, 1) is one represented by existing species at the present day, and belongs to the Perch family.

The genus *Osmeroides*, again (fig. 208, 2), is supposed to be related to the living Smelts (*Osmerus*), and, therefore, to belong to the Salmon tribe.

No remains of *Amphibians* have hitherto been detected in any part of the Cretaceous series; but *Reptiles* are extremely numerous, and belong to very varied types. As regards the great extinct groups of Reptiles which characterize the Mesozoic period as a whole, the huge "Enaliosaurs" or "Sea-Lizards" are still represented by the *Ichthyosaur* and the *Plesiosaur*. Nearly allied to the latter of these is the *Elasmosaurus* of the American Cretaceous, which combined the long tail of the Ichthyosaur with the long neck of the Plesiosaur. The length of this monstrous Reptile could not have been less than fifty feet, the neck consisting of over sixty vertebræ and measuring over twenty feet in length. The extraordinary Flying Reptiles of the Jurassic are likewise well represented in the Cretaceous rocks by species of the genus *Pterodactylus* itself, and these later forms are much more gigantic in their dimensions than their predecessors. Thus some of the Cretaceous *Pterosaurs* seem to have had a spread of wing of from twenty to forty-five feet, more than realizing the "Dragons" of fable in point of size. The most remarkable, however, of the Cretaceous *Pterosaurs* are the forms which have recently been described by Professor Marsh under the generic title of *Pteranodon*. In these singular forms—so far only known as American—the animal possessed a skeleton in all respects similar to that of the typical Pterodactyles, except that the jaws are completely destitute of teeth. There is, therefore, the strongest probability that the jaws were encased in a horny sheath, thus coming to resemble the beak of a Bird. Some of the recognized species of *Pteranodon* are very small; but the skull of one species (*P. longiceps*) is not less than a yard in length, and there are portions of the skull of another species which would indicate a length of four feet for the cranium. These measurements would point to dimensions larger than those of any other known Pterosaurs.

The great Mesozoic order of the *Deinosaurs* is largely represented in the Cretaceous rocks, partly by genera which previously existed in the Jurassic period, and partly by entirely new types. The great delta-deposit of the Wealden, in the Old World, has yielded the remains of various of these huge terrestrial Reptiles, and very many others have been found in the Cretaceous deposits of North America. One of the most celebrated of the Cretaceous Deinosaurs is the *Iguanodon,* so called from the curious resemblance of its teeth to those of the existing but comparatively diminutive *Iguana.* The teeth (fig. 209) are soldered to the inner face of the jaw, instead of being sunk in distinct sockets; and they have the form of somewhat flattened prisms, longitudinally ridged on the outer surface, with an obtusely triangular crown, and having the enamel crenated on one or both sides. They present the extraordinary feature that the crowns became worn down flat by mastication, showing that the *Iguanodon* employed its teeth in actually chewing and triturating the vegetable matter on which it fed. There can therefore be no doubt but that the *Iguanodon,* in spite of its immense bulk, was an herbivorous Reptile, and lived principally on the foliage of the Cretaceous forests amongst which it dwelt. Its size has been variously estimated at from thirty to fifty feet, the thigh-bone in large examples

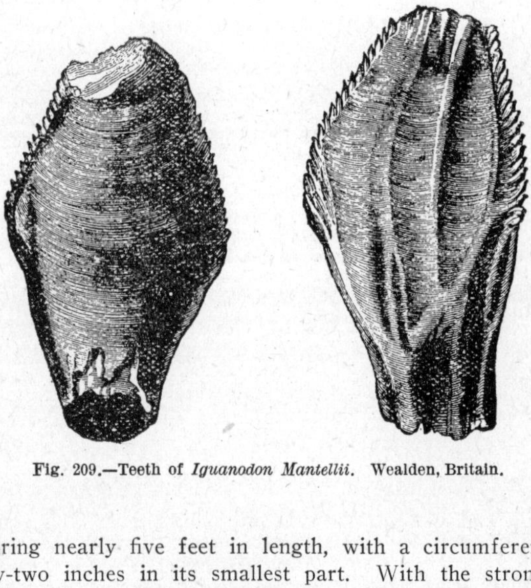

Fig. 209.—Teeth of *Iguanodon Mantellii.* Wealden, Britain.

measuring nearly five feet in length, with a circumference of twenty-two inches in its smallest part. With the strong and massive hind-limbs are associated comparatively weak and small fore-limbs; and there seems little reason to doubt that the *Iguanodon* must have walked temporarily or permanently upon its hind-limbs, after the manner of a Bird. This conjecture is further supported by the occurrence in the strata which contain the bones of the *Iguanodon* of gigantic three-toed footprints, disposed *singly* in a double track. These prints have undoubtedly been produced by some animal walking on two legs; and they can hardly, with any probability, be ascribed to any other than this enormous Reptile. Closely allied to the

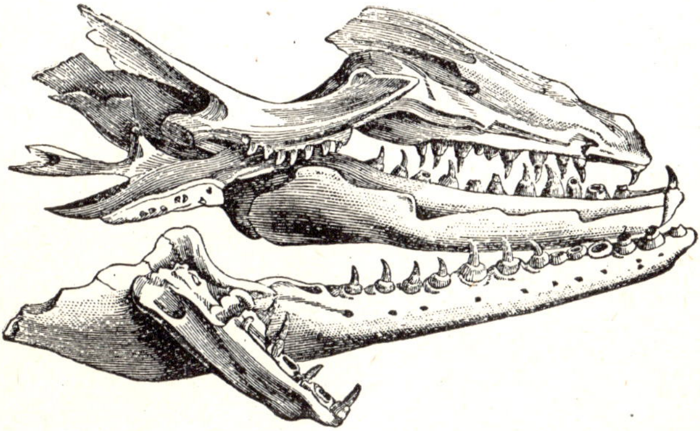

Fig. 210.—Skull of *Mosasaurus Camperi*, greatly reduced. Maestricht Chalk.

Iguanodon is the *Hadrosaurus* of the American Cretaceous, the length of which is estimated at twenty-eight feet. *Iguanodon* does not appear to have possessed any integumentary skeleton; but the great *Hylæosaurus* of the Wealden seems to have been furnished with a longitudinal crest of large spines running down the back, similar to that which is found in the comparatively small Iguanas of the present day. The *Megalosaurus* of the Oolites continued to exist in the Cretaceous period; and, as we have previously seen, it was carnivorous in its habits. The American *Lælaps* was also carnivorous, and, like the Megalosaur, which it very closely resembles, appears to have walked upon its hind-legs, the fore-limbs being disproportionately small.

Another remarkable group of Reptiles, exclusively confined to the Cretaceous series, is that of the *Mosasauroids*, so called from the type-genus *Mosasaurus.* The first species of *Mosasaurus* known to science was the *M. Camperi* (fig. 210), the skull of which—six feet in length—was discovered in 1780 in the Maestricht Chalk at Maestricht. As this town stands on the river Meuse, the name of *Mosasaurus* ("Lizard of the Meuse") was applied to this immense Reptile. Of late years the remains of a large number of Reptiles more or less closely related to *Mosasaurus,* or absolutely belonging to it have been discovered in the Cretaceous deposits of North America, and have been described by Professors Cope and Marsh. All the known forms of this group appear to have been of large size—one of them, *Mosasaurus princeps,* attaining the length of seventy-five or eighty feet, and thus rivalling the largest of existing Whales in its dimensions. The teeth in the "Mosasauroids" are long, pointed, and slightly curved; and instead of being sunk in distinct sockets, they are firmly amalgamated with the jaws, as in modern Lizards. The palate also carried teeth, and the lower jaw was so constructed as to allow of the mouth being opened to an immense width, somewhat as in the living Serpents. The body was long and snake-like, with a very long tail, which is laterally compressed, and must have served as a powerful swimming-apparatus. In addition to this, both pairs of limbs have the bones connecting them with the trunk greatly shortened; whilst the digits were enclosed in the integuments, and constituted paddles, closely resembling in structure the "flippers" of Whales and Dolphins. The neck is sometimes moderately long, but oftener very short, as the great size and weight of the head would have led one to anticipate. Bony plates seem in some species to have formed an at any rate partial covering to the skin; but it is not certain that these integumentary appendages were present in all. Upon the whole, there can be no doubt but that the Mosasauroid Reptiles—the true "Sea-serpents" of the Cretaceous period—were essentially aquatic in their habits, frequenting the sea, and only occasionally coming to the land.

The "Mosasauroids" have generally been regarded as a greatly modified group of the Lizards (*Lacertilia*). Whether this reference be correct or not—and recent investigations render it dubious—the Cretaceous rocks have yielded the remains of small Lizards not widely removed from existing forms. The recent order of the *Chelonians* is also represented in the Cretaceous rocks, by forms closely resembling living types. Thus the fresh-water deposits of the Wealden have yielded examples of the "Terrapins" or "Mud-Turtles" (*Emys*); and the marine Cretaceous strata have been found to contain the remains of various species of Turtles, one of which is here figured (fig. 211). No true Serpents (*Ophidia*) have as yet been detected in the Cretaceous rocks; and this order does not appear to have come into existence till the Tertiary period. Lastly, true Crocodiles are known to have existed in considerable numbers in the Cretaceous period. The oldest of these occur in the fresh-water deposits of the Wealden; and they differ from the existing forms of the group in the fact that the bodies of the vertebræ, like those of the Jurassic Crocodiles, are bi-concave, or hollowed out at both ends. In the Greensand of North America, however, occur the remains of Crocodiles which agree with all the living species in having the bodies of the vertebræ in the region of the back hollowed out in front and convex behind.

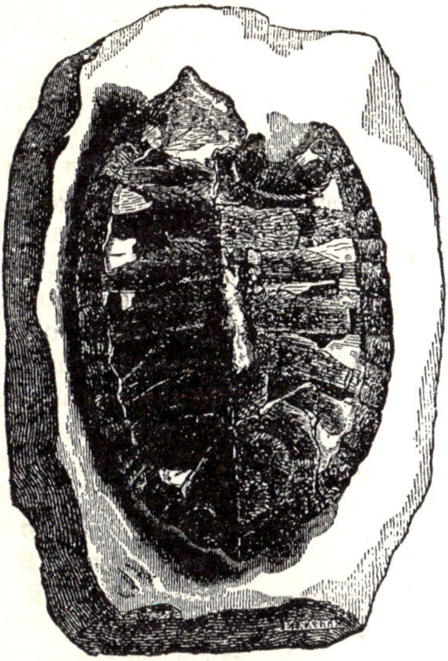

Fig. 211.—Carapace of *Chelone Benstedi*. Lower Chalk. (After Owen.)

Birds have not hitherto been shown, with certainty, to have existed in Europe during the Cretaceous period, except in a few instances in which fragmentary remains belonging to this class have been discovered. The Cretaceous deposits of North America have, however, been shown by Professor Marsh to contain a considerable number of the remains of Birds, often in a state of excellent preservation. Some of these belong to swimming or Wading Birds, differing in no point of special interest from modern birds of similar habits. Others, however, exhibit such extraordinary peculiarities that they merit more than a passing notice. One of the forms in question constitutes the genus *Ichthyornis* of Marsh, the type-species of which (*I. dispar*) was about as large as a Pigeon. In two remarkable respects, this singular Bird differs from all known living members of the class. One of these respects concerns the jaws, both of which exhibit the Reptilian character of being armed with numerous small pointed *teeth* (fig. 212, *a*), sunk in distinct sockets. No existing bird possesses teeth; and this character forcibly recalls the Bird-like Pterosaurs, with their toothed jaws. *Ichthyornis,* however, possessed fore-limbs constructed strictly on the type of the "wing" of the living Birds; and it cannot, therefore, be separated from this class. Another extraordinary peculiarity of *Ichthyornis* is, that the bodies of the *vertebræ* (fig. 213, *c*) were *bi-concave,* as is the case with many extinct Reptiles and almost all Fishes, but as does not occur in any living Bird. There can be little doubt that *Ichthyornis* was aquatic in its habits, and that it lived principally upon fishes; but its powerful wings at the same time indicate that it was capable of prolonged flight. The tail of *Ichthyornis* has, unfortunately, not been discovered; and it is at present impossible to say whether this resembled the tail of existing Birds, or whether it was elongated and composed of separate vertebræ, as in the Jurassic *Archæopteryx.*

Still more wonderful than *Ichthyornis* is the marvellous bird described by Marsh under the name of *Hesperornis regalis.* This presents us with a gigantic diving bird, somewhat resembling the existing "Loons" (*Colymbus*), but agreeing with *Ichthyornis* in having the jaws furnished with conical, recurved, pointed teeth (fig. 212, *b*). Hence these forms are grouped together in a new sub-class, under the name of *Odontornithes* or "Toothed Birds." The teeth of *Hesperornis* (fig. 212, *d*) resemble those of *Ichthyornis* in their general form; but instead of being sunk in distant sockets, they are simply implanted in a deep continuous groove in the bony substance of the jaw. The front of the upper jaw does not carry teeth, and was probably encased in a horny beak. The breast-bone is entirely destitute of a central ridge or keel, and the wings are minute and quite rudimentary; so that *Hesperornis,* unlike *Ichthyornis,* must have been wholly deprived of the power of flight, in this respect approaching the existing Penguins. The tail consists of about twelve vertebræ, of which the last three or four are amalgamated to form a flat terminal mass, there being at the same time clear indications that the tail was capable of up and down movement in a vertical plane, this probably fitting it to serve as a swimming-paddle or rudder. The legs were powerfully constructed, and the feet were adapted to assist the bird in rapid motion through the water. The known remains of *Hesperornis regalis* prove it to have been a swimming and diving bird, of larger dimensions than any of the aquatic members of the class of Birds with which we are acquainted at the present day. It appears to have stood between five and six feet high, and its inability to fly is fully compensated for by the numerous adaptations of its structure to a

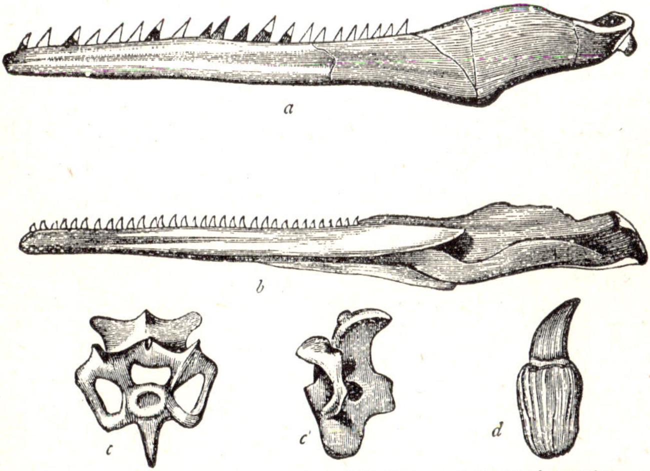

Fig. 212.—Toothed Birds (*Odontornithes*) of the Cretaceous Rocks of America. *a*, Left lower jaw of *Ichthyornis dispar*, slightly enlarged ; *b*, Left lower jaw of *Hesperornis regalis*, reduced to nearly one-fourth of the natural size ; *c*, Cervical vertebra of *Ichthyornis dispar*, front view, twice the natural size ; *c'*, Side view of the same ; *d*, Tooth of *Hesperornis regalis*, enlarged to twice the natural size. (After Marsh.)

watery life. Its teeth prove it to have been carnivorous in its habits, and it probably lived upon fishes. It is a curious fact that two Birds agreeing with one another in the wholly abnormal character of possessing teeth, and in other respects so entirely different, should, like *Ichthyornis* and *Hesperornis*, have lived not only in the some geological period, but also in the same geographical area; and it is equally curious that the area inhabited by these toothed Birds should at the same time have been tenanted by winged and bird-like Reptiles belonging to the toothed genus *Pterodactylus* and the toothless genus *Pteranodon*.

No remains of *Mammals*, finally, have as yet been detected in any sedimentary accumulation of Cretaceous age.

Literature.

The following list comprises of the more important works and memoirs which may be consulted with reference to the Cretaceous strata and their fossil contents:—

(1) 'Memoirs of the Geological Survey of Great Britain.'
(2) 'Geology of England and Wales.' Conybeare and Phillips.
(3) 'Geology of Yorkshire,' vol. ii. Phillips.
(4) 'Geology of Oxford and the Thames Valley.' Phillips.
(5) 'Geological Excursions through the Isle of Wight.' Mantell.
(6) 'Geology of Sussex.' Mantell.
(7) 'Report on Londonderry,' &c. Portlock.
(8) 'Recherches sur le Terrain Crétacé Supérieur de l'Angleterre et de l'Irlande.' Barrois.
(9) "Geological Survey of Canada"—'Report of Progress, 1872-73.'
(10) 'Geological Survey of California.' Whitney.
(11) 'Geological Survey of Montana, Idaho, Wyoming, and Utah.' Hayden and Meek.
(12) 'Report on Geology,' &c. (British North America Boundary Commission). G. M. Dawson.
(13) 'Manual of Geology.' Dana.
(14) 'Lethæa Rossica.' Eichwald.
(15) 'Petrefacta Germaniæ.' Goldfuss.
(16) 'Fossils of the South Downs.' Mantell.
(17) 'Medals of Creation.' Mantell.
(18) 'Mineral Conchology.' Sowerby.
(19) 'Lethæa Geognostica.' Bronn.
(20) 'Malacostracous Crustacea of the British Cretaceous Formation' (Palæontographical Society). Bell.
(21) 'Brachiopoda of the Cretaceous Formation' Palæontographical Society). Davidson.
(22) 'Corals of the Cretaceous Formation' (Palæontographical Society). Milne-Edwards and Haime.

(23) 'Supplement to the Fossil Corals' (Palæontographical Society). Martin Duncan.

(24) 'Echinodermata of the Cretaceous Formation' (Palæontographical Society). Wright.

(25) 'Monograph of the Belemnitidæ' (Palæontographical Society). Phillips.

(26) 'Monograph of the Trigoniæ' (Palæontographical Society). Lycett.

(27) 'Fossil Cirripedes' (Palæontographical Society). Darwin.

(28) 'Fossil Mollusca of the Chalk of Britain' (Palæontographical Society). Sharpe.

(29) 'Entomostraca of the Cretaceous Formation' (Palæontographical Society). Rupert Jones.

(30) 'Monograph of the Fossil Reptiles of the Cretaceous Formation' (Palæontographical Society). Owen.

(31) 'Manual of Palæontology.' Owen.

(32) 'Synopsis of Extinct Batrachia and Reptilia.' Cope.

(33) "Structure of the Skull and Limbs in Mosasauroid Reptiles"—'American Journ. Sci. and Arts, 1872.' Marsh.

(34) "On Odontornithes"—'American Journ. Sci. and Arts, 1875.' Marsh.

(35) 'Ossemens Fossiles.' Cuvier.

(36) 'Catalogue of Ornithosauria.' Seeley.

(37) 'Paléontologie Française.' D'Orbigny.

(38) 'Synopsis des Echinides fossiles.' Desor.

(39) 'Cat. Raisonné des Echinides.' Agassiz and Desor.

(40) "Echinoids"—'Decades of the Geol. Survey of Britain.' E. Forbes.

(41) 'Paléontologie Française.' Cotteau.

(42) 'Versteinerungen der Böhmischen Kreide-formation.' Reuss.

(43) "Cephalopoda, Gasteropoda, Pelecypoda, Brachiopoda, &c., of the Cretaceous Rocks of India"—'Palæontologica Indica,' ser. i., iii., v., vi., viii. Stoliczka.

(44) "Cretaceous Reptiles of the United States"—'Smithsonian Contributions to Knowledge,' vol. xiv. Leidy.

(45) 'Invertebrate Cretaceous, and Tertiary Fossils of the Upper Missouri Country.' 1876. Meek.

CHAPTER XVIII.

THE EOCENE PERIOD.

Before commencing the study of the subdivisions of the Kainozoic series, there are some general considerations to be noted. In the first place, there is in the Old World a complete and entire physical break between the rocks of the Mesozoic and Kainozoic periods. In no instance in Europe are Tertiary strata to be found resting conformably upon any Secondary rock. The Chalk has invariably suffered much erosion and denudation before the lowest Tertiary strata were deposited upon it. This is shown by the fact that the actually eroded surface of the Chalk can often be seen; or, failing this, that we can point to the presence of the chalk-flints in the Tertiary strata. This last, of course, affords unquestionable proof that the Chalk must have been subjected to enormous denudation prior to the formation of the Tertiary beds, all the chalk itself having been removed, and nothing left but the flints, while these are all rolled and rounded. In the continent of North America, on the other hand, the lowest Tertiary strata have been shown to graduate downwards conformably with the highest Cretaceous beds, it being a matter of difficulty to draw a precise line of demarcation between the two formations.

In the second place, there is a marked break in the *life* of the Mesozoic and Kainozoic periods. With the exception of a few *Foraminifera,* and one *Brachiopod* (the latter doubtful), no Cretaceous *species* is known to have survived the Cretaceous period; while several characteristic *families,* such as the *Ammonitidæ, Belemnitidæ,* and *Hippuritidæ,* died out entirely with the close of the Cretaceous rocks. In the Tertiary rocks, on the other hand, not only are all the animals and plants more or less like existing types, but we meet with a constantly-increasing number of *living species* as we pass from the bottom of the Kainozoic series to the top. Upon this last fact is founded the modern classification of the Kainozoic rocks, propounded by Sir Charles Lyell.

The absence in strata of Tertiary age of the chambered Cephalopods, the Belemnites, the *Hippurites,* the *Inocerami,* and the diversified types of Reptiles which form such conspicuous features in the Cretaceous fauna, render the palæontological break between the Chalk and the Eocene one far too serious to be overlooked. At the same time, it is to be remembered that the evidence afforded by the explorations carried out of late years as to the animal life of the deep sea, renders it certain that the extinction of marine forms of life at the close of the Cretaceous period was far less extensive than had been previously assumed. It is tolerably certain, in fact, that we may look upon some of the inhabitants of the depths of our existing oceans as the direct, if modified, descendants of animals which were in existence when the Chalk was deposited.

It follows from the general want of conformity between the Cretaceous and Tertiary rocks, and still more from the great difference in life, that the Cretaceous and Tertiary periods are separated, in the Old World at any rate, by an enormous lapse of unrepresented time. How long this interval may have been, we have no means of judging exactly, but it very possibly was as long as the whole Kainozoic epoch itself. Some day we shall doubtless find, at some part of the earth's surface, marine strata which were deposited during this period, and which will contain fossils intermediate in character between the organic remains which respectively characterize the Secondary and Tertiary periods. At present, we have only slight traces of such deposits—as, for instance, the Maestricht beds, the Faxöe Limestone, and the Pisolitic Limestone of France.

Classification of the Tertiary Rocks.—The classification of the Tertiary rocks is a matter of unusual difficulty, in consequence of their occurring in disconnected basins, forming a series of detached areas, which hold no relations of superposition to one another. The order, therefore, of the Tertiaries in point of time, can only be determined by an appeal to fossils; and in such determination Sir Charles Lyell proposed to take as the basis of classification the *proportion of living or existing species of Mollusca which occurs in each stratum or group of strata.* Acting upon this principle, Sir Charles Lyell divides the Tertiary series into four groups:—

I. The *Eocene* formation (Gr. *eos,* dawn; *kainos,* new), containing the smallest proportion of existing species, and being, therefore, the oldest division. In this classification, only the *Mollusca* are taken into account; and it was found that of these about three and a half per cent were identical with existing species.

II. The *Miocene* formation (Gr. *meion*, less; *kainos*, new), with more recent species than the Eocene, but *less* than the succeeding formation, and less than one-half the total number in the formation. As before, only the Mollusca are taken into account, and about 17 per cent of these agree with existing species.

III. The *Pliocene* formation (Gr. *Pleion*, more; *kainos*, new) with generally *more* than half the species of shells identical with existing species—the proportion of these varying from 35 to 50 per cent in the lower beds of this division, up to 90 or 95 per cent in its higher portion.

IV. The *Post-Tertiary Formations*, in which *all the shells belong to existing species*. This, in turn, is divided into two minor groups—the *Post-Pliocene* and *Recent Formations*. In the *Post-Pliocene* formations, while all the *Mollusca* belong to existing species, most of the *Mammals* belong to extinct species. In the *Recent* period, the quadrupeds, as well as the shells, belong to living species.

The above, with some modifications, was the original classification proposed by Sir Charles Lyell for the Tertiary rocks, and now universally accepted. More recent researches, it is true, have somewhat altered the proportions of existing species to extinct, as stated above. The general principle, however, of an increase in the number of living species, still holds good; and this is as yet the only satisfactory basis upon which it has been proposed to arrange the Tertiary deposits.

Eocene Formation.

The Eocene rocks are the lowest of the Tertiary series, and comprise all those Tertiary deposits in which there is only a small proportion of existing *Mollusca*—from three and a half to five per cent. The Eocene rocks occur in several basins in Britain, France, the Netherlands, and other parts of Europe, and in the United States. The subdivisions which have been established are extremely numerous, and it is often impossible to parallel those of one basin with those of another. It will be sufficient, therefore, to accept the division of the Eocene formation into three great groups—Lower, Middle, and Upper

Eocene—and to consider some of the more important beds comprised under these heads in Europe and in North America.

I. EOCENE OF BRITAIN. (1.) LOWER EOCENE.—The base of the Eocene series in Britain is constituted by about 90 feet of light-colored, sometimes argillaceous sands (*Thanet Sands*), which are of marine origin. Above these, or forming the base of the formation where these are wanting, come mottled clays and sands with lignite (*Woolwich and Reading series*), which are estuarine or fluvio-marine in origin. The highest member of the Lower Eocene of Britain is the "*London Clay,*" consisting of a great mass of dark-brown or blue clay, sometimes with sandy beds, or with layers of "septaria," the whole attaining a thickness of from 200 to as much as 500 feet. The London Clay is a purely marine deposit, containing many marine fossils, with the remains of terrestrial animals and plants; all of which indicate a high temperature of the sea and tropical or subtropical condition of the land.

(2.) MIDDLE EOCENE.—The inferior portion of the Middle Eocene of Britain consists of marine beds, chiefly consisting of sand, clays, and gravels, and attaining a very considerable thickness (*Bagshot and Bracklesham beds*). The superior portion of the Middle Eocene of Britain, on the other hand, consists of deposits which are almost exclusively fresh-water or brackish-water in origin (*Headon and Osborne series*).

The chief Continental formations of Middle Eocene age are the "Calcaire grossier" of the Paris basin, and the "Nummulitic Limestone" of the Alps.

(3.) UPPER EOCENE.—If the Headon and Osborne beds of the Isle of Wight be placed in the Middle Eocene, the only British representatives of the upper Eocene are the *Bembridge* beds. These strata consist of limestones, clays, and marls, which have for the most part been deposited in fresh or brackish water.

II. EOCENE BEDS OF THE PARIS BASIN.—The Eocene strata are very well developed in the neighborhood of Paris, where they occupy a large area or basin scooped out of the Chalk. The beds of this area are partly marine, partly freshwater in origin; and the following table (after Sir Charles Lyell) shows their subdivisions and their parallelism with the English series:—

General Table of French Eocene Strata.

UPPER EOCENE.

French Subdivisions.	*English Equivalents.*
A. 1. Gypseous series of Montmatre.	1. Bembridge series.
A. 2. Calcaire silicieux, or Travertin Inférieur.	2. Osborne and Headon series.
A. 3. Grès de Beauchamp, or Sables Moyens.	3. White sand and clay of Barton Cliff, Hants.

MIDDLE EOCENE.

B. 1. Calcaire Grossier.	1. Bagshot and Bracklesham beds.
B. 2. Soissonnais Sands, or Lits Coquilliers.	2. Wanting.

LOWER EOCENE.

C. 1. Argile de Londres at base of Hill of Cassel, near Dunkirk.	1. London clay.
C. 2. Argile plastique and lignite.	2. Plastic clay and sand with lignite (Woolwich and Reading series).
C. 3. Sables de Bracheux.	3. Thanet sands.

III. Eocene Strata of the United States.—The lowest member of the Eocene deposits of North America is the so-called "*Lignitic Formation,*" which is largely developed in Mississippi, Tennessee, Arkansas, Wyoming, Utah, Colorado, and California, and sometimes attain a thickness of several thousand feet. Stratigraphically, this formation exhibits the interesting point that it graduates downwards insensibly and conformably into the Cretaceous, whilst it is succeeded *unconformably* by strata of Middle Eocene age. Lithologically, the series consists principally of sands and clays, with beds of lignite and coal, and its organic remains show that it is principally of fresh-water origin with a partial intermixture of marine beds. These marine strata of the "Lignitic formation" are of special interest, as showing such a commingling of Cretaceous and Tertiary types of life, that it is impossible to draw any rigid line in this region between the Mesozoic and Kainozoic systems. Thus the marine beds of the Lignitic series contain such characteristic Cretaceous forms as *Inoceramus* and *Am-*

monites, along with a great number of Univalves of a distinctly Tertiary type (Cones, Cowries, &c.) Upon the whole, therefore, we must regard this series of deposits as affording a kind of transition between the Cretaceous and the Eocene, holding in some respects a position which may be compared with that held by the Purbeck beds in Britain as regards the Jurassic and Cretaceous.

The Middle Eocene of the United States is represented by the *Claiborne* and *Jackson* beds. The *Claiborne series* is extensively developed at Claiborne, Alabama, and consists of sands, clays, lignites, marls, and impure limestones, containing marine fossils along with numerous plant-remains. The *Jackson series* is represented by lignitic clays and marls which occur at Jackson, Mississippi. Amongst the more remarkable fossils of this series are the teeth and bones of Cretaceans of the genus *Zeuglodon.*

Strata of Upper Eocene age occur in North America at Vicksburg, Mississippi, and are known as the *Vicksburg series.* They consist of lignites, clays, marls, and limestones. Fresh-water deposits of Eocene age are also largely developed in parts of the Rocky Mountain region. The most remarkable fossils of these beds are Mammals, of which a large number of species have been already determined.

Life of the Eocene Period.

The fossils of the Eocene deposits are so numerous that nothing more can be attempted here than to give a brief and general sketch of the life of the period, special attention being directed to some of the more prominent and interesting types, amongst which—as throughout the Tertiary series—the Mammals hold the first place. It is not uncommon, indeed, to speak of the Tertiary period as a whole under the name of the "Age of Mammals," a title at least as well deserved as that of "Age of Reptiles" applied to the Mesozoic, or "Age of Molluscs" applied to the Palæozoic epoch.

As regards the *plants* of the Eocene, the chief point to be noticed is, that the conditions which had already set in with the commencement of the Upper Cretaceous, are here continued, and still further enforced. The *Cycads* of the Secondary period, if they have not totally disappeared, are exceedingly rare; and the *Conifers,* losing the predominance which they enjoyed in the Mesozoic, are now relegated to a subordinate though well-defined place in the terrestrial vegetation. The great majority of the Eocene plants are referable to the groups of the Angiospermous Exogens and the Monocotyledons; and the vegetation of the period, upon the whole, approximates closely to that now existing upon the earth. The plants of the European Eocene are, however, in the main most closely allied to forms which are now characteristic of tropical or sub-tropical regions. Thus, in the London Clay are found numerous fruits of Palms (*Nipadites,* fig. 213), along with various other plants, most of which indicate a warm climate as prevailing in the south of England at the commencement of the Eocene period. In the Eocene strata of North America occur numerous plants belonging to existing types—such as Palms, Conifers, the Magnolia, Cinnamon, Fig, Dog-wood, Maple, Hickory, Poplar, Plane, &c. Taken as a whole, the Eocene flora of North America is nearly related to that of the Miocene strata of Europe, as well as that now existing in the American area. We may conclude, therefore, that "the forests of the American Eocene resembled those of the European Miocene, and even of modern America" (Dana).

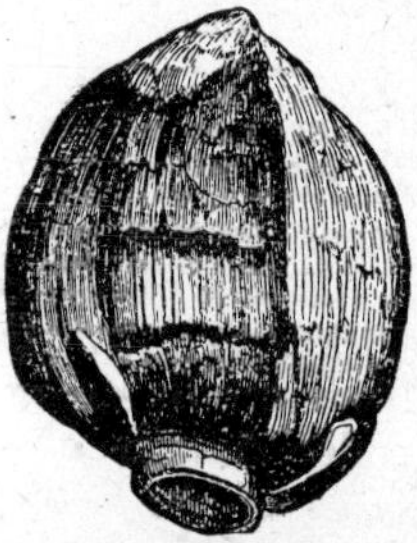

Fig. 213.—*Nipadites ellipticus*, **the fruit of a fossil Palm. London Clay, Isle of Sheppey.**

As regards the *animals* of the Eocene period, the *Protozoans* are represented by numerous *Foraminifera,* which reach here their maximum of development, both as regards the size of individuals and the number of generic types. Many of the Eocene Foraminifers are of small size; but even these not uncommonly form whole rock-masses. Thus, the so-called "Miliolite Limestone" of the Paris basin, largely used as a building-stone, is almost wholly composed of the shells of a small species of *Miliola.* The most remarkable, however, of the many members of this group of animals which flourish in Eocene times, are the "Nummulites" (*Nummulina*), so called from their resemblance in shape to coins (Lat. *nummus,* a coin). The Nummulites are amongst the largest of all known *Foraminifera,* sometimes attaining a size of three inches in circumference; and their internal structure is very complex (fig. 214). Many species are known, and they are particularly characteristic of the Middle and Upper of these periods—their place being sometimes taken by *Orbitoides*, a form very similar to the Nummulite in external appearance, but differing in its internal details. In the Middle Eocene, the remains of Nummulites are found in vast numbers in very widely-spread and easily-recognized formation known as the "Nummulitic Limestone" (fig. 10). According to Sir Charles Lyell, "the Nummulitic Limestone of the Swiss Alps rises to more than 10,000 feet above the level of the sea, and attains here and in other mountain-chains a thickness of several thousand feet. It may be said to play a far more conspicuous part than any other Tertiary group in the solid framework of the earth's crust, whether in Europe, Asia, or Africa. It occurs in Algeria and Morocco, and has been traced from Egypt, where it was largely quarried of old for the building of the Pyramids, into Asia Minor, and across Persia by Bagdad to the mouths of the Indus. It has been observed not only in Cutch, but in the mountain-ranges

1 2 3

Fig. 214.—*Nummulina lævigata.* Middle Eocene.

which separate Scinde from Persia, and which form the passes leading to Cabul; and it has been followed still further eastward into India, as far as Eastern Bengal and the frontiers of China." The shells of Nummulites have been found at an elevation of 16,500 feet above the level of the sea in Western Thibet; and the distinguished and philosophical geologist just quoted, further remarks, that "when we have once arrived at the conviction that the Nummulitic formation occupies a middle and upper place in the Eocene series, we are struck with the comparatively modern date to which some of the greatest revolutions in the physical geography of Europe, Asia, and Northern Africa must be referred. All the mountain-chains—such as the Alps, Pyrenees, Carpathians, and Himalayas—into the composition of whose central and loftiest parts the Nummulitic strata enter bodily, could have had no existence till after the Middle Eocene period. During that period, the sea prevailed where these chains now rise; for Nummulites and their accompanying Testacea were unquestionably inhabitants of salt water."

The *Cœlenterates* of the Eocene are represented principally by *Corals,* mostly of types identical with or nearly allied to those now in existence. Perhaps the most characteristic group of these is that of the *Turbinolidæ,* comprising a number of simple "cup-corals," which probably lived in moderately deep water. One of the forms belonging to this family is here figured (fig. 215). Besides true Corals, the Eocene deposits have yielded the remains of the "Sea-pens" (*Pennatulidæ*) and the branched skeletons of the "Sea-shrubs" (*Gorgonidæ*).

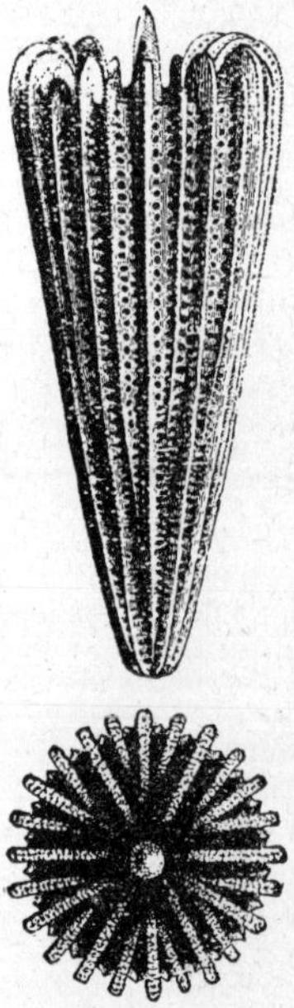

Fig. 215.—*Turbinolia sulcata*, viewed from one side, and from above. Eocene.

The *Echinoderms* are represented principally by Sea-urchins, and demand nothing more than mention. It is to be observed, however, that the great group of the Sea-lilies (*Crinoids*) is now verging on extinction, and is but very feebly represented.

Amongst the *Mollusca,* the *Polyzoans* and *Brachiopods* also require no special mention, beyond the fact that the latter are greatly reduced in numbers, and belong principally to the existing genera *Terebratula* and *Rhynchonella.* The Bivalves (*Lamellibranchs*) and the Univalves (*Gasteropods*) are exceedingly numerous, and almost all the principal existing genera are now represented; though less than five per cent of the Eocene *species* are identical with those now living. It is difficult to make any selection from the many Bivalves which are known in deposits of this age; but species of *Cardita, Crassatella, Leda, Cyrena, Mactra, Cardium, Psammobia,* &c., may be mentioned as very characteristic. The *Cardita planicosta* here figured (fig. 216) is not only very abundant in the Middle Eocene, but is very widely distributed, ranging from Europe to the Pacific coast of North America. The *Univalves* of the Eocene are extremely numerous, and generally beautifully preserved. The majority of them belong to that great section of the *Gasteropods* in which the mouth of the shell is notched or produced into a canal (when the shell is said to be "siphonostomatous")—this section including the carnivorous and most highly-organized groups of the class. Not only is this the case, but a large number of the Eocene Univalves belong to types which now attain their maximum of development in the warmer regions of the globe. Thus we find numerous species

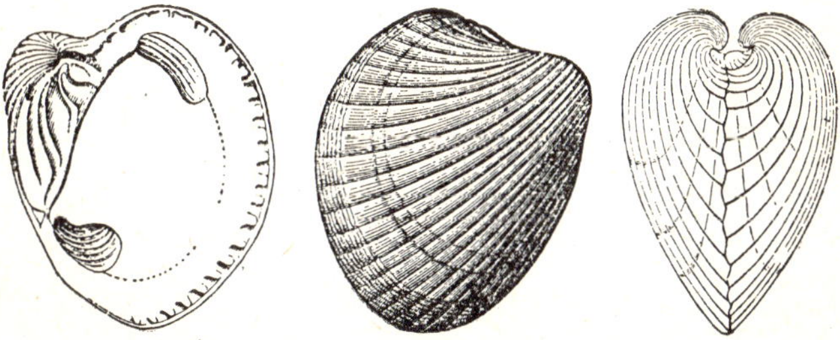

Fig. 216.—*Cardita planicosta.* Middle Eocene.

of Cones (*Conus*), Volutes (*Voluta*), Cowries (*Cypræa*, fig. 218), Olives and Rice-shells (*Oliva*), Mitre-shells (*Mitra*), Trumpet-shells (*Triton*), Auger-shells (*Terebra*), and Fig-shells (*Pyrula*). Along with these are many forms of *Pleurotoma, Rostellaria*, Spindle-shells (*Fusus*), Dog-whelks (*Nassa*), *Murices,* and many round-mouthed ("holostomatous") species, belonging to such

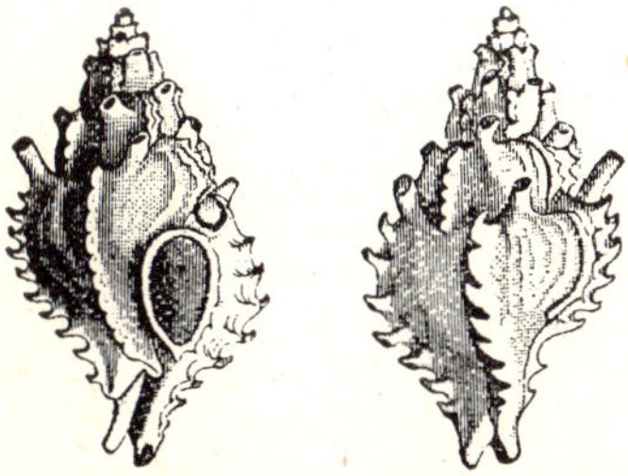

Fig. 217.—*Typhis tubifer*, a "siphonostomatous" Univalve. Eocene.

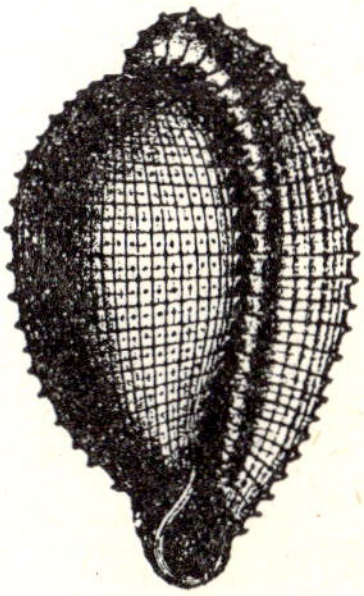

Fig. 218. *Cypræa elegans*. Eocene.

genera as *Turritella, Nerita, Natica, Scalaria,* &c. The genus *Cerithium* (fig. 219), most of the living forms of which are found in warm regions, inhabiting fresh or brackish waters, undergoes a vast development in the Eocene period, where it is represented by an immense number of specific forms, some of which attain very large dimensions. In the Eocene strata of the Paris basin alone, nearly one hundred and fifty species of this genus have been detected. The more strictly fresh-water deposits of the Eocene period have also yielded numerous remains of Univalves such as are now proper to rivers and lakes, together with the shells of true Land-snails. Amongst these may be mentioned numerous species of *Limnæa* (fig. 220), *Physa* (fig. 221), *Melania, Paludina, Planorbis, Helix, Bulimus,* and *Cyclostoma* (fig. 222).

Fig. 219.—*Cerithium hexagenum.* Eocene.

With regard to the *Cephalopods,* the chief point to be noticed is, that all the beautiful and complex forms which peculiarly characterized the Cretaceous period have here disappeared. We no longer meet with a single example of the Turrilite, the Baculite, the Hamite, the Scaphite, or the Ammonite. The only exception to this statement is the occurrence of one species of Ammonite in the so-called "Lignitic Formation" of North America; but the beds containing this may possibly be rather referable to the Cretaceous—and this exception does not affect the fact that the *Ammonitidæ*, as a family, had become extinct before the Eocene strata were deposited. The ancient genus *Nautilus* still survives, the sole representative of the once mighty order of the Tetrabranchiate Cephalopods.

Fig. 220.—*Limnæa pyramidalis.* **Eocene.**

Fig. 221.—*Physa columnaris.* **Eocene.**

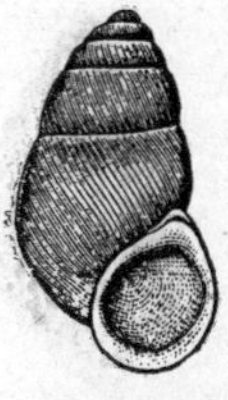

Fig. 222.—*Cyclostoma Arnoudii.* **Eocene.**

In the order of the *Dibranchiates,* we have a like phenomenon to observe in the total extinction of the great family of the "Belemnites." No form referable to this group has hitherto been found in any Tertiary stratum; but the internal skeletons of Cuttle-fishes (such as *Belosepia*) are not unknown.

Remains of *Fishes* are very abundant in strata of Eocene age, especially in certain localities. The most famous depot for the fossil fishes of this period is the limestone of Monte Bolca, near Verona, which is interstratified with beds of volcanic ashes, the whole being referable to the Middle Eocene. The fishes here seem to have been suddenly destroyed by a volcanic eruption, and are found in vast numbers. Agassiz

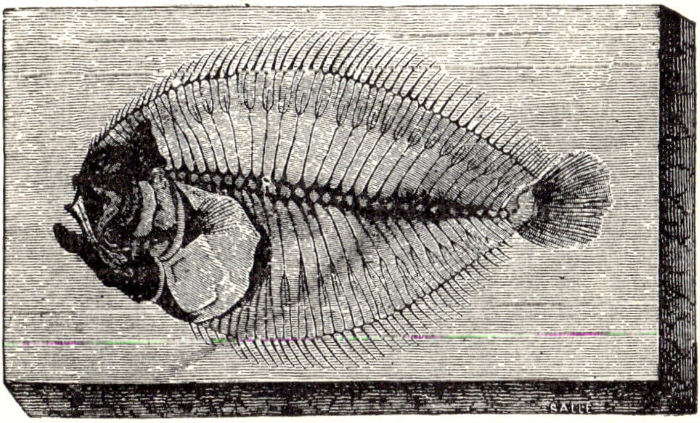

Fig. 223.—*Rhombus minimus*, a small fossil Turbot from the Eocene Tertiary, Monte Bolca.

has described over one hundred and thirty species of Fishes from this locality, belonging to seventy-seven genera. All the *species* are extinct; but about one-half of the *genera* are represented by living forms. The great majority of the Eocene Fishes belong to the order of the "Bony Fishes" (*Teleosteans*), so that in the main the forms of Fishes characterizing the Eocene are similar to those which predominate in existing seas. In addition to the above, a few *Ganoids* and a large number of *Placoids* are known to occur in the Eocene rocks. Amongst the latter are found numerous teeth of true Sharks, such as *Otodus* (fig. 224) and *Carcharodon*. The pointed and serrated teeth of the latter sometimes attain a length of over half a foot, indicating that these predaceous fishes attained gigantic dimensions; and it is interesting to note that teeth, in external appearance very similar to those of the early Tertiary genus *Carcharodon,* have been dredged

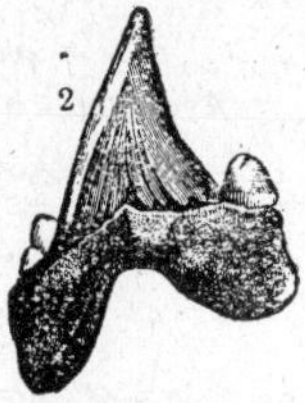

Fig. 224.—Tooth of *Otodus obliquus.* Eocene.

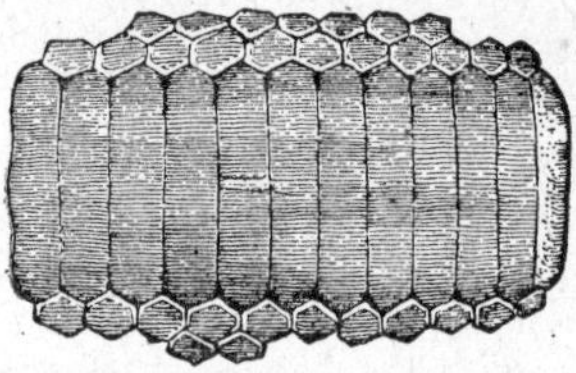

Fig. 225.—Flattened dental plates of a Ray (*Myliobatis Edwardsii*). Eocene.

from great depths during the recent expedition of the Challenger. There also occur not uncommonly the flattened teeth of Rays (fig. 225), consisting of flat bony pieces placed close together, and forming "a kind of mosaic pavement on both the upper and lower jaws." (Owen).

In the class of the *Reptiles,* the disappearance of the characteristic Mesozoic types is as marked a phenomenon as the introduction of new forms. The Ichthyosaurs, the Plesiosaurs, the Pterosaurs, and the Mosasaurs of the Mesozoic, find no representatives in the Eocene Tertiary; and the same is true of the Deinosaurs, if we except a few remains from the doubtfully-situated "Lignitic formation" of the United States. On the other hand, all the modern orders of Reptiles are known to have existed during the Eocene period. The *Chelonians* are represented by true marine Turtles, by "Terrapins" (*Emydidæ*), and by "Soft Tortoises" (*Trionycidæ*). The order of the Snakes and Serpents (*Ophidia*) makes its appearance here for the first time under several forms—all of which, however, are referable to the non-venomous group of the "Constricting Serpents" (*Boidæ*). The oldest of these is the *Palæophis toliapicus* of the London Clay of Sheppey, first made known to science by the researches of Professor Owen. The nearly-allied *Palæophis typhæus* of the Eocene beds of Bracklesham appears to have been a Boa-constrictor-like Snake of about twenty feet in length. Similar Python-like Snakes (*Palæophis, Dinophis,* &c.) have been described from the Eocene deposits of the United States. True Lizards

(*Lacertilians*) are found in some abundance in the Eocene deposits,—some being small terrestrial forms, like the common European lizards of the present day; whilst others equal or exceed the living Monitors in size. Lastly, the modern order of the *Crocodilia* is largely represented in Eocene times, by species belonging to all the existing genera, together with others referable to extinct types. As pointed out by Owen, it is an interesting fact that in the Eocene rocks of the south-west of England, there occur fossil remains of all the three living types of Crocodilians—namely, the Gavials, the true Crocodiles, and the Alligators (fig. 226)—though at the present day these forms are all geographically restricted in their range, and are never associated together.

Almost all the existing orders of *Birds*, if not all, are represented in the Eocene deposits by remains often very closely allied to existing types. Thus, amongst the Swimming Birds (*Natatores*) we find examples of forms allied to the living Pelicans and Mergansers; amongst the Waders (*Grallatores*) we have birds resembling the Ibis (the *Numenius gypsorum* of the Paris basin); amongst the Running Birds (*Cursores*) we meet with the great *Gastornis Parisiensis*, which equalled the African Ostrich in height, and the still more

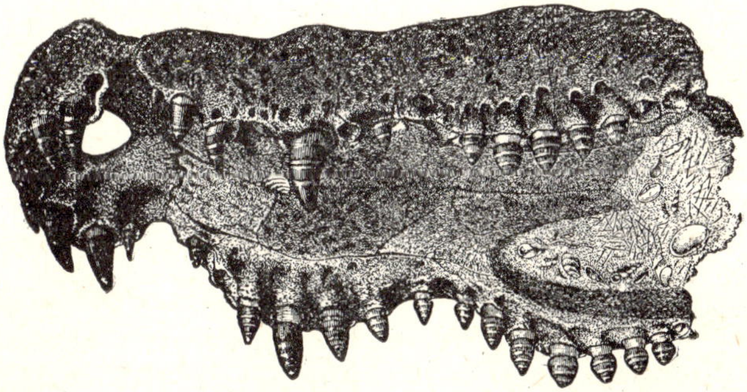

Fig. 226.—Upper jaw of Alligator. Eocene Tertiary. Isle of Wight.

gigantic *Dasornis Londinensis;* remains of a Partridge represent the Scratching Birds (*Rasores*); the American Eocene has yielded the bones of one of the Climbing Birds (*Scansores*), apparently referable to the Woodpeckers; the *Protornis Glarisiensis* of the Eocene Schists of Glaris is the oldest known example of the Perching Birds (*Insessores*); and the

Birds of Prey (*Raptores*) are represented by Vultures, Owls, and Hawks. The toothed Birds of the Upper Cretaceous are no longer known to exist; but professor Owen has recently described from the London Clay the skull of a very remarkable Bird, in which there is, at any rate, an approximation to the structure of *Ichthyornis* and *Hesperornis*. The bird in question has been named the *Odontopteryx toliapicus*, its generic title being derived from the very remarkable characters of its jaws. In this singular form (fig. 227) the margins of both jaws are furnished with tooth-like denticulations, which differ from true teeth in being actually portions of the bony

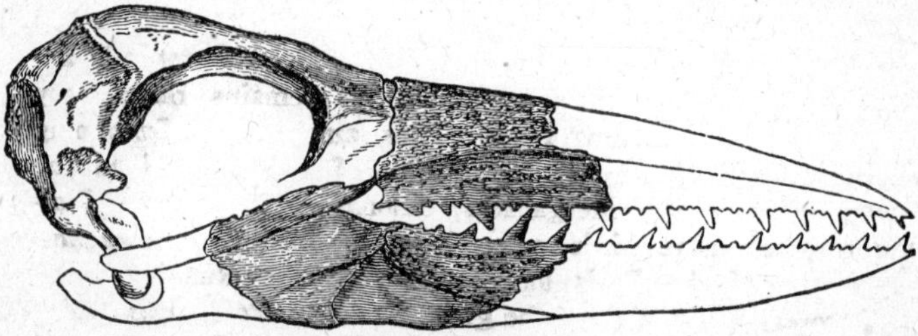

Fig. 227.—Skull of *Odontopteryx toliapicus*, restored. (After Owen.)

substance of the jaw itself, with which they are continuous, and which were probably encased by extensions of the horny sheath of the bill. These tooth-like processes are of two sizes, the larger ones being comparable to canines; and they are all directed forwards, and have a triangular or compressed conical form. From a careful consideration of all the discovered remains of this bird, Professor Owen concludes that "*Odontopteryx* was a warm-blooded feathered biped, with wings; and further, that it was web-footed and a fish-eater, and that in the catching of its slippery prey it was assisted by this Pterosauroid armature of its jaws." Upon the whole, *Odontopteryx* would appear to be most nearly related to the family of the Geese (*Anserinæ*) or Ducks (*Anatidæ*); but the extension of the bony substance of the jaws into tooth-like processes is an entirely unique character, in which it stands quite alone.

The known *Mammals* of the Mesozoic period, as we have seen, are all of small size; and with one not unequivocal exception, they appear to be referable to the order of the

Pouched Quadrupeds (*Marsupials*), almost the lowest group of the whole class of the *Mammalia*. In the Eocene rocks, on the other hand, numerous remains of Quadrupeds have been brought to light, representing most of the great Mammalian orders now in existence upon the earth, and in many cases indicating animals of very considerable dimensions. We are, in fact, in a position to assert that the majority of the great groups of Quadrupeds with which we are familar at the present day were already in existence in the Eocene period, and that their ancient root-stocks were even in this early time separated by most of the fundamental differences of structure which distinguish their living representatives. At the same time, there are some amongst the Eocene quadrupeds which have a "generalized" character, and which may be regarded as structural types standing midway between groups now sharply separated from one another.

The order of the *Marsupials*—including the existing Kangaroos, Wombats, Opossums, Phalangers, &c.—is poorly represented in deposits of Eocene age. The most celebrated example of this group is the *Didelphys gypsorum* of the Gypseous beds of Montmartre, near Paris, an Opossum very nearly allied to the living Opossums of North and South America.

No member of the *Edentates* (Sloths, Ant-eaters, and Armadillos) has hitherto been detected in any Eocene deposit. The aquatic order of the *Sirenians* (Dugongs and Manatees), with their fish-like bodies and tails, paddle-shaped fore-limbs, and wholly deficient hind-limbs, are represented in strata of this age by remains of the ancient "Sea-Cows," to which the name of *Halitherium* has been applied. Nearly allied to the preceding is the likewise aquatic order of the Whales and Dolphins (*Cetaceans*), in which the body is also fish-like, the hind-limbs are wanting, the fore-limbs are converted into powerful "flippers" or swimming-paddles, and the terminal extremity of the body is furnished with a horizontal tail-fin. Many existing Cetaceans (such as the Whalebone Whales) have no true teeth; but others (Dolphins, Porpoises, Sperm Whales) possess simple conical teeth. In strata of Eocene age, however, we find a singular group of Whales constituting the genus *Zeuglodon* (fig. 228), in which the teeth differed from those of all existing forms in being of two kinds,—the front ones being conical incisors, whilst the back teeth or molars have serrated triangular crowns, and are inserted in the jaw by two roots. Each molar (fig. 228, A) looks as if it were composed of two separate teeth united on the one side by their crowns; and it is this peculiarity which is expressed by the generic name (Gr. *zeugle*, a yoke; *odous*, tooth). The best-known species of the genus is the *Zeuglodon cetoides* of Owen, which attained a length of seventy feet. Remains of these gigantic Whales are very common in the "Jackson Beds" of the Southern United States. So common are they that, according to Dana,

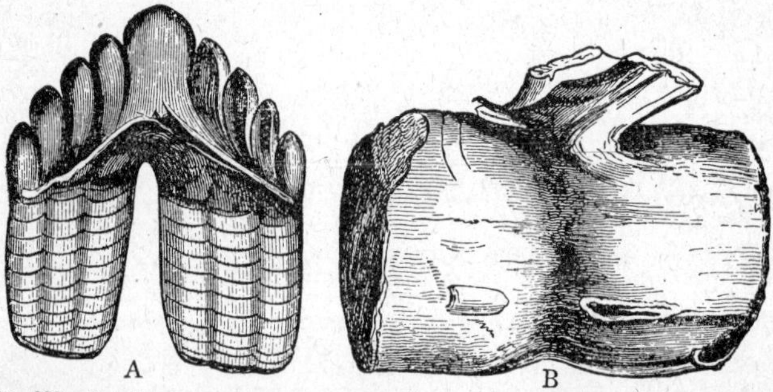

Fig. 228.—*Zeuglodon cetoides*. A, Molar tooth of the natural size; B, Vertebra, reduced in size. From the Middle Eocene of the United States. (After Lyell.)

"the large vertebræ, some of them a foot and a half long and a foot in diameter, were formerly so abundant over the country, in Alabama, that they were used for making walls, or were burned to rid the fields of them."

The great and important order of the Hoofed Quadrupeds (*Ungulata*) is represented in the Eocene by examples of both of its two principal sections—namely, those with an uneven number of toes (one or three) on the foot (*Perissodactyle Ungulates*), and those with an even number of toes (two or four) to each foot (*Artiodactyle Ungulates*). Amongst the Odd-toed Ungulates, the living family of the Tapirs (*Tapiridæ*) is represented by the genus *Coryphodon* of Owen. Nearly related to the preceding are the species of *Palæotherium*, which have a historical interest as being amongst the first of the Tertiary Mammals investigated by the illustrious Cuvier. Several species of *Palæothere* are known, varying greatly in size, the smallest being little bigger than a hare, whilst the largest must have equalled a good-sized horse in its dimensions. The species of *Palæotherium* appear to have agreed with the existing Tapirs in possessing a lengthened and flexible nose, which formed a short proboscis or trunk (fig. 229), suitable as an instrument for stripping off the foliage of trees—the char-

Fig. 229.—Outline of *Palæotherium magnum*, restored. Upper Eocene, Europe. (After Cuvier.)

acters of the molar teeth showing them to have been strictly herbivorous in their habits. They differ, however, from the Tapirs, amongst other characters, in the fact that both the fore and the hind feet possessed three toes each; whereas in the latter there are four toes on each fore-foot, and the hind-feet alone are three-toed. The remains of *Palæotheria* have been found in such abundance in certain localities as to show that these animals roamed in great herds over the fertile plains of France and the south of England during the later portion of the Eocene period. The accompanying illustration (fig. 229) represents the notion which the great Cuvier was induced by his researches to form as to the outward appearance of *Palæotherium magnum*. Recent discoveries, however, have rendered it probable that this restoration is in some important respects inaccurate. Instead of being bulky, massive, and more or less resembling the living Tapirs in form, it would rather appear that *Palæotherium magnum* was in reality a slender, graceful, and long-necked animal, more closely resembling in general figure a Llama, or certain of the Antelopes.

The singular genus *Anchitherium* forms a kind of transition between the *Palæotheria* and the true Horses (*Equidæ*). The Horse (fig. 230, D) possesses but one fully-developed toe to each foot, this being terminated by a single broad hoof, and representing the *middle* toe—the *third* of the typical five-fingered or five-toed limb of Quadrupeds in general. In addition, however, to this fully-developed toe, each foot in the horse carries two rudimentary toes which are concealed beneath the skin, and are known as the "splint-bones." These are respectively the *second* and *fourth* toes, in an aborted condition; and the *first* and *fifth* toes are wholly wanting. In *Hipparion* (fig. 230, C), the foot is essentially like that of the modern Horses, except that the second and fourth toes no longer are mere "splint-bones," hidden beneath the skin; but have now little hoofs, and hang freely, but uselessly, by the side of the great middle toe, not being sufficiently developed to reach the ground. In *Anchitherium*, again (fig. 230, B), the foot is three-toed, like that of *Hipparion;* but the two lateral toes (the second and fourth) are so far developed that they now reach the ground. The *first* digit (thumb or great toe) is still wanting; as also is the *fifth* digit

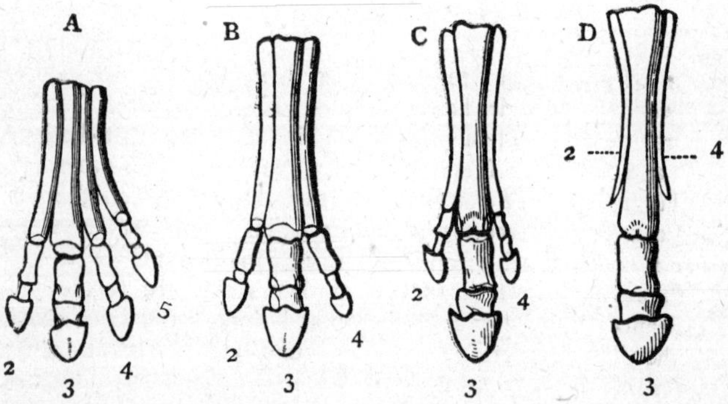

Fig. 230.—Skeleton of the foot in various forms belonging to the family of the *Equidæ*, A, Foot of *Orohippus*, Eocene; B, Foot of *Anchitherium*, Upper Eocene and Lower Miocene; C, Foot of *Hipparion*, Upper Miocene and Pliocene; D, Foot of Horse (*Equus*), Pliocene and Recent. The figures indicate the numbers of the digits in the typical five-fingered hand of Mammals. (After Marsh.)

(little finger or little toe). Lastly, the Eocene rocks have yielded in North America the remains of a small Equine quadruped, to which Marsh has given the name of *Orohippus*. In this singular form—which was not larger than a fox—the foot (fig. 230, A) carries *four* toes, all of which are hoofed and touch the ground, but of which the *third* toe is still the largest.

The *first* toe (thumb or great toe) is still wanting; but in this ancient representative of the Horses, the *fifth* or "little" toe appears for the first time. As all the above-mentioned forms succeed one another in point of time, it may be regarded as probable that we shall yet be able to point, with some certainty, to some still older example of the *Equidæ,* in which

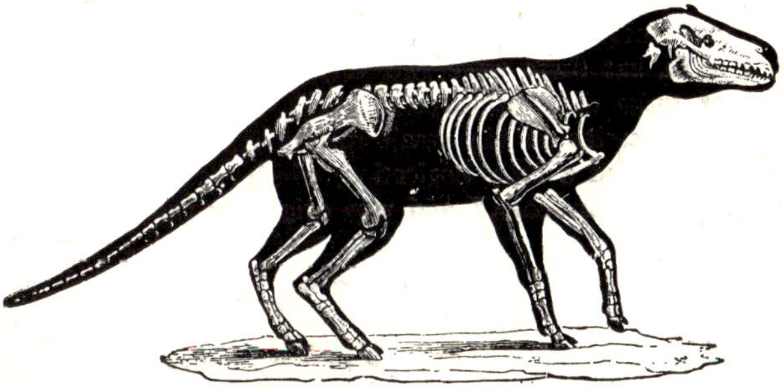

Fig. 231.—*Anoplotherium commune.* Eocene Tertiary, France. (After Cuvier.)

the first digit is developed, and the foot assumes its typical five-fingered condition.

Passing on to the Even-toed or *Artiodactyle Ungulates,* no representative of the *Hippopotamus* seems yet to have existed, but there are several forms (*Chæropotamus, Hyopotamus,* &c.) more or less closely allied to the Pigs (*Suida*); and the singular group of the *Anoplotheridæ* may be regarded as forming a kind of transition between the Swine and the Ruminants. The *Anoplotheria* (fig. 231) were slender in form, the largest not exceeding a donkey in size, with long tails, and having the feet terminated by two hoofed toes each, sometimes with a pair of small accessory hoofs as well. The teeth exhibit the peculiarity that they are arranged in a continuous series, without any gap or interval between the molars and the canines; and the back teeth, like those of all the Ungulates, are adapted for grinding vegetable food, their crowns resembling in form those of the true Ruminants. The genera *Dichobune* and *Xiphodon,* of the Middle and Upper Eocene, are closely related to *Anoplotherium,* but are more slender and deer-like in form. No example of the great Ruminant group of the Ungulate Quadrupeds has as yet been detected in deposits of Eocene age.

Whilst true Ruminants appear to be unknown, the Eocene strata of North America have yielded to the researches of Professor Marsh examples of an extraordinary group (*Dinocerata*), which may be considered as in some respects intermediate between the Ungulates and the Proboscideans. In *Dinoceras* itself (fig. 232) we have a large animal, equal in dimensons to the living Elephants, which it further resembles in the structure of the massive limbs, except that there are only four toes to each foot. The upper jaw was devoid of front teeth, but there were two very large canine teeth, in the form of tusks directed perpendicularly downwards; and there was also a series of six small molars on each. Each upper jaw-bone carried a bony projection, which was probably of the nature of a "horn-core," and was originally sheathed in horn.

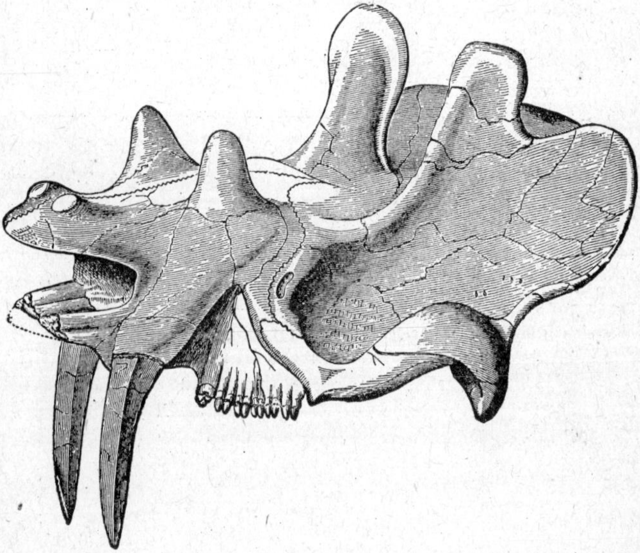

Fig. 232.—Skull of *Dinoceras mirabilis*, greatly reduced. Eocene, North America. (After Marsh.)

Two similar, but smaller, horn-cores are carried on the nasal bones; and two much larger projections, also probably of the nature of horn-cores, were carried upon the forehead. We may thus infer that *Dinoceras* possessed three pairs of horns, all of which resembled the horns of the Sheep and Oxen in consisting of a central bony "core," surrounded by a horny sheath. The nose was not prolonged into a proboscis or "trunk," as in the existing Elephants; and the tail was short and slender. Many forms of the *Dinocerata* are known; but all these singular and gigantic quadrupeds appear to have been confined to the North American continent, and to be restricted to the Eocene period.

The important order of the Elephants (*Proboscidea*) is also not known to have come into existence during the Eocene period. On the other hand, the great order of the Beasts of Prey (*Carnivora*) is represented in Eocene strata by several formations belonging to different types. Thus the *Arctocyon* pre-

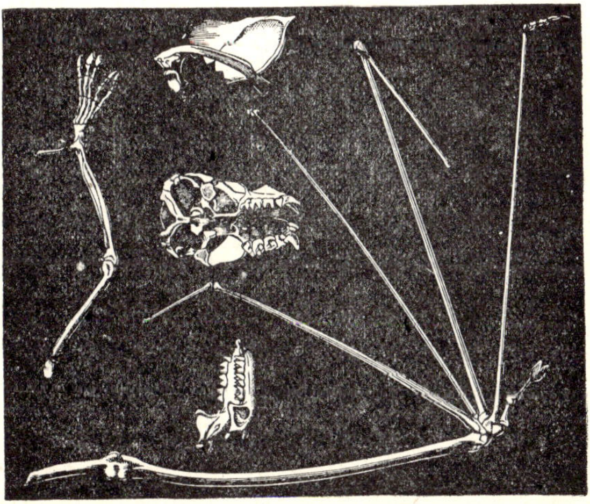

Fig. 233.—Portion of the skeleton of *Vespertilio Parisiensis*. Eocene Tertiary, France.

sents us with an Eocene Carnivore more or less closely allied to the existing Racoons; the *Palæonyctis* appears to be related to the recent Civet-cats; the genus *Hyænodon* is in some respects comparable to the living Hyænas; and the *Canis Parisiensis* of the gypsum-bearing beds of Montmartre may perhaps be allied to the Foxes.

The order of the Bats (*Cheiroptera*) is represented in Eocene strata of the Paris basin (Gypseous series of Montmartre) by the *Vespertilio Parisiensis* (fig. 233), an insect-eating Bat very similar to some of the existing European forms. Lastly, the Eocene deposits have yielded more or less satisfactory evidence of the existence in Europe at this period of examples of the orders of the Gnawing Mammals (*Rodentia*), the Insect-eating Mammals (*Insectivora*), and the Monkeys (*Quadrumana*).*

CHAPTER XIX.

THE MIOCENE PERIOD.

The Miocene rocks comprise those Tertiary deposits which contain less than about 35 per cent of existing species of shells (*Mollusca*), and more than 5 per cent—or those deposits in which the proportion of living shells is less than of extinct species. They are divisible into a *Lower Miocene* (*Oligocene*) and an *Upper Miocene* series.

In *Britain,* the Miocene rocks are very poorly developed, one of their leading developments being at Bovey Tracy in Devonshire, where there occur sands, clays, and beds of lignite or imperfect coal. These strata certain numerous plants, amongst which are Vines, Figs, the Cinnamon-tree, Palms, and many Conifers, especially those belonging to the genus *Sequoia* (the "Red-woods"). These Bovey Tracy lignites are of Lower Miocene age, and they are lacustrine in origin. Also of Lower Miocene age are the so-called "Hempstead Beds" of Yarmouth in the Isle of Wight. These attain a thickness of less than 200 feet, and are shown by their numerous fossils to be principally a true marine formation. Lastly, the Duke of Argyll, in 1851, showed that there existed at Ardtun, in the island of Mull, certain Tertiary strata containing numerous remains of plants; and these also are now regarded as belonging to the Lower Miocene.

* A short list of the more important works relating to the Eocene rocks and fossils will be given after all the Tertiary deposits have been treated of.

In *France,* the Lower Miocene is represented in Auvergne, Cantal, and Velay, by a great thickness of nearly horizontal strata of sands, sandstone, clays, marls, and limestones, the whole of fresh-water origin. The principal fossils of these lacustrine deposits are *Mammalia,* of which the remains occur in great abundance. In the valley of the Loire occur the typical European deposits of Upper Miocene age. These are known as the "Faluns," from a provincial term applied to shelly sands, employed to spread upon soils which are deficient in lime; and the Upper Miocene is hence sometimes spoken of as the "Falunian" formation. The Faluns occur in scattered patches, which are rarely more than 50 feet in thickness, and consist of sands and marls. The fossils are chiefly marine; but there occur also land and fresh-water shells, together with the remains of numerous Mammals. About 25 per cent of the shells of the Faluns are identical with existing species. The sands, limestones, and marls of the Department of Gers, near the base of the Pyrenees, rendered famous by the number of Mammalian remains exhumed from them by M. Lartet, also belong to the age of the Faluns.

In *Switzerland,* between the Alps and the Jura, there occurs a great series of Miocene deposits, known collectively as the "Molasse," from the soft nature of a greenish sandstone, which constitutes one of its chief members. It attains a thickness of many thousands of feet, and rises into lofty mountains, some of which—as the Rigi—are more than 6000 feet in height. The middle portion of the Molasse is of marine origin, and is shown by its fossils to be of the age of the Faluns; but the lower and upper portions of the formation are mainly or entirely of fresh-water origin. The Lower Molasse (of Lower Miocene age) has yielded about 500 species of plants, mostly of tropical or sub-tropical forms. The Upper Molasse has yielded about the same number of plants, with about 900 species of Insects, such as wood-eating Beetles, Water-beetles, White Ants, Dragon-flies, &c.

In *Belgium,* strata of both Lower and Upper Miocene age are known,—the former (*Rupelian Clays*) containing numerous marine fossils; whilst the latter (*Bolderberg Sands*) have yielded numerous shells corresponding with those of the Faluns.

In *Austria,* Miocene strata are largely developed, marine beds belonging to both the Lower and Upper division of the formation occurring extensively in the Vienna basin. The well-known Brown Coals of Radaboj, in Croatia, with numerous plants and insects, are also of Lower Miocene age.

In *Germany*, deposits belonging to both the Lower and Upper division of the Miocene formation are extensively developed. To the former belong the marine strata of the Mayence basin, and the marine *Rupelian Clay* near Berlin; whilst a celebrated group of strata belonging to the Upper Miocene occurs near Epplesheim, in Hesse-Darmstadt, and is well known for the number of its Mammalian remains.

In *Greece*, at Pikermé, near Athens, there occurs a celebrated deposit of Upper Miocene age, well known to palæontologists through the researches of M. M. Wagner, Roth, and Gaudry upon the numerous Mammalia which it contains. In *Italy*, also, strata of both Lower and Upper Miocene age are well developed in the neighborhood of Turin.

In the *Siwâlik Hills*, in India, at the southern foot of the Himalayas, occurs a series of Upper Miocene strata, which have become widely celebrated through the researches of Dr. Falconer and Sir Proby Cautley upon the numerous remains of Mammals and Reptiles which they contain. Beds of corresponding age, with similar fossils, are known to occur in the island of Perim in the Gulf of Cambay.

Lastly, Miocene deposits are found in *North America*, in New Jersey, Maryland, Virginia, Missouri, California, Oregon, &c., attaining a thickness of 1500 feet or more. They consist principally of clays, sands, and sandstones, sometimes of marine and sometimes of fresh-water origin. Near Richmond, in Virginia, there occurs a remarkable stratum, wrongly called "Infusorial Earth," which is occasionally 30 feet in thickness, and consists almost wholly of the siliceous envelopes of certain low forms of plants (Diatoms), along with the spicules of Sponges and other siliceous organisms (see fig. 16). The *White River Group* of Hayden occurs in the Upper Missouri region, and is largely exposed over the barren and desolate district known as the "Mauvaises Terres." They have a thickness of 1000 feet or more, and contain numerous remains of Mammals. They are of lacustrine origin, and are believed to be of the age of the Lower Miocene. Upon the whole, about from 15 to 30 per cent of the *Mollusca* of the American Miocene are identical with existing species.

In addition to the regions previously enumerated, Miocene strata are known to be developed in *Greenland, Iceland, Spitzbergen,* and in other areas of less importance.

The *life* of the Miocene period is extremely abundant, and, from the nature of the deposits of this age, also extremely varied in its character. The marine beds of the formation have yielded numerous remains of both Vertebrate and Invertebrate sea-animals; whilst the fresh-water deposits contain the skeletons of such shells, fishes, &c., as now inhabit rivers or lakes. Both the marine and the lacustrine beds have been shown to contain an enormous number of plants, the latter more particularly; whilst the Brown Coals of the formation are made up of vegetable matter little altered from its original condition. The remains of air-breathing animals, such as Insects, Reptiles, Birds, and Mammals, are also abundantly found, more especially in the fresh-water beds.

The *plants* of the Miocene period are extraordinarily numerous, and only some of the general features of the vegetation of this epoch can be indicated here. Our chief sources of information as to the Miocene plants are the Brown Coals of Germany and Austria, the Lower and Upper Molasse of Switzerland, and the Miocene strata of the Arctic regions. The lignites of Austria have yielded very numerous plants, chiefly of a tropical character—one of the most noticeable forms being a Palm of

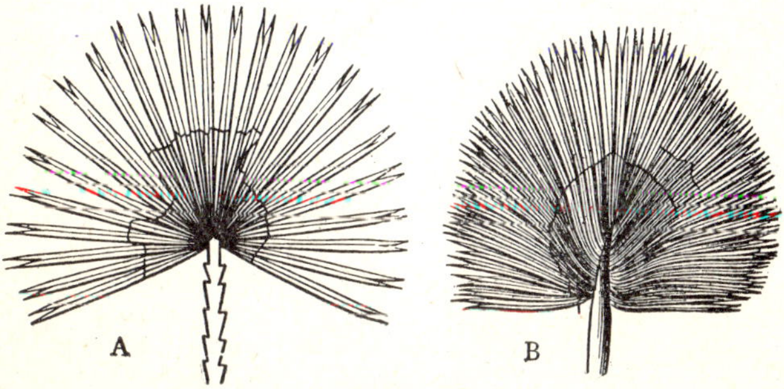

Fig. 234.—Miocene Palms. A, *Chamærops Helvetica* ; B, *Sabal major*. Lower Miocene of Switzerland and France.

the genus *Sabal* (fig. 234, B), now found in America. The plants of the Lower Miocene of Switzerland are also mostly of a tropical character, but include several forms now found in North America, such as a Tulip-tree (*Liriodendron*) and a Cypress (*Taxodium*). Amongst the more remarkable forms from these beds may be mentioned Fan-Palms (*Chamærops*, fig. 234, A), numerous tropical ferns, and two species of Cinnamon. The plant-remains of the Upper Molasse of Switzerland indicate an extraordinarily rank and luxuriant vegetation, composed mainly of plants which now live in warm countries. Among the commoner plants of this formation may be enumerated many species of Maple (*Acer*), Plane-trees (*Platanus*

Fig. 235.—*Platanus aceroides*, an Upper Miocene Plane-tree. *a*, Leaf; b, The core of a bundle of fruits; *c*, A single fruit.

Fig. 236. — *Cinnamomum polymorphum*. *a*, Leaf; *b*, Flower. Upper Miocene.

fig. 235), Cinnamon-trees (fig. 236), and other members of the *Lauraceæ*, many species of *Proteaceæ* (*Banksia, Grevillea,* &c.), several species of Sarsaparilla (*Smilax*), Palms, Cypresses, &c.

In Britain, the Lower Miocene strata of Bovey Tracy have yielded remains of Ferns, Vines, Fig, Cinnamon, *Proteaceæ*, &c., along with numerous Conifers. The most abundant of these last is a gigantic pine—the *Sequoia Couttsiæ*—which is very nearly allied to the huge *Sequoia* (*Wellingtonia*) *gigantea* of California. A nearly-allied form (*Sequoia Langsdorffi*) has been detected in the leaf-bed of Ardtun, in the Hebrides.

In Greenland, as well as in other parts of the Arctic regions, Miocene strata have been discovered which have yielded a great number of plants, many of which are identical with species found in the European Miocene. Amongst these plants are found many trees, such as Conifers, Beeches, Oaks, Maples, Plane-trees, Walnuts, Magnolias, &c., with numerous shrubs, ferns, and other smaller plants. With regard to the Miocene flora of the Arctic regions, Sir Charles Lyell remarks that "more than thirty species of Coniferæ have been found, including several Sequoias (allied to the gigantic Wellingtonia of California), with species of *Thujopsis* and *Salisburia*, now peculiar to Japan. There are also beeches, oaks, planes, poplars, maples, walnuts, limes, and even a magnolia, two cones of which have recently been obtained, proving that this splendid evergreen not only lived but ripened its fruit within the Arctic circle. Many of the limes, planes, and oaks were large-leaved species; and both flowers and fruits, besides immense quantities of leaves, are in many cases preserved. Among the shrubs are many evergreens, as *Andromeda,* and two extinct genera, *Daphnogene* and *M'Clintockia,* with fine leathery leaves, together with hazel, blackthorn, holly, logwood, and hawthorn. A species of Zamia (*Zamites*) grew in the swamps, with *Potamogeton, Sparganium,* and *Menyanthes;* while ivy and vines twined around the forest-trees, and the broad-level ferns grew beneath their shade. Even in Spitzbergen, as far north as lat. 78° 56′, no less than ninety-five species of fossil plants have been obtained, including *Taxodium* of two species, hazel, poplar, alder, beech, plane-tree, and lime. Such a vigorous growth of trees within 12° of the pole, where now a dwarf willow and a few herbaceous plants form the only vegetation, and where the ground is covered with almost perpetual snow and ice, is truly remarkable."

Taking the Miocene flora as a whole, Dr. Heer concludes from his study of about 3000 plants contained in the European Miocene alone, that the Miocene plants indicate tropical or sub-tropical conditions, but that there is a striking intermixture of forms which are at present found in countries widely removed from one another. It is impossible to state with certainty how many of the Miocene plants belong to existing species, but it appears that the larger number are extinct. According to Heer, the American types of plants are most largely represented in the Miocene flora, next those of Europe and Asia, next those of Africa, and lastly those of Australia. Upon the whole, however, the Miocene flora of Europe is mostly nearly allied to the plants which we now find inhabiting the warmer parts of the United States; and this has led to the suggestion that in Miocene times the Atlantic Ocean was dry land, and that a migration of American plants to Europe was thus permitted. This view is borne out by the fact that the Miocene plants of Europe are most nearly allied to the living plants of the eastern or Atlantic seaboard of the United States, and also by the occurrence of a rich Miocene flora in Greenland. As regards Greenland,

Dr. Heer has determined that the Miocene plants indicate a temperate climate in that country, with a mean annual temperature at least 30° warmer than it is at present.

The present limit of trees is the isothermal which gives the mean temperature of 50° Fahr. in July, or about the parallel of 67° N. latitude. In Miocene times, however, the Limes, Cypresses, and Plane-trees reach the 79th degree of latitude, and the Pines and Poplars must have ranged even further north than this.

The *Invertebrate Animals* of the Miocene period are very numerous, but they belong for the most part to existing types, and they can only receive scanty consideration here. The little shells of *Foraminifera* are extremely abundant in some beds, the genera being in many cases such as now flourish abundantly in our seas. The principal forms belong to the genera *Textularia* (fig. 237), *Robulina, Glandulina, Polystomella, Amphistegina,* &c. *Corals* are very abundant, in many instances forming regular "reefs;" but all the more important groups are in existence at the present day. The Red Coral (*Corallium*), so largely sought after as an ornamental material, appears for the first time in deposits of this age. Amongst the *Echinoderms,* we meet with Heart-Urchins (*Spatangus*), Cake-Urchins (*Scutella,* fig. 238), and various other forms, the majority of which are closely allied to forms now in existence.

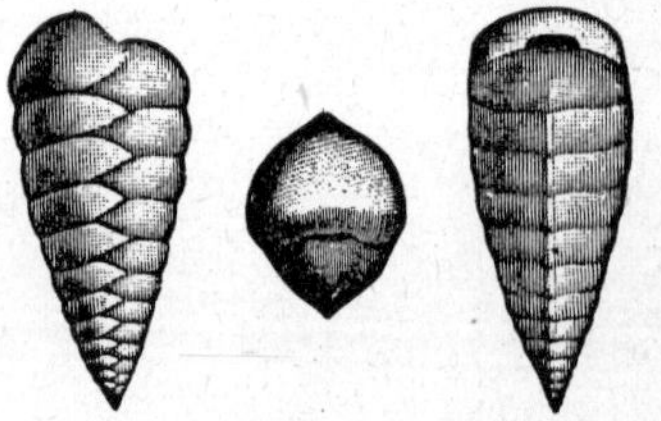

Fig. 237.—*Textularia Meyeriana*, greatly enlarged. Miocene Tertiary.

Numerous Crabs and Lobsters represent the *Crustacea;* but the most important of the Miocene Articulate Animals are the *Insects.* Of these, more than thirteen hundred species have been determined by Dr. Heer from the Miocene strata of Switzerland alone. They include almost all the existing orders of insects, such as numerous and varied forms of Beetles (*Coleoptera*), Forest-bugs (*Hemiptera*), Ants (*Hymenoptera*), Flies (*Diptera*), Termites and Dragon-flies (*Neuroptera*), Grasshoppers (*Orthoptera*), and Butterflies (*Lepidoptera*). One of the latter, the well-known *Vanessa Pluto* of the Brown Coals of Croatia, even exhibits the pattern of the wing, and to some extent its original coloration; whilst the more durablyconstructed insects are often in a state of exquisite preservation.

The *Mollusca* of the Miocene period are very numerous, but call for little special comment. Upon the whole, they are generically very similar to the Shell-fish of the present day; whilst, as before stated, from fifteen to thirty per cent of the

Fig. 238.—Different views of *Scutella subrotunda*, a Miocene "Cake-Urchin" from the south of France.

species are identical with those now in existence. So far as the European area is concerned, the Molluscs indicate a decidedly hotter climate than the present one, though they have not such a distinctly tropical character as is the case with the Eocene shells. Thus we meet with many Cones, Volutes, Cowries, Olive-shells, Fig-shells, and the like, which are decidedly indicative of a high temperature of the sea. *Polyzoans* are abundant, and often attain considerable dimensions; whilst *Brachiopods*, on the other hand, are few in number. *Bivalves* and *Univalves* are extremely plentiful; and we meet here with the shells of Winged-Snails (*Pteropods*), belonging to such existing genera as *Hyalea* (fig. 239) and *Cleodora*. Lastly, the *Cephalopods* are represented both by the chambered shells of *Nautili* and by the internal skeletons of Cuttle-fishes (*Spirulirostra.*)

Fig. 239.—Different views of the shell of *Hyalea Orbignyana*, a Miocene Pteropod.

The *Fishes* of the Miocene period are very abundant, but of little special importance. Besides the remains of Bony Fishes, we meet in the marine deposits of this age with numerous pointed teeth belonging to different kinds of Sharks. Some of the genera of these—such as *Carcharodon* (fig. 241), *Oxyrhina* (fig. 240), *Lamna*, and *Galeocerdo*—are very widely distributed, ranging through both the Old and New Worlds; and some of the species attain gigantic dimensions.

Amongst the *Amphibians* we meet with distinctly modern types, such as Frogs (*Rana*) and Newts or Salamanders. The most celebrated of the latter is the famous *Andrias Scheuchzeri* (fig. 242), discovered in the year 1725 in the fresh-water Miocene deposits of Œningen, in Switzerland. The skeleton indicates an animal nearly five feet in length; and it was originally described by Scheuchzer, a Swiss physician, in a dissertation published in 1731, as the remains of one of the human beings who were in existence at the time of the Noachian Deluge. Hence he applied to it the name of *Homo diluvii testis*. In reality, however, as shown by Cuvier, we have here the skeleton of a huge Newt, very closely allied to the Giant Salamander (*Menopoma maxima*) of Java.

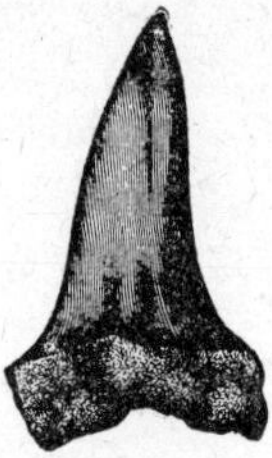

Fig. 240. — Tooth of *Oxyrhina xiphodon*. Miocene.

Fig, 241. — Tooth of *Carcharodon productus*. Miocene.

The remains of *Reptiles* are far from uncommon in the Miocene rocks, consisting principally of Chelonians and Crocodilians. The Land-tortoises (*Testudinidæ*) make their first appearance during this period. The most remarkable form of this group is the huge *Colossochelys Atlas* of the Upper Miocene deposits of the Siwâlik Hills in India, described by Dr. Falconer and Sir Proby Cautley. Far exceeding any living Tortoise in its dimensions, this enormous animal is estimated as having had a length of about twenty feet, measured from the tip of the snout to the extremity of the tail, and to have stood upwards of seven feet high. All the details of its organization, however, prove that it must have been "strictly a land animal, with herbivorous habits, and probably of the most inoffensive nature." The accomplished palæontologist just quoted, shows further that some of the traditions of the Hindoos would render it not improbable that this colossal Tortoise had survived into the earlier portion of the human period.

Of the *Birds* of the Miocene period it is sufficient to remark that though specifically distinct, they belong, so far as known, wholly to existing groups, and therefore present no points of special palæontological interest.

The *Mammals* of the Miocene are very numerous, and only

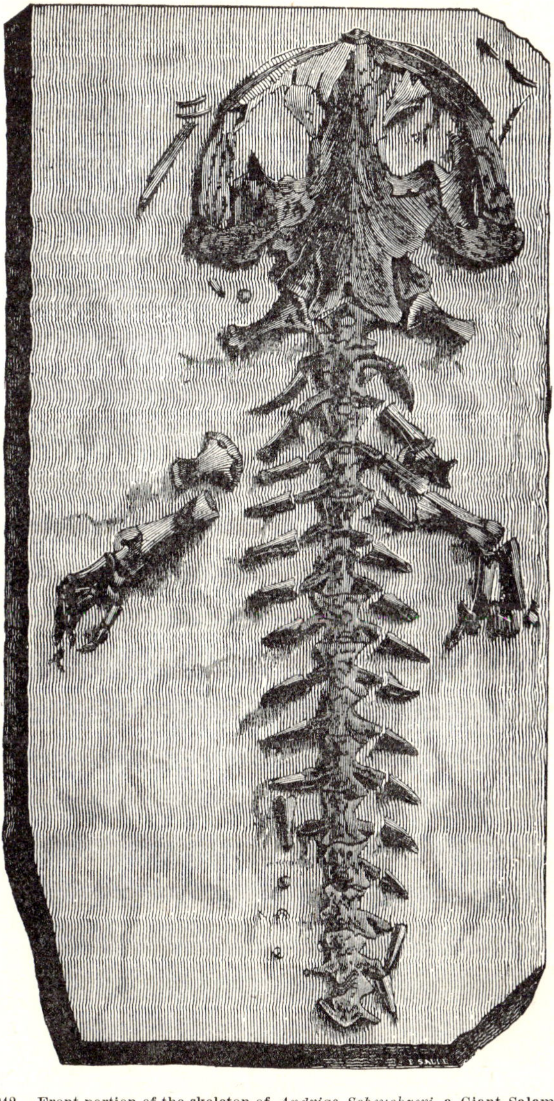

Fig. 242.—Front portion of the skeleton of *Andrias Scheuchzeri*, a Giant Salamander from the Miocene Tertiary of Œningen, in Switzerland. Reduced in size.

the more important forms can be here alluded to. Amongst the *Marsupials,* the Old World still continued to possess species of Opossum (*Didephys*), allied to the existing American forms. The *Edentates* (Sloths, Armadillos, and Ant-eaters), at the present day mainly South American, are represented by two large European forms. One of these is the large *Macrotherium giganteum* of the Upper Miocene of Gers in Southern France, which appears to have been in many respects allied to the existing Scaly Ant-eaters or Pangolins, at the same time that the disproportionately long fore-limbs would indicate that it possessed the climbing habits of the Sloths. The other is the still more gigantic *Ancylotherium Pentelici* of the Upper Miocene of Pikermé, which seems to have been as large as, or larger than, the Rhinoceros, and which must have been terrestrial in its habits. This conclusion is further borne out by the comparative equality of length which subsists between the fore and hind limbs, and is not affected by the curvature and crookedness of the claws, this latter feature being well marked in such existing terrestrial Edentates as the Great Ant-eater.

The aquatic *Sirenians* and *Cetaceans* are represented in Miocene times by various forms of no special importance. Amongst the former, the previously existing genus *Halitherium* continued to survive, and amongst the latter we meet with remains of Dolphins and of Whales of the "Zeuglodont" family. We may also note here the first appearance of true "Whalebone Whales," two species of which, resembling the living "Right Whale" of Arctic seas, and belonging to the same genus (*Balæna*), have been detected in the Miocene beds of North America.

The great order of the *Ungulates* or Hoofed Quadrupeds is very largely developed in strata of Miocene age, various new types of this group making their appearance here for the first time, whilst some of the characteristic genera of the preceding period are still represented under new shapes. Amongst the Odd-toed or "Perissodactyle" Ungulates, we meet for the first time with representatives of the family *Rhinocerida* comprising only the existing Rhinoceroses. In India in the Upper Miocene beds of the Sikâlik Hills, and in North America, several species of Rhinoceros have been detected, agreeing with the existing forms in possessing three toes to each foot, and in having one or two solid fibrous "horns" carried upon the front of the head. On the other hand, the forms of this group which distinguish the Miocene deposits of Europe appear to have been for the most part hornless, and to have resembled the Tapirs in having three-toed hind-feet, but four-toed fore-feet.

The family of the Tapirs is represented, both in the Old and New Worlds, by species of the genus *Lophiodon*, some of which were quite diminutive in point of size, whilst others attained the dimensions of a horse. Nearly allied to this family, also, is the singular group of quadrupeds which Marsh has described from the Miocene strata of the United States under the name of *Brontotheridæ*. These extraordinary animals, typified by *Brontotherium* (fig. 243) itself, agree with the

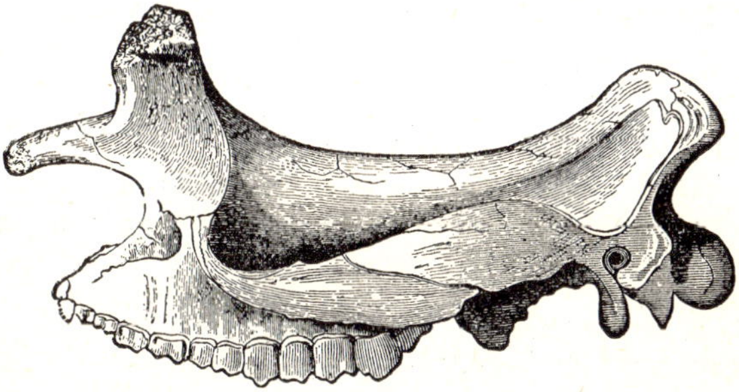

Fig. 243.—Skull of *Brontotherium ingens*. Miocene Tertiary, United States. (After Marsh.)

existing Tapirs of South America and the Indian Archipelago in having the fore-feet four-toed, whilst the hind-feet are three-toed; and a further point of resemblance is found in the fact (as shown by the form of the nasal bones) that the nose was long and flexible, forming a short movable proboscis or trunk, by means of which the animal was enabled to browse on shrubs or trees. They differ, however, from the Tapirs, not only in the apparent presence of a long tail, but also in the possession of a pair of very large "horn-cores," carried upon the nasal bones, indicating that the animal possessed horns of a similar structure to those of the "Hollow-horned" Ruminants (*e. g.*, Sheep and Oxen). *Brontotherium gigas* is said to be nearly as large as an Elephant, whilst *B. ingens* appears to have attained dimensions still more gigantic. The well-known genus *Titanotherium* of the American Miocene would also appear to belong to this group.

The family of the Horses (*Equidæ*) appears under various forms in the Miocene, but the most important and best known of these is *Hipparion.* In this genus the general conformation of the skeleton is extremely similar to that of the existing Horses, and the external appearance of the animal must have been very much the same. The foot of *Hipparion,* however, as has been previously mentioned, differed from that of the Horse in the fact that whilst both possess the middle toe greatly developed and enclosed in a broad hoof, the former, in addition, possessed two lateral toes, which were sufficiently developed to carry hoofs, but were so far rudimentary that they hung idly by the side of the central toe without touching the ground (see fig. 230). In the Horse, on the other hand, these lateral toes, though present, are not only functionally useless, but are concealed beneath the skin. Remains of the *Hipparion* have been found in various regions in Europe and in India; and from the immense quantities of their bones found in certain localities, it may be safely inferred that these Middle Tertiary ancestors of the Horses lived, like their modern representatives, in great herds, and in open grassy plains or prairies.

Amongst the Even-toed or *Artiodactyle* Ungulates, we for the first time meet with examples of the *Hippopotamus,* with its four-toed feet, its massive body, and huge tusk-like lower canine teeth. The Miocene deposits of Europe have not hitherto yielded any remains of *Hippopotamus;* but several species have been detected in the Upper Miocene of the Siwâlik Hills by Dr. Falconer and Sir Proby Cautley. These ancient Indian forms, however, differ from the existing *Hippopotamus amphibius* of Africa in the fact that they possessed six incisor teeth in each jaw (fig. 244), whereas the latter has only four.

Amongst the other Even-toed Ungulates, the family of the Pigs (*Suida*) is represented by true Swine (*Sus Erymanthius*), Peccaries (*Dicotyles antiquus*), and by forms which, like the great *Elotherium* of the American Miocene, have no representative at the present day. The Upper Miocene of India has yielded examples of the Camels. Small Musk-deer (*Amphitragulus* and *Dremotherium*) are known to have existed in France and Greece; and the true Deer (*Cervidæ*), with their solid bony antlers, appear for the first time here in the person of species allied to the living Stags (*Cervus*), accompanied by the extinct genus *Dorcatherium.* The Giraffes (*Camelopardalidæ*), now confined to Africa, are known to have lived in India and Greece; and the allied *Helladotherium,* in some respects intermediate between the Giraffes and the Antelopes, ranged over Southern Europe from Attica to France. The great group of the "Hollow-horned" Ruminants (*Cavicornia*), lastly, came into existence in the Miocene period; and though the typical families of the Sheep and Oxen are apparently wanting, there are true Antelopes, together with forms which, if systematic-

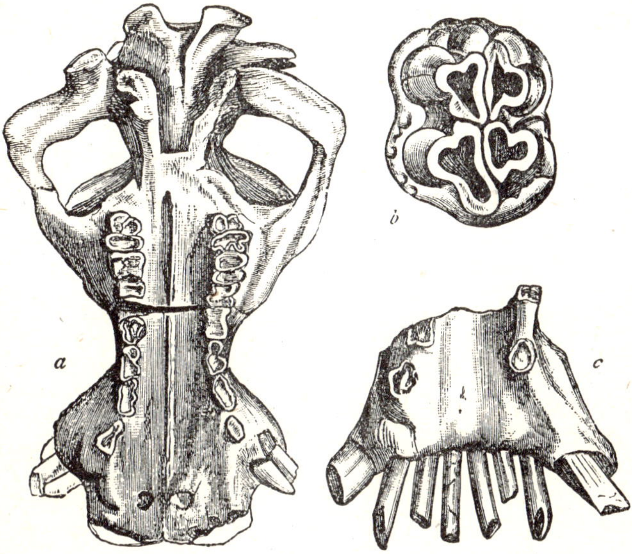

Fig. 244.—*a*, Skull of *Hippopotamus Sivalensis*, viewed from below, one-eighth of the natural size; *b*, Molar tooth of the same, showing the surface of the crown, one-half of the natural size; *c*, Front of the lower jaw of the same, showing the six incisors and the tusk-like canines, one-eighth of the natural size. Upper Miocene, Siwâlik Hills. (After Falconer and Cautley.)

ally referable to the *Antilopidæ*, nevertheless are more or less clearly transitional between this and the family of the Sheep and Goats. Thus the *Palæoreas* of the Upper Miocene of Greece may be regarded as a genuine Antelope; but the *Tragoceras* of the same deposit is intermediate in its characters between the typical Antelopes and the Goats. Perhaps the most remarkable, however, of these Miocene Ruminants is the *Sivatherium giganteum* (fig. 245) of the Siwâlik Hills, in India. In this extraordinary animal there were two pairs of horns, supported by bony "horn-cores," so that there can be no hesitation in referring *Sivatherium* to the Cavicorn Ruminants. If all these horns had been simple, there would have been no difficulty in considering *Sivatherium* as simply a gigantic four-horned Antelope, essentially similar to the living *Antilope* (*Tetraceros*) *quadricornis* of India. The hinder pair of horns, however, is not only much larger than the front pair, but each possesses two branches or snags—a peculiarity not to be paralleled amongst any existing Antelope, save the abnormal Prongbuck (*Antilocapra*) of North America. Dr. Murie, however, in an admirable memoir on the structure and relationships of *Sivatherium*, has drawn attention to the fact that the Prong-

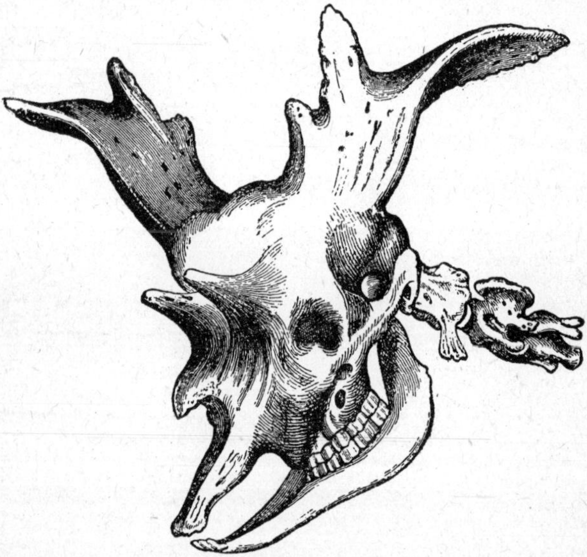

Fig. 245.—Skull of *Sivatherium giganteum*, reduced in size. Miocene, India. (After Murie.)

buck sheds the *sheath* of its horns annually, and has suggested that this may also have been the case with the extinct form. This conjecture is rendered probable, amongst other reasons, by the fact that no traces of a horny sheath surrounding the horn-cores of the Indian fossil have been as yet detected. Upon the whole, therefore, we may regard the elephantine *Sivatherium* as being most nearly allied to the Prongbuck of Western America, and thus as belonging to the family of the Antelopes.

It is to the Miocene period, again, to which we must refer the first appearance of the important order of the Elephants and their allies (*Proboscideans*), all of which are characterized by their elongated trunk-like noses, the possession of five toes to the foot, the absence of canine teeth, the development of two or more of the incisor teeth into long tusks, and the adaptation of the molar teeth to a vegetable diet. Only three generic groups of this order are known—namely, the extinct *Deinotherium,* the equally extinct *Mastodons,* and the *Elephants;* and all these three types are known to have been in existence as early as the Miocene period, the first of them being exclusively confined to deposits of this age. Of the three, the genus *Deinotherium* is much the most abnormal in its characters; so much so, that good authorities regard it as being one of the Sea-cows (*Sirenia*)—though this view has been rendered untenable by the discovery of limb-bones which can hardly belong to any other animal, and which are distinctly Proboscidean in type. The most celebrated skull of the Deinothere (fig. 246) is one which was exhumed from the Upper Miocene deposits of Epplesheim, in Hesse-Darmstadt, in the year 1836. This skull was four and a half feet in length, and indicated an animal larger than any existing species of Elephant. The upper jaw is destitute of incisor or canine teeth, but is furnished on each side with five molars, which are opposed to a corresponding series of grinding teeth in the lower jaw. No canines are present in the lower jaw; but the front portion of the jaw is abruptly bent downwards, and carries two huge tusk-like incisor teeth, which are curved downwards and backwards, and the use of which is rather problematical. Not only does the Deinothere occur in Europe, but remains belonging to this genus have also been detected in the Siwâlik Hills, in India.

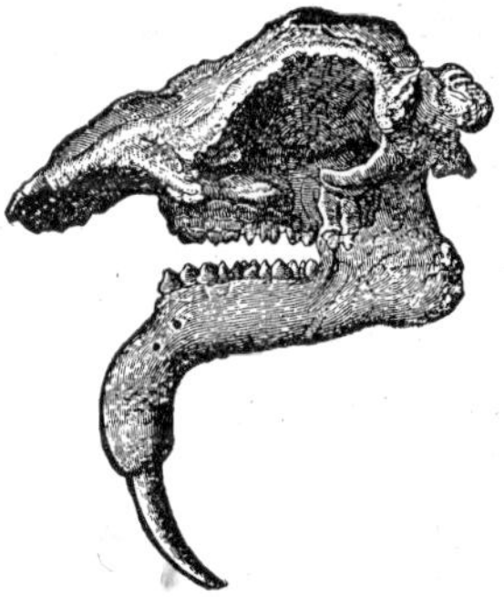

Fig. 246.—Skull of *Deinotherium giganteum*, greatly reduced. From the Upper Miocene of Germany.

The true Elephants (*Elephas*) do not appear to have existed during the Miocene period in Europe, but several species have been detected in the Upper Miocene deposits of the

Siwâlik Hills, in India. The fossil forms, though in all cases specifically, and in some cases even sub-generically, distinct, agree with those now in existence in the general conformation of their skeleton, and in the principal characters of their dentition. In all, the canine teeth are wanting in both jaws; and there are no incisor teeth in the lower jaw, whilst there are two incisors in the front of the upper jaw, which are developed into two huge "tusks." There are six molar teeth on each side of both the upper and lower jaw, but only

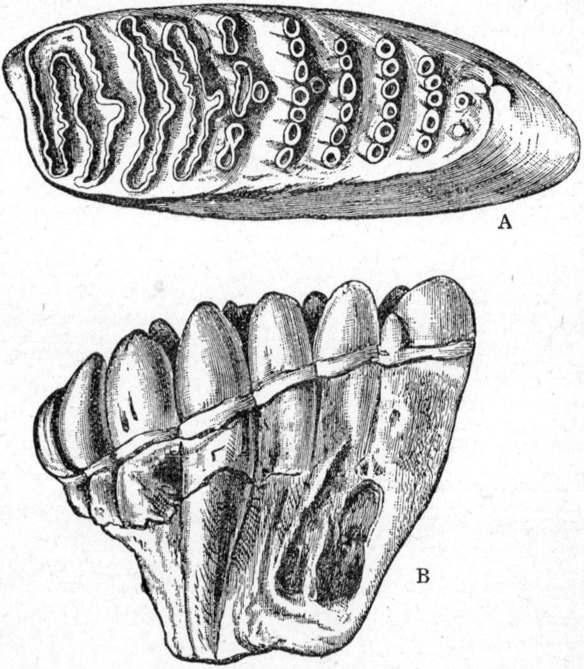

Fig. 247.—A, Molar tooth of *Elephas planifrons*, one-third of the natural size, showing the grinding surface—from the Upper Miocene of India; B, Profile view of the last upper molar of *Mastodon Sivalensis*, one-third of the natural size—from the Upper Miocene of India. (After Falconer.)

one, or at most a part of two, is in actual use at any given time; and as this becomes worn away, it is pushed forward and replaced by its successor behind it. The molars are of very large size, and are each composed of a number of transverse plates of enamel united together by ivory; and by the process of mastication, the teeth become worn down to a flat surface, crossed by the enamel-ridges in varying patterns. These patterns are different in the different species of Elephants, though constant for each; and they constitute one of the most readily available means of separating the fossil forms from one another. Of the seven Miocene Elephants of India, as judged by the characters of the molar teeth, two are allied to the existing Indian Elephant, one is related to the living African Elephant, and the remaining four are in some respects intermediate between the true Elephants and the Mastodons.

The *Mastodons,* lastly, though quite elephantine in their general characters, possess molar teeth which have their crowns furnished with conical eminences or tubercles placed in pairs (fig. 247, B), instead of having the approximately flat surface characteristic of the grinders of the Elephants. As in the latter, there are two upper incisor teeth, which grow permanently during the life of the animal, and which constitute great tusks; but the Mastodons, in addition, often possess two lower incisors, which in some cases likewise grow into small tusks. Three species of *Mastodon* are known to occur in the Upper Miocene of the Siwâlik Hills of India; and the Miocene deposits of the European area have yielded the remains of four species, of which the best known are the *M. longirostris* and the *M. angustidens.*

Whilst herbivorous Quadrupeds, as we have seen, were extremely abundant during Miocene times, and often attained gigantic dimensions, Beasts of Prey (*Carnivora*) were by no means wanting, most of the principal existing families of the order being represented in deposits of this age. Thus, we find aquatic Carnivores belonging to both the living groups of the Seals and Walruses; true Bears are wanting, but their place is filled by the closely-allied genus *Amphicyon,* of which various species are known; Weasels and Otters were not unknown, and the *Hyænictis* and *Ictitherium* of the Upper Miocene of Greece are apparently intermediate between the Civet-cats and the Hyænas; whilst the great Cats of subsequent periods are more than adequately represented by the huge "Sabre-toothed Tiger" (*Machairodus*), with its immense trenchant and serrated canine teeth.

Amongst the *Rodent* Mammals, the Miocene rocks have yielded remains of Rabbits, Porcupines (such as the *Hystrix primigenius* of Greece), Beavers, Mice, Jerboas, Squirrels, and Marmots. All the principal living groups of this order were therefore differentiated in Middle Tertiary times.

The *Cheiroptera* are represented by small insect-eating Bats; and the order of the Insectivorous Mammals is represented by Moles, Shrew-mice, and Hedgehogs.

Lastly, the Monkeys (*Quadrumana*) appear to have existed during the Miocene period under a variety of forms, remains of these animals having been found both in Europe and in India; but no member of this order has as yet been detected in the Miocene Tertiary of the North American continent. Amongst the Old World Monkeys of the Miocene, the two most interesting are the *Pliopithecus* and *Dryopithecus* of France.

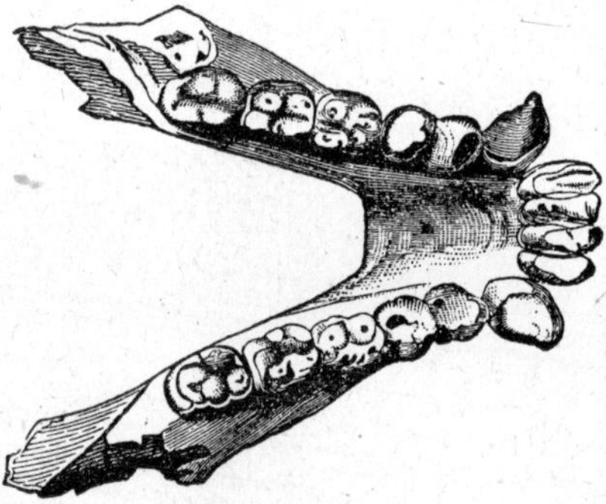

Fig. 248.—Lower jaw of *Pliopithecus antiquus*. Upper Miocene, France.

The former of these (fig. 248) is supposed to have been most nearly related to the living *Semnopitheci* of Southern Asia, in which case it must have possessed a long tail. The *Mesopithecus* of the Upper Miocene of Greece is also one of the lower Monkeys, as it is most closely allied to the existing Macaques. On the other hand, the *Dryopithecus* of the French Upper Miocene is referable to the group of the "Anthropoid Apes," and is most nearly related to the Gibbons of the present day, in which the tail is rudimentary and there are no cheek-pouches. *Dryopithecus* was, also, of large size, equalling Man in stature, and apparently living amongst the trees and feeding upon fruits.

CHAPTER XX.

THE PLIOCENE PERIOD.

The highest division of the Tertiary deposits is termed the *Pliocene* formation, in accordance with the classification proposed by Sir Charles Lyell. The Pliocene formations contain from 45 to 95 per cent of existing species of *Mollusca,* the remainder belonging to extinct species. They are divided by Sir Charles Lyell into two divisions, the Older Pliocene and Newer Pliocene.

The Pliocene deposits of Britain occur in Suffolk, and are known by the name of "Crags," this being a local term used for certain shelly sands, which are employed in agriculture. Two of these Crags are referable to the Older Pliocene, viz., the White and Red Crags,—and one belongs to the Newer Pliocene, viz., the Norwich Crag.

The *White* or *Coralline Crag* of Suffolk is the oldest of the Pliocene deposits of Britain, and is an exceedingly local formation occurring in but a single small area, and having a maximum thickness of not more than 50 feet. It consists of soft sands, with occasional intercalations of flaggy limestone. Though of small extent and thickness, the Coralline Crag is of importance from the number of fossils which it contains. The name "Coralline" is a misnomer; since there are few true Corals, and the so-called "Corals" of the formation are really *Polyzoa,* often of very singular forms. The shells of the Coralline Crag are mostly such as inhabit the seas of temperate regions; but there occur some forms usually looked upon as indicating a warm climate.

The *Upper* or *Red Crag* of Suffolk—like the Coralline Crag —has a limited geographical extent and a small thickness, rarely exceeding 40 feet. It consists of quartzose sands, usually deep red or brown in color, and charged with numerous fossils.

Altogether more than 200 species of shells are known from the Red Crag, of which 60 per cent are referable to existing species. The shells indicate, upon the whole, a temperate or even cold climate, decidedly less warm than that indicated by the organic remains of the Coralline Crag. It appears, therefore, that a gradual refrigeration was going on during the Pliocene period, commencing in the Coralline Crag, becoming intensified in the Red Crag, being still more severe in the Norwich Crag, and finally culminating in the Arctic cold of the Glacial period.

Besides the *Mollusca,* the Red Crag contains the ear-bones of Whales, the teeth of Sharks and Rays, and remains of the Mastodon, Rhinoceros, and Tapir.

The *Newer Pliocene* deposits are represented in Britain by the *Norwich Crag,* a local formation occurring near Norwich. It consists of incoherent sands, loams, and gravels, resting in detached patches, from 2 to 20 feet in thickness, upon an eroded surface of Chalk. The Norwich Crag contains a mixture of marine, land, and fresh-water shells, with remains of fishes and bones of mammals; so that it must have been deposited at a local sea-deposit near the mouth of an ancient river. It contains altogether more than 100 marine shells, of which 89 per cent belong to existing species. Of the Mammals, the two most important are an Elephant (*Elephas meridionalis*), and the characteristic Pliocene Mastodon (*M. Arvernensis*), which is hitherto the only Mastodon found in Britain.

According to the most recent views of high authorities, certain deposits—such as the so-called "Bridlington Crag" of Yorkshire, and the "Chillesford beds" of Suffolk—are to be also included in the Newer Pliocene, upon the ground that they contain a small proportion of extinct shells. Our knowledge, however, of the existing Molluscan fauna, is still so far incomplete, that it may reasonably be doubted if these supposed extinct forms have actually made their final disappearance, whilst the strata in question have a strong natural connection with the "Glacial deposits," as shown by the number of Arctic Mollusca which they contain. Here, therefore, these beds will be included in the Post-Pliocene series, in spite of the fact that some of their species of shells are not known to exist at the present day.

The following are the more important Pliocene deposits which have been hitherto recognized out of Britain:—

1. In the neighborhood of Antwerp occur certain "crags,"

which are the equivalent of the White and Red Crag in part. The lowest of these contains less than 50 per cent, and the highest 60 per cent, of existing species of shells, the remainder being extinct.

2. Bordering the chain of the Apennines, in Italy, on both sides is a series of low hills made up of Tertiary strata, which are known as the Sub-Apennine beds. Part of these is of Miocene age, part is Older Pliocene, and a portion is Newer Pliocene. The Older Pliocene portion of the Sub-Apennines consists of blue or brown marls, which sometimes attain a thickness of 2000 feet.

3. In the valley of the Arno, above Florence, are both Older and Newer Pliocene strata. The former consist of blue clays and lignites, with an abundance of plants. The latter consist of sands and conglomerates, with remains of large Carnivorous Mammals, Mastodon, Elephant, Rhinoceros, Hippopotamus, &c.

4. In Sicily, Newer Pliocene strata are probably more largely developed than anywhere else in the world, rising somtimes to a height of 3000 feet above the sea. The series consists of clays, marls, sands, and conglomerates, capped by a compact limestone, which attains a thickness of from 700 to 800 feet. The fossils of these beds belong almost entirely to living species, one of the commonest being the Great Scallop of the Mediterranean (*Pecten Jacobœus*).

5. Occupying an extensive area round the Caspian, Aral, and Azof Seas, are Pliocene deposits known as the "Aralo-Caspian" beds. The fossils in these beds are partly fresh-water, partly marine, and partly intermediate in character, and they are in great part identical with species now inhabiting the Caspian. The entire formation appears to indicate the former existence of a great sheet of brackish water, forming an inland sea, like the Caspian, but as large as, or larger than, the Mediterranean.

6. In the United States, strata of Pliocene age are found in North and South Carolina. They consist of sands and clays, with numerous fossils, chiefly *Molluscs* and *Echinoderms*. From 40 to 60 per cent of the fossils belong to existing species. On the Loup Fork of the river Platte, in the Upper Missouri region, are strata which are also believed to be referable to the Pliocene period, and probably to its upper division. They are from 300 to 400 feet thick, and contain land-shells,

with the bones of numerous Mammals, such as Camels, Rhinoceroses, Mastodons, Elephants, the Horse, Stag, &c.

As regards the *life* of the Pliocene period, there are only two classes of organisms to which our attention need be directed—namely, the Shell-fish and the Mammals. So far as the former are concerned, we have to note in the first place that the introduction of new species of animals upon the globe went on rapidly during this period. In the Older Pliocene deposits, the number of shells of existing species is only from 40 to 60 per cent; but in the Newer Pliocene the proportion of living forms rises to as much as from 80 to 95 per cent. Whilst the Molluscs thus become rapidly modernized, the Mammals still all belong to extinct species, though modern generic types gradually supersede the more antiquated forms of the Miocene. In the second place, there is good evidence to show that the Pliocene period was one in which the climate of the northern hemisphere underwent a gradual refrigeration. In the Miocene period, there is evidence to show that Europe possessed a climate very similar to that now enjoyed by the Southern United States, and certainly very much warmer than it is at present. The presence of Palm-trees upon the land, and of numerous large Cowries, Cones, and other shells of warmer regions in the sea, sufficiently proves this. In the Older Pliocene deposits, on the other hand, northern forms predominate amongst the Shells, though some of the types of hotter regions still survive. In the Newer Pliocene, again, the Molluscs are such as almost exclusively inhabit the seas of temperate or even cold regions; whilst if we regard deposits like the "Bridlington Crag" and "Chillesford beds" as truly referable to this period, we meet at the close of this period with shells such as nowadays are distinctively characteristic of high latitudes. It might be thought that the occurrence of Quadrupeds such as the Elephant, Rhinoceros, and Hippopotamus, would militate against this generalization, and would rather support the view that the climate of Europe and the United States must have been a hot one during the later portion of the Pliocene period. We have, however, reason to believe that many of these extinct Mammals were more abundantly furnished with hair, and more adapted to withstand a cool temperature, than any of their living congeners. We have also to recollect that many of these large herbivorous quadrupeds may have been, and indeed probably were, more or less migratory in their habits;

and that whilst the winters of the later portion of the Pliocene period were cold, the summers might have been very hot. This would allow of a northward migration of such terrestrial animals during the summer-time, when there would be an ample supply of food and a suitably high temperature, and a southward recession towards the approach of winter.

The chief palæontological interests of the Pliocene deposits, as of the succeeding Post-Pliocene, center round the Mammals of the period; and amongst the many forms of these we may restrict our attention to the orders of the Hoofed Quadrupeds (*Ungulates*), the *Proboscideans*, the *Carnivora*, and the *Quadrumana*. Almost all the other Mammalian orders are more

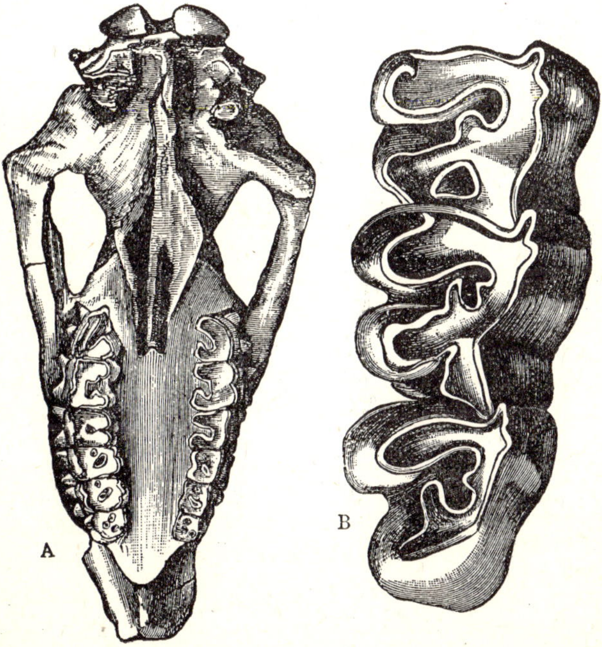

Fig. 249.—A, Under surface of the skull of *Rhinoceros Etruscus*, one-seventh of the natural size—Pliocene, Italy ; B. Crowns of the three true molars of the upper jaw, left side, of *Rhinoceros megarhinus* (*R. leptorhinus*, Falconer), one-half of the natural size—Pliocene, France. (After Falconer.)

or less fully represented in Pliocene times, but none of them attains any special interest till we enter upon the Post-Pliocene.

Amongst the Odd-toed Ungulates, in addition to the remains of true Tapirs (*Tapirus Arvernensis*), we meet with the bones of several species of Rhinoceros, of which the *Rhinoceros Etruscus* and *R. megarhinus* (fig. 249) are the most important. The former of these (fig. 249, A) derives its specific name from its abundance in the Pliocene deposits of the Val d'Arno, near Florence, and though principally Pliocene in its distribution, it survived into the earlier portion of the Post-Pliocene period. *Rhinoceros Etruscus* agreed with the existing African forms in having two horns placed one behind the other, the front one being the longest; but it was comparatively slight and slender in its build, whilst the nostrils were separated by an incomplete bony partition. In the *Rhinoceros megarhinus* (fig. 249, B), on the other hand, no such partition exists between the nostrils, and the nasal bones are greatly developed in size. It was a two-horned form, and is found associated with *Elephas meridionalis* and *E. antiquus* in the Pliocene deposits of the Val d'Arno, near, Florence. Like the preceding, it survived, in diminished numbers, into the earlier portion of the Post-Pliocene period.

The Horses (*Equidæ*) are represented, both in Europe and America, by the three-toed Hipparions, which survive from the Miocene, but are now verging upon extinction. For the first time, also, we meet with genuine Horses (*Equus*), in which each foot is provided with a single complete toe only, encased in a single broad hoof. One of the American species of this period (the *Equus excelsus*) quite equalled the modern Horse in stature; and it is interesting to note the occurrence of indigenous horses in America at such a comparatively late geological epoch, seeing that this continent certainly possessed none of these animals when first discovered by the Spaniards.

Amongst the Even-toed Ungulates, we may note the occurrence of Swine (*Suida*), of forms allied to the Camels (*Camelidæ*), and of various kinds of Deer (*Cervidæ*); but the most interesting Pliocene Mammals belonging to this section is the great *Hippopotamus major* of Britain and Europe. This well-known species is very closely allied to the living *Hippopotamus amphibius* of Africa, from which it is separated only by its larger dimensions, and by certain points connected with the conformation of the skeleton. It is found very abundantly in the Pliocene deposits of Italy and France, associated with the remains of the Elephant, Mastodon, and Rhinoceros, and it survived into the earlier portion of the Post-Pliocene period.

During this last-mentioned period, it extended its range northwards, and is found associated with the Reindeer, the Bison, and other northern animals. From this fact it has been inferred, with great probability, that the *Hippopotamus major* was furnished with a long coat of hair and fur, thus differing from its nearly hairless modern representative, and resembling its associates, the Mammoth and the Woolly Rhinoceros.

Passing on to the Pliocene Proboscideans, we find that the great *Deinotheria* of the Miocene have now wholly disappeared, and the sole representatives of the order are Mastodons and Elephants. The most important member of the former group is the *Mastodon Arvernensis* (fig. 250), which ranged widely over Southern Europe and England, being generally associated with remains of the *Elephas meridionalis, E. antiquus, Rhinoceros megarhinus,* and *Hippopotamus major.* The lower jaw

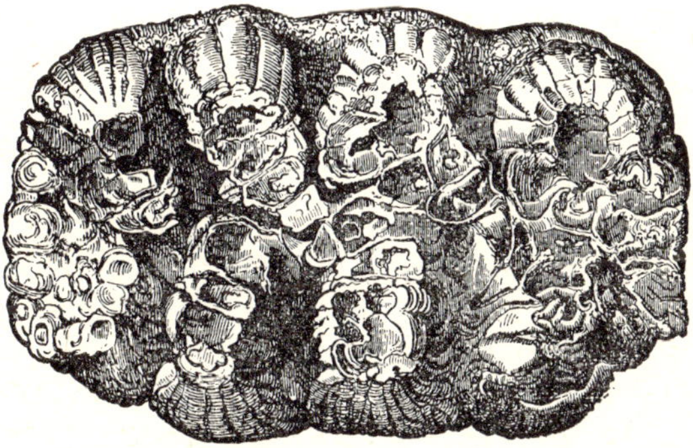

Fig. 250.—Third milk-molar of the left side of the upper jaw of *Mastodon Arvernensis*, showing the grinding surface. Pliocene.

seems to have been destitute of incisor teeth; but the upper incisors are developed into great tusks, which sometimes reach a length of nine feet, and which have the simple curvature of the tusks of the existing Elephants. Amongst the Pliocene Elephants the two most important are the *Elephas meridionalis* and the *Elephas antiquus.* Of these, the *Elephas meridionalis* (fig. 251) is found abundantly in the Pliocene deposits of Southern Europe and England, and also survived into the earlier portion of the Post-Pliocene period. Its molar teeth are of the type of those of the existing African Elephant, the spaces enclosed by the transverse enamel-plates being more or less lozenge-shaped, whilst the curvature of the tusks is simple. The *Elephas antiquus* (fig. 252) is very generally associated with the preceding, and it survived to an even later stage of the Post-Pliocene period. The molar teeth are of the type of the existing Indian Elephant, with comparatively thin enamel-ridges, placed closer together than in the African type; whilst the tusks were nearly straight.

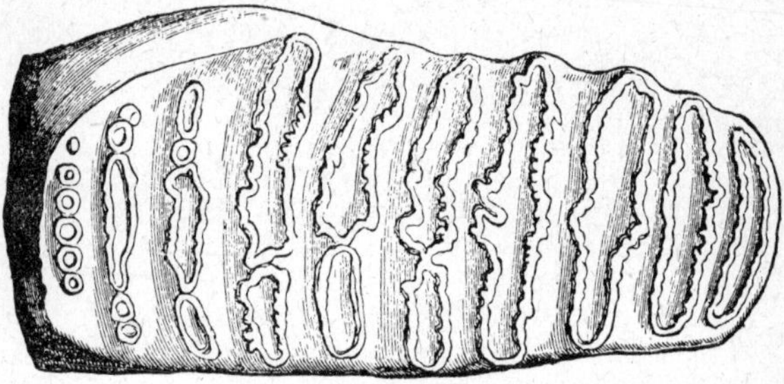

Fig. 251.—Molar tooth of *Elephas meridionalis*, one-third of the natural size. Pliocene and Post-Pliocene.

Amongst the Pliocene *Carnivores,* we meet with true Bears (*Ursus Arvernensis*), Hyænas (such as *Hyæna Hipparionum*), and genuine Lions (such as the *Felis augustus* of North America); but the most remarkable of the beasts of prey of this period is the great "Sabre-toothed Tiger" (*Machairodus*),

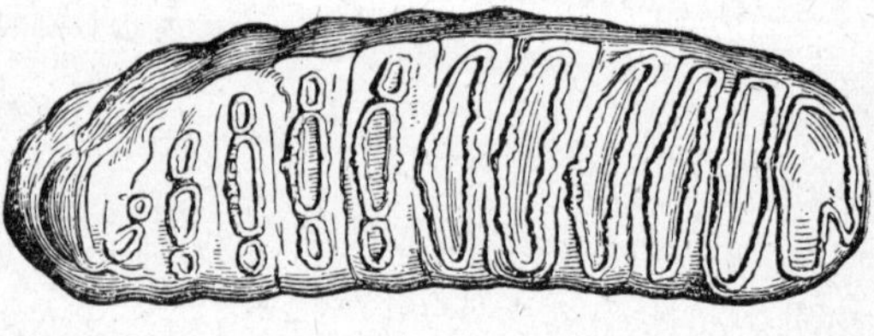

Fig. 252.—Molar tooth of *Elephas antiquus*, one-third of the natural size. Pliocene and Post-Pliocene.

species of which existed in the earlier Miocene, and survived to the later Post-Pliocene. In this remarkable form we are presented with perhaps the most highly carnivorous type of all known beasts of prey. Not only are the jaws shorter in proportion even than those of the great Cats of the present day, but the canine teeth (fig. 253) are of enormous size, greatly flattened so as to assume the form of a poignard, and

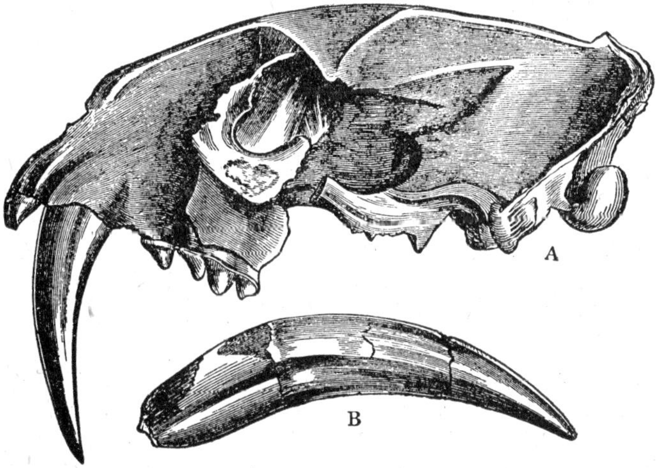

Fig. 253.—A, Skull of *Machairodus cultridens*, without the lower jaw, reduced in size ; B, Canine tooth of the same, one-half the natural size. Pliocene, France.

having their margins finely serrated. Apart from the characters of the skull, the remainder of the skeleton, so far as known, exhibits proofs that the Sabre-toothed Tiger was extraordinarily muscular and powerful, and in the highest degree adapted for a life of rapine. Species of *Machairodus* must have been as large as the existing Lion; and the genus is not only European, but is represented both in South America and in India, so that the geographical range of these predaceous beasts must have been very extensive.

Lastly, we may note that Pliocene deposits of Europe have yielded the remains of Monkeys (*Quadrumana*), allied to the existing *Semnopitheci* and Macaques.

Literature.

The following list comprises a small selection of some of the more important and readily accessible works and memoirs relating to the Tertiary rocks and their fossils. With few exceptions, foreign works relating to the Tertiary strata of the continent of Europe or their organic remains have been omitted:—

(1) 'Elements of Geology.' Lyell.
(2) 'Students' Elements of Geology.' Lyell.
(3) 'Manual of Palæontology.' Owen.
(4) 'British Fossil Mammals and Birds.' Owen.
(5) 'Traité de Paléontologie.' Pictet.
(6) 'Cours Elémentaire de Paléontologie.' D'Orbigny.
(7) "Probable Age of the London Clay," &c.—'Quart. Journ. Geol. Soc.,' vol. iii. Prestwich.
(8) 'Structure and Probable Age of the Bagshot Sands'—Ibid., vol. iii. Prestwich.
(9) 'Tertiary Formations of the Isle of Wight'—Ibid., vol. ii. Prestwich.
(10) 'Structure of the Strata between the London Clay and the Chalk,' &c.—Ibid., vols. vi., viii., and x. Prestwich.
(11) 'Correlation of the Eocene Tertiaries of England, France, and Belgium'—Ibid., vol. xxvii. Prestwich.
(12) 'On the Fluvio-marine Formations of the Isle of Wight'—Ibid., vol. ix. Edward Forbes.
(13) 'Newer Tertiary Deposits of the Sussex Coast'—Ibid., vol. xiii. Godwin-Austen.
(14) 'Kainozoic Formations of Belgium'—Ibid., vol. xxii. Godwin-Austen.
(15) 'Tertiary Strata of Belgium and French Flanders'—Ibid., vol. viii. Lyell.
(16) 'On Tertiary Leaf-beds in the Isle of Mull'—Ibid., vol. vii. The Duke of Argyll.
(17) 'Newer Tertiaries of Suffolk and their Fauna'—Ibid., vol. xxvi. Ray Lankester.
(18) 'Lower London Tertiaries of Kent'—Ibid., vol. xxii. Whitaker.
(19) "Guide to the Geology of London"—'Mem. Geol. Survey.' Whitaker.
(20) 'Memoirs of the Geological Survey of Great Britain.'
(21) 'Introductory Outline of the Geology of the Crag District' (Supplement to Crag Mollusca, Palæontographical Society). S. V. Wood, jun., and F. W. Harmer.
(22) "Tertiary Fluvio-marine Deposits of the Isle of Wight." Edward Forbes. Edited by Godwin-Austen; with Descriptions of the Fossils by Morris, Salter, and Rupert Jones—'Memoirs of the Geological Survey.'
(23) 'Geological Excursions round the Isle of Wight.' Mantell.
(24) 'Catalogue of British Fossils.' Morris.

(25) 'Catalogue of Fossils in the Museum of Practical Geology.' Etheridge.
(26) 'Monograph of the Crag Polyzoa' (Palæontographical Society). Busk.
(27) 'Monograph of the Tertiary Brachiopoda' (Ibid.) Davidson.
(28) 'Monograph of the Tertiary Malacostracous Crustacea' (Ibid.) Bell.
(29) 'Monograph of the Tertiary Corals' (Ibid.) Milne-Edwards and Haime.
(30) 'Supplement to the Tertiary Corals' (Ibid.) Martin Duncan.
(31) 'Monograph of the Eocene Mollusca' (Ibid.) Fred. E. Edwards.
(32) 'Monograph of the Eocene Mollusca' (Ibid.) Searles V. Wood.
(33) 'Monograph of the Crag Mollusca' (Ibid.) Searles V. Wood.
(34) 'Monograph of the Tertiary Entomostraca' (Ibid.) Rupert Jones.
(35) 'Monograph of the Foraminifera of the Crag' (Ibid.) Rupert Jones, Parker, and H. B. Brady.
(36) 'Monograph of the Radiaria of the London Clay' (Ibid.) Edward Forbes.
(37) 'Monograph of the Cetacea of the Red Crag' (Ibid.) Owen.
(38) 'Monograph of the Fossil Reptiles of the London Clay' (Ibid.) Owen and Bell.
(39) "On the Skull of a Dentigerous Bird from the London Clay of Sheppey"—'Quart. Journ. Geol. Soc.,' vol. xxix. Owen.
(40) 'Ossemens Fossiles.' Cuvier.
(41) 'Fauna Antiqua Sivalensis.' Falconer and Sir Proby Cautley.
(42) 'Palæontological Memoirs.' Falconer.
(43) 'Animaux Fossiles et Géologie de l'Attique.' Gaudry.
(44) "Principal Characters of the Dinocerata"—'American Journ. of Science and Arts,' vol. xi. Marsh.
(45) 'Principal Characters of the Brontotheridæ' (Ibid.) Marsh.
(46) 'Principal Characters of the Tillodontia' (Ibid.) Marsh.
(47) "Extinct Vertebrata of the Eocene of Wyoming"—'Geological Survey of Montana,' &c., 1872. Cope.
(48) "Ancient Fauna of Nebraska"—'Smithsonian Contributions to Knowledge,' vol. vi. Leidy.
(49) 'Manual of Geology.' Dana.
(50) "Palæontology and Evolution" (Presidential Address to the Geological Society of London, 1870)—'Quart. Journ. Geol. Soc.,' vol. xxvi. Huxley.
(51) 'Mineral Conchology.' Sowerby.
(52) 'Description des Coquilles Fossiles,' &c. Deshayes.
(53) 'Description des Coquilles Tertiaries de Belgique.'
(54) 'Fossilen Polypen des Wiener Tertiär-beckens.' Reuss.

(55) 'Palæontologische Studien über die älteren Tertiärschichten der Alpen.' Reuss.
(56) 'Land- und Süss-wasser Conchylien der Vorwelt.' Sandberger.
(57) 'Flora Tertiaria Helvetica.' Heer.
(58) 'Flora Fossilis Arctica.' Heer.
(59) 'Recherches sur le Climat et la Végétation du Pays Tertiaire.' Heer.
(60) 'Fossil Flora of Great Britain.' Lindley and Hutton.
(61) 'Fossil Fruits and Seeds of the London Clay.' Bowerbank.
(62) "Tertiary Leaf-beds of the Isle of Mull"—'Quart. Journn. Geol. Soc.,' vol vii. Edward Forbes.
(63) 'The Geology of England and Wales.' Horace B. Woodward.

CHAPTER XXI.

THE QUATERNARY PERIOD.

THE POST-PLIOCENE PERIOD.

Later than any of the Tertiary formations are various detached and more or less superficial accumulations, which are generally spoken of as the *Post-Tertiary formations,* in accordance with the nomenclature of Sir Charles Lyell—or as the *Quaternary formations,* in accordance with the general usage of Continental geologists. In all these formations we meet with no *Mollusca* except such as are now alive—with the partial and very limited exception of some of the oldest deposits of this period, in which a few of the shells occasionally belong to the species not known to be in existence at the present day. Whilst the *Shell-fish* of the Quaternary deposits are, generally speaking, identical with existing forms, the *Mammals* are sometimes referable to living, sometimes to extinct species. In accordance with this, the Quaternary formations are divided into two groups: (1) The *Post-Pliocene,* in which the shells are almost invariably referable to existing species, but *some of the Mammals are extinct;* and (2) the *Recent,* in which *the shells and the Mammals alike belong to existing species.* The *Post-Pliocene* deposits are often spoken of as the *Pleistocene* formations (Gr. *pleistos,* most; *kainos,* new or recent), in allusion to the fact that the great majority of the living beings of this period belong to the species characteristic of the "new" or Recent period.

The *Recent* deposits, though of the highest possible interest, do not properly concern the palæontologist strictly so-called, but the zoologist, since they contain the remains of none but existing animals. They are "Pre-historic," but they belong entirely to the existing terrestrial order. The *Post-Pliocene* deposits, on the other hand, contain the remains of various extinct Mammals; and though Man undoubtedly existed in, at any rate, the later portion of this period, if not throughout the whole of it, they properly form part of the domain of the palæontologist.

The Post-Pliocene deposits are extremely varied, and very widely distributed; and owing to the mode of their occurrence, the ordinary geological tests of age are in their case but very partially available. The subject of the classification of these deposits is therefore an extremely complicated one; and as regards the age of even some of the most important of them, there still exists considerable difference of opinion. For our present purpose, it will be convenient to adopt a classification of the Post-Pliocene deposits founded on the relations which they bear in time to the great "Ice-age" or "Glacial period;" though it is not pretended that our present knowledge is sufficient to render such a classification more than a provisional one.

In the early Tertiary period, as we have seen, the climate of the northern hemisphere, as shown by the Eocene animals and plants, was very much hotter than it is at present—partaking, indeed, of a sub-tropical character. In the Middle Tertiary or Miocene period, the temperature, though not so high, was still much warmer than that now enjoyed by the northern hemisphere; and we know that the plants of temperate regions at this time flourished within the Arctic circle. In the later Tertiary or Pliocene period, again, there is evidence that the northern hemisphere underwent a further progressive diminution of temperature; though the climate of Europe generally seems at the close of the Tertiary period to have been if anything warmer, or at any rate not colder, than it is at the present day. With the commencement of the Quaternary period, however, this diminution of temperature became more decided; and beginning with a temperate climate, we find the greater portion of the northern hemisphere to become gradually subjected to all the rigours of intense Arctic cold. All the mountainous regions of Northern and Central Europe, of Britain, and of North America, became the nurseries of huge ice-streams, and large areas of the land appear to have been covered with a continuous ice-sheet. The Arctic conditions of this, the well-known "Glacial period," relaxed more than once, and were more than once re-established with lesser intensity. Finally, a gradual but steadily progressive amelioration of temperature took place; the ice slowly gave way, and ultimately disappeared altogether; and the climate once more became temperate, except in high northern latitudes.

The changes of temperature sketched out above took place slowly and gradually, and occupied the whole of the Post-Pliocene period. In each of the three periods marked out by these changes—in the early temperate, the central cold, and the later temperate period—certain deposits were laid down over the surface of the northern hemisphere; and these deposits collectively constitute the Post-Pliocene formations. Hence we may conveniently classify all the accumulations of this age under the heads of (1) *Pre-Glacial* deposits, (2) *Glacial* deposits, and (3) *Post-Glacial* deposits, according as they were formed before, during, or after the "Glacial period." It cannot by any means be asserted that we can definitely fix the precise relations in time of all the Post-Pliocene deposits to the Glacial period. On the contrary, there are some which hold a very disputed position as regards this point; and there are others which do not admit of definite allocation in this manner at all, in consequence of their occurrence in regions where no "Glacial Period" is known to have been established. For our present purpose, however, dealing as we shall have to do principally with the northern hemisphere, the above classification, with all its defects, has greater advantages than any other that has been yet proposed.

I. Pre-Glacial Deposits.—The chief pre-glacial deposit of Britain is found on the Norwalk coast, reposing upon the Newer Pliocene (Norwich Crag), and consists of an ancient land-surface which is known as the "Cromer Forest-bed."

This consists of an ancient soil, having embedded in it the stumps of many trees, still in an erect position, with remains of living plants, and the bones of recent extinct quadrupeds. It is overlaid by fresh-water and marine beds, all the shells of which belong to existing species, and it is finally surmounted by true "glacial drift." While all the shells and plants of the Cromer Forest-bed and its associated strata belong to existing species, the Mammals are partly living, partly extinct. Thus we find the existing Wolf (*Canis lupus*), Red Deer (*Cervus elaphus*), Roebuck (*Cervus capreolus*), Mole (*Talpa Europæa*), and Beaver (*Castor fiber*), living in western England side by side with the *Hippopotamus major, Elephas antiquus, Elephas meridionalis, Rhinoceros Etruscus,* and *R. megarhinus* of the Pliocene period, which are not only extinct, but imply an at any rate moderately warm climate. Besides the above, the Forest-bed has yielded the remains of several extinct species of Deer, of the great extinct Beaver (*Trogontherium Cuvieri*), of the Caledonian Bull or "Urus" (*Bos primigenius*), and of a Horse (*Equus fossilis*), little if at all distinguishable from the existing form.

The so-called "Bridlington Crag" of Yorkshire, and the "Chillesford Beds" of Suffolk, are probably to be regarded as also belonging to this period; though many of the shells which they contain are of an Arctic character, and would indicate that they were deposited in the commencement of the Glacial period itself. Owing, however, to the fact that a few of the shells of these deposits are known to occur in a living condition, these, and some other similar accumulations, are sometimes considered as referable to the Pliocene period.

II. Glacial Deposits.—Under this head is included a great series of deposits which are widely spread over both Europe and America, and which were formed at a time when the climate of these countries was very much colder than it is at present, and approached more or less closely to what we see at the present day in the Arctic regions. These deposits are known by the general name of the *Glacial deposits,* or by the more specialized names of the Drift, the Northern Drift, the Boulder-clay, the Till, &c.

These glacial deposits are found in Britain as far south as the Thames, over the whole of Northern Europe, in all the more elevated portions of Southern and Central Europe, and over the whole of North America, as far south as the 39th parallel. They generally occur as sands, clays, and gravels, spread in widely-extended sheets over all the geological formations alike, except the most recent, and are commonly spoken of under the general term of "Glacial drift." They vary much in their exact nature in different districts, but they universally consist of one, or all, of the following members:—

1. *Unstratified* clays,, or loams, containing numerous angular or sub-angular blocks of stone, which have often been transported for a greater or less distance from their parent rock, and which often exhibit polished, grooved, or striated surfaces. These beds are what is called *Boulder-clay,* or *Till.*

2. Sands, gravels, and clays, often more or less regularly *stratified,* but containing erratic blocks, often of large size, and with their edges *unworn,* derived from considerable distances from the place where they are now found. In these beds it is not at all uncommon to find fossil shells; and these, though of existing species, are generally of an Arctic character, comprising a greater or less number of forms which are now exclusively found in the icy waters of the Arctic seas. These beds are often spoken of as "Stratified Drift."

3. *Stratified* sands and gravels, in which the pebbles are *worn* and rounded, and which have been produced by a re-arrangement of ordinary glacial beds by the sea. These beds are commonly known as "Drift-gravels," or "Regenerated Drift."

Some of the last-mentioned of these are doubtless post-glacial; but, in the absence of fossils, it is often impossible to arrive at a positive opinion as to the precise age of superficial accumulations of this nature. It is also the opinion of high authorities that a considerable number of the so-called "cave-deposits," with the bones of extinct Mammals, truly belong to the Glacial period, being formed during warm intervals when the severity of the Arctic cold had become relaxed. It is further believed that some, at any rate, of the so-called "high-level" river-gravels and "brick-earths" have likewise been deposited during mild or warm intervals in the great age of ice; and in two or three instances this has apparently been demonstrated—deposits of this nature, with the bones of extinct animals and the implements of man, having been shown to be overlaid by true Boulder-clay.

The fossils of the undoubted Glacial deposits are principally shells, which are found in great numbers in certain localities, sometimes with *Foraminifera,* the bivalved cases of Ostracode Crustaceans, &c. Whilst some of the shells of the "Drift" are such as now live in the seas of temperate regions, others, as previously remarked, are such as are now only known to live in the seas of high latitudes; and these therefore afford unquestionable evidence of cold conditions. Amongst these Arctic forms of shells which characterize the Glacial beds may be mentioned *Pecten Islandicus* (fig. 254), *Pecten Grœnlandicus, Scalaria Grœnlandica, Leda truncata, Astarte borealis, Tellina proxima, Natica clausa,* &c.

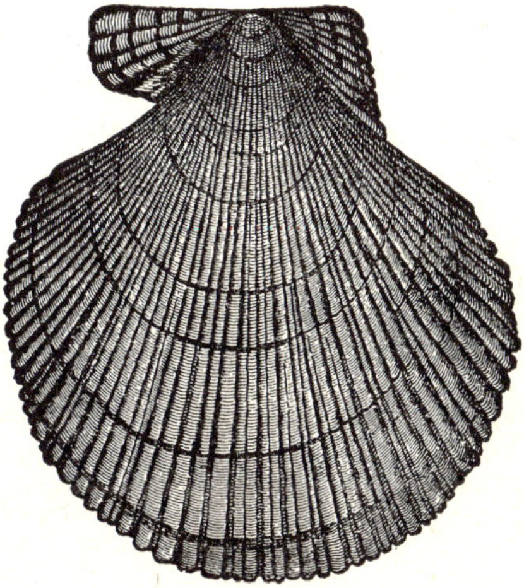

Fig. 254.—Left valve of *Pecten Islandicus*. Glacial and Recent.

III. Post-Glacial Deposits.—As the intense cold of the Glacial period became gradually mitigated, and temperate conditions of the climate were once more re-established, various deposits were formed in the northern hemisphere, which are found to contain the remains of extinct Mammals, and which, therefore, are clearly of Post-Pliocene age. To these deposits the general name of *Post-Glacial formations* is given; but it is obvious that, from the nature of the case, and with our present limited knowledge, we cannot draw a rigid line of demarcation between the deposits formed toward the close of the Glacial period, or during warm "Interglacial" periods, and those laid down after the ice had fairly disappeared. Indeed it is extremely improbable that any such rigid line of demarcation should ever have existed; and it is far more likely that the Glacial and Post-Glacial periods, and their corresponding deposits, shade into one another by an imperceptible gradation. Accepting this reservation, we may group together, under the general head of "Post-Glacial Deposits," most of the so-called "Valley-gravels," "Brick-earths," and "Cave-deposits," together with some "raised beaches" and various deposits of peat. Though not strictly within the compass of this work, a few words may be said here as to the origin and mode of formation of the Brick-earths, Valley-gravels, and Cave-deposits, as the subject will thus be rendered more clearly intelligible.

Every river produces at the present day beds of fine mud and loam, and accumulations of gravel, which it deposits at various parts of its course—the gravel generally occupying the lowest position, and the finer sands and mud coming above. Numerous deposits of a similar nature are found in most countries in various localities, and at various heights above the present channels of our rivers. Many of these fluviatile (Lat. *fluvius,* a river) deposits consist of fine loam, worked for brick-making, and known as "Brick-earths;" and they have yielded the remains of numerous extinct Mammals, of which the Mammoth (*Elephas primigenius*) is the most abundant. In the valley of the Rhine these fluviatile loams (known as "Loess") attain a thickness of several hundred feet, and contain land and fresh-water shells of existing species. With these occur the remains of Mammals, such as the Mammoth and Woolly Rhinoceros. Many of these Brick-earths are undoubtedly Post-Glacial, but others seem to be clearly "inter glacial;" and instances have recently been brought forward in which deposits of Brick-earth containing bones and shells of fresh-water Molluscs have been found to be overlaid by regular unstratified boulder-clay.

The so-called "Valley-gravels," like the Brick-earths, are fluviatile deposits, but are of a coarser nature, consisting of sands and gravels. Every river gives origin to deposits of this kind at different points along the course of its valley; and it is not uncommon to find that there exist in the valley of a single river two or more sets of these gravel-beds, formed by the river itself, but formed at times when the river ran at different levels, and therefore formed at different periods. These different accumulations are known as the "high-level" and "low-level" gravels; and a reference to the accompanying diagram will explain the origin and nature of these deposits (fig. 255). When a river begins to occupy a particular line of drainage, and to form its own channel, it will deposit fluviatile sands and gravels along its sides. As it goes on deepening the bed or valley through which it flows, it will deposit other fluviatile strata at a lower level besides its new bed. In this way have arisen the terms "high-level" and "low-level" gravels. We find, for instances, a modern river flowing through a valley which it has to a great extent or entirely formed itself; by the side of its immediate channel

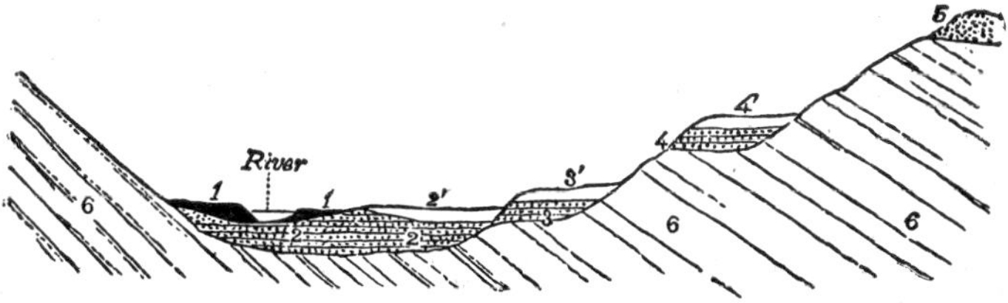

Fig. 255.—Recent and Post-Pliocene Alluvial Deposits. 1, Peat of the recent period; 2, Gravel of the modern river; 2′, Loam of the modern river; 3, Lower-level valley-gravel with bones of extinct Mammals (Post-Pliocene); 3′, Loam of the same age as 3; 4, Higher-level valley-gravel (Post-Pliocene); 4′, Loam of the same age as 4; 5, Upland gravel of various kinds (often glacial drift); 6, Older rocks. (After Sir Charles Lyell.)

we may find gravels, sand, and loam (fig. 255, 2 2′) deposited by the river flowing in its present bed. These are *recent* fluviatile or alluvial deposits. At some distance from the present bed of the river, and at a higher level, we may find other sands and gravels, quite like the recent ones in character and origin, but formed at a time when the stream flowed at a higher level, and before it had excavated its valley to its present depth. These (fig. 255, 3 3′) are the so-called "*low-level* gravels" of a river. At a still higher level, and still farther removed from the present bed of the river, we may find another terrace, composed of just the same materials as the lower one, but formed at a still earlier period, when the excavation of the valley had proceeded to a much less extent. These (fig. 255, 4 4′) are the so-called "*high-level* gravels" of a river, and there may be one or more terraces of these.

The important fact to remember about these fluviatile deposits is this—that here the ordinary geological rule is reversed. The high-level gravels are, of course, the highest, so far as their actual elevation above the sea is concerned; but geologically the lowest, since they are obviously much older than the low-level gravels, as these are than the recent gravels. How much older the high-level gravels may be than the low-level ones, it is impossible to say. They occur at heights varying from 10 to 100 feet above the present river-channels, and they are therefore older than the recent gravels by the time required by the river to dig out its own bed to this depth. How long this period may be, our data do not enable us to determine accurately; but if we are to calculate from the observed rate of erosion of the actually existing rivers, the period between the different valley-gravels must be a very long one.

The lowest or recent fluviatile deposits which occur beside the bed of the present river, are referable to the Recent period, as they contain the remains of none but living Mammals. The two other sets of gravels are Post-Pliocene, as they contain the bones of extinct Mammals, mixed with land and fresh-water shells of existing species. Among the more important extinct Mammals of the low-level and high-level valley-gravels may be mentioned the *Elephas antiquus,* the Mammoth (*Elephas primigenius*), the Woolly Rhinoceros (*R. tichorhinus*), the Hippopotamus, the Cave-lion, and the Cave-bear. Along with these are found unquestionable traces of the existence of Man, in the form of rude flint implements of undoubted human workmanship.

The so-called "Cave-deposits," again, though exhibiting peculiarities due to the fact of their occurrence in caverns or fissures in the rocks, are in many respects essentially similar to the older valley-gravels. Caves, in the great majority of instances, occur in limestone. When this is not the case, it will generally be found that they occur along lines of sea-coast, or along lines which can be shown to have anciently formed the coast-line. There are many caves, however, in the making of which it can be shown that the sea has had no hand; and these are most of the caves of limestone districts. These owe their origin to the solvent action upon lime of water holding carbonic acid in solution. The rain which falls upon a limestone district absorbs a certain amount of carbonic acid from the air, or from the soil. It then percolates through the rock, generally along the lines of jointing so characteristic of limestones, and in its progress it dissolves and carries off a certain quantity of carbonate of lime. In this way, the natural joints and fissures in the rock are widened, as can be seen at the present day in any or all limestone districts. By a continuance of this action for a sufficient length of time, caves may ultimately be produced. Nothing, also, is commoner in a limestone district than for the natural drainage to take the line of some fissure, dissolving the rock in its course. In this way we constantly meet in limestone districts with springs issuing from the limestone rock—sometimes as large rivers—the waters of which are charged with carbonate of lime, obtained by the solution of the sides of the fissure through which the waters have flowed. By these and similar actions, every district in which limestones are extensively developed will be found to exhibit a number of natural caves, rents, or fissures. The first element, therefore, in the production of cave-deposits, is the existence of a period in which limestone rocks were largely dissolved, and caves were formed in consequence of the then existing drainage taking the lime of some fissure.

Secondly, there must have been a period in which various deposits were accumulated in the caves thus formed. These cavern-deposits are of various nature, consisting of mud, loam, gravel, or breccias of different kinds. In all cases, these materials have been introduced into the cave at some period subsequent to, or contemporaneous with, the formation of the cave. Sometimes the cave communicates with the surface by a fissure through which sand, gravel, &c., may be washed by rains or by floods from some neighboring river. Sometimes the cave has been the bed of an ancient stream, and the deposits have been formed as are fluviatile deposits at the surface. Or, again, the river has formerly flowed at a greater elevation than it does at present, and the cave has been filled with fluviatile deposits by the river at a time prior to the excavation of its bed to the present depth (fig. 256). In this last case, the cave-deposits obviously bear exactly the same relation in point of antiquity to recent deposits, as do the low-level and high-level valley-gravels to recent river-gravels. In any case, it is necessary for the physical geography of the district to change to some extent, in order that the cave-deposits should be preserved. If the materials have been introduced by a fissure, the cave will probably become ultimately filled to the roof, and the aperture of admission thus blocked up. If a river has flowed through the cave, the surface configuration of the district must be altered so far as to divert the river into a new channel. And if the cave is placed in the side of a river-valley, as in fig. 256, the river must have excavated its channel to such a depth that it can no longer wash out the contents of a cave even in high floods.

If the cave be entirely filled, the included deposits generally get more or less completely cemented together by the percolation through them of water holding carbonate of lime in solution. If the cave is only partially filled, the dropping of water from the roof holding lime in solution, and its subsequent

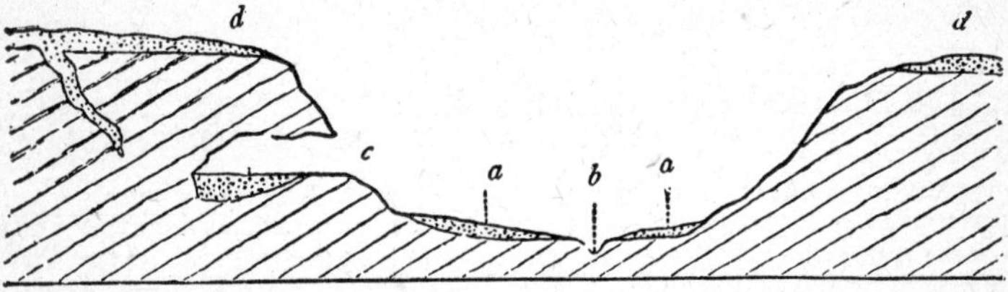

Fig. 256.—Diagrammatic section across a river-valley and cave. *a a*, Recent valley-graves near the channel (*b*) of the existing river; *c*, Cavern, party filled with cave-earth; *d d*, High-level gravels, filling fissures in the limestone, which perhaps communicate in some instances with the cave, and form a channel by which materials of various kinds were introduced into it; *e e*, Inclined beds of limestone.

evaporation, would lead to the formation over the deposits below of a layer of stalagmite, perhaps several inches, or even feet, in thickness. In this way cave-deposits, with their contained remains, may be hermetically sealed up and preserved without injury for an altogether indefinite period of time

In all caves in limestone in which deposits containing bones are found, we have then evidence of three principal sets of changes. (1.) A period during which the cave was slowly hollowed out by the percolation of acidulated water; (2.) A period in which the cave became the channel of an engulfed river, or otherwise came to form part of the general drainage-system of the district; (3.) A period in which the cave was inhabited by various animals.

As a typical example of a cave with fossiliferous Post-Pliocene deposits, we may take Kent's Cavern, near Torquay, in which a systematic and careful examination has revealed the following sequence of accumulations in descending order:—

(*a*) Large blocks of limestone, which lie on the floor of the cave, having fallen from the roof, and which are sometimes cemented together by stalagmite.

(*b*) A layer of black mould, from three to twelve inches thick, with human bones, fragments of pottery, stone and bronze implements, and the bones of animals now living in Britain. This, therefore, is a *recent* deposit.

(*c*) A layer of stalagmite, from sixteen to twenty inches thick, but sometimes as much as five feet, containing the bones of Man, together with those of extinct Post-Pliocene Mammals.

(*d*) A bed of red cave-earth, sometimes four feet in thickness, with numerous bones of extinct Mammals (Mammoth, Cave-bear, &c.), together with human implements of flint and horn.

(*e*) A second bed of stalagmite, in places twelve feet in thickness, with bones of the Cave-bear.

(*f*) A red-loam and cave-breccia, with remains of the Cave-bear and human implements.

The most important Mammals which are found in cave-deposits in Europe generally, are the Cave-bear, the Cave-lion, the Cave-hyæna, the Reindeer, the Musk-ox, the Glutton, and the Lemming—of which the first three are probably identical with existing forms, and the remainder are certainly so—together with the Mammoth and the Woolly Rhinoceros, which are undoubtedly extinct. Along with these are found the implements, and in some cases the bones, of Man himself, in such a manner as to render it absolutely certain that an early race of men was truly contemporaneous in Western Europe with the animals above mentioned.

IV. Unclassified Post-Pliocene Deposits.—Apart from any of the afore-mentioned deposits, there occur other accumulations—sometimes superficial, sometimes in caves—which are found in regions where a "Glacial period" has not been fully demonstrated, or where such did not take place; and which, therefore, are not amenable to the above classification. The most important of these are known to occur in South America and Australia ; and though their numerous extinct Mammalia place their reference to the Post-Pliocene period beyond doubt, their relations to the glacial period and its deposits in the northern hemisphere have not been precisely determined.

CHAPTER XXII.

THE POST-PLIOCENE PERIOD.—Continued.

As regards the *life* of the Post-Pliocene period, we have, in the first place, to notice the effect produced throughout the northern hemisphere by the gradual supervention of the Glacial period. Previous to this the climate must have been temperate or warm-temperate; but as the cold gradually came on, two results were produced as regards living beings of the area thus affected. In the first place, all those Mammals which, like the Mammoth, the Woolly Rhinoceros, the Lion, the Hyæna, and the Hippopotamus, require, at any rate, moderately warm conditions, would be forced to migrate southwards to regions not affected by the new state of things. In the second place, Mammals previously inhabiting higher latitudes, such as the Reindeer, the Musk-ox, and the Lemming, would be enabled by the increasing cold to migrate southwards, and to invade provinces previously occupied by the Elephant and the Rhinoceros. A precisely similar, but more slowly-executed process, must have taken place in the sea, the northern Mollusca moving southwards as the arctic conditions of the Glacial period became established, whilst the forms proper to temperate seas receded. As regards the readily locomotive Mammals, also, it is probable that this process was carried on repeatedly in a partial manner, the southern and northern forms alternately fluctuating backwards and forwards over the same area, in accordance with the fluctuations of temperature which have been shown by Mr. James Geikie to have characterized the Glacial period as a whole. We can thus readily account for the intermixture which is sometimes found of northern and southern types of Mammalia in the same deposits, or in deposits apparently synchronous, and within a single district. Lastly, at the final close of the arctic cold of the Glacial period, and the re-establishment of temperate conditions over the northern hemisphere, a reversal of the original process took place—the northern Mammals retiring within their ancient limits, and the southern forms pressing northwards and reoccupying their original domains.

The *Invertebrate* animals of the Post-Pliocene deposits require no further mention—all the known forms, except a few of the shells in the lowest beds of the formation, being identical with species now in existence upon the globe. The only point of importance in this connection has been previously noticed—namely, that in the true Glacial deposits themselves a considerable number of the shells belong to northern or Arctic types.

As regards the *Vertebrate* animals of the period, no extinct forms of Fishes, Amphibians, or Reptiles are known to occur, but we meet with both extinct Birds and extinct Mammals. The remains of the former are of great interest, as indicating the existence during Post-Pliocene times, at widely remote points of the southern hemisphere, of various wingless, and for the most part gigantic, Birds. All the great wingless Birds of the order *Cursores* which are known as existing at the present day upon the globe, are restricted to regions which are either wholly or in great part south of the equator. Thus the true Ostriches are African; the Rheas are South American; the Emeus are Australian; the Cassowaries are confined to Northern Australia, Papua, and the Indian Archipelago; the species of *Apteryx* are natives of New Zealand; and the Dodo and Solitaire (wingless, though probably not true *Cursores*), both of which have been exterminated within historical times, were inhabitants of the islands of Mauritius and Rodriguez, in the Indian Ocean. In view of these facts, it is noteworthy that, so far as known, all the Cursorial Birds of the Post-Pliocene period should have been confined to the same hemisphere as that inhabited by the living representatives of the order. It is still further interesting to notice that the extinct forms in question are only found in geographical provinces which are now, or have been within historical times, inhabited by similar types. The greater number of the remains of these have been discovered in New Zealand, where there now live several species of the curious wingless genus *Apteryx;* and they have been referred by Professor Owen to several generic groups, of which *Dinornis* is the most important (fig 257). Fourteen species of *Dinornis* have been described by the distinguished palæontologist just mentioned, all of them being large wingless birds of the type of the existing Ostrich, having enormous powerful hind-limbs adapted for running, but with the wings wholly rudimentary, and the breast-bone devoid of the keel or ridge which characterizes this bone in all birds which fly. The largest species is the *Dinornis giganteus,* one of the most gigantic of living or fossil birds, the shank (*tibia*) measuring a yard in length, and the total height being at least ten feet. Another species, the *Dinornis elephantopus* (fig. 257), though not standing more than about six feet in height, was of an even more ponderous construction—"the framework

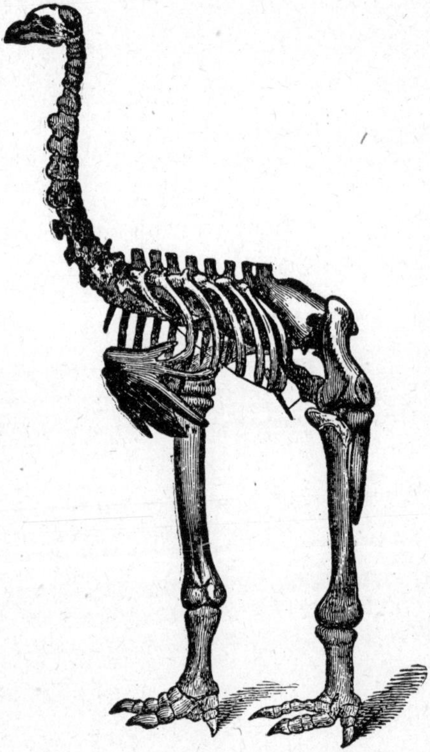

Fig. 257.—Skeleton of *Dinornis elephantopus*, greatly reduced. Post-Pliocene, New Zealand. (After Owen.)

of the skeleton being the most massive of any in the whole class of Birds," whilst "the toe-bones almost rival those of the Elephant" (Owen). The feet in *Dinornis* were furnished with three toes, and are of interest as presenting us with an undoubted Bird big enough to produce the largest of the footprints of the Triassic Sandstones of Connecticut. New Zealand has now been so far explored, that it seems questionable if it can retain in its recesses any living example of *Dinornis;* but it is certain that species of this genus were alive during the human period, and survived up to quite a recent date. Not only are the bones very numerous in certain localities, but they are found in the most recent and superficial deposits, and they still contain a considerable proportion of animal matter; whilst in some instances bones have been found with the feathers attached, or with the horny skin of the legs still adhering to them. Charred bones have been found in connection with native "ovens;" and the traditions of the Maories contain circumstantial accounts of gigantic wingless Birds, the "Moas," which were hunted both for their flesh and their plumage. Upon the whole, therefore, there can be no doubt but that the Moas of New Zealand have been exterminated at quite a recent period—perhaps within the last century— by the unrelenting pursuit of Man,—a pursuit which their wingless condition rendered them unable to evade.

In Madagascar, bones have been discovered of another huge wingless Bird, which must have been as large as, or larger than, the *Dinornis giganteus,* and which has been described under the name of *Æpiornis maximus.* With the bones have been found eggs measuring from thirteen to fourteen inches in diameter, and computed to have the capacity of three Ostrich eggs. At least two other small species of *Æpiornis* have been described by Grandidier and Milne-Edwards as occurring in Madagascar; and they consider the genus to be so closely allied to the *Dinornis* of New Zealand, as to prove that these regions, now so remote, were at one time united by land. Unlike New Zealand, where there is the *Apteryx,* Madagascar is not known to possess any living wingless Birds; but in the neighboring island of Mauritius the wingless Dodo (*Didus ineptus*) has been exterminated less than three hundred years ago; and the little island of Rodriguez, in the same geographical province, has in a similar period lost the equally wingless Solitaire (*Pezophaps*), both of these, however, being generally referred to the *Rasores.*

The *Mammals* of the Post-Pliocene period are so numerous, that in spite of the many points of interest which they present, only a few of the more important forms can be noticed here, and that but briefly. The first order that claims our attention is that of the *Marsupials,* the headquarters of which at the present day is the Australian province. In Oolitic times Europe possessed its small Marsupials, and similar forms existed in the same area in the Eocene and Miocene periods; but if size be any criterion, the culminating point in the history of the order was attained during the Post-Pliocene period in Australia. From deposits of this age there has been disentombed a whole series of remains of extinct, and for the most part gigantic, examples of this group of Quadrupeds. Not to speak of Wombats and Phalangers, two forms stand out prominently as representatives of the Post-Pliocene animals of Australia. One of these is *Diprotodon* (fig. 258), representing, with many differences, the well-known modern group of the Kangaroos. In its teeth, *Diprotodon* shows itself to be closely allied to the

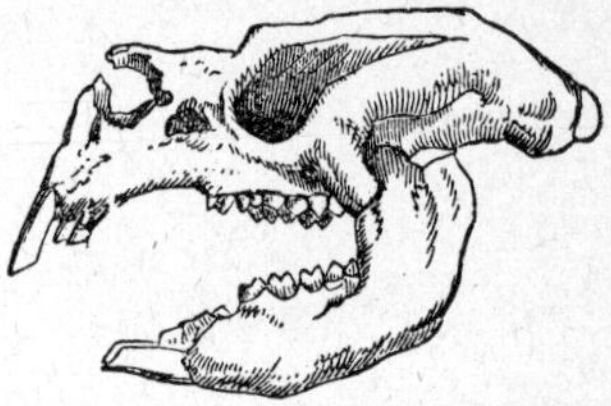

Fig. 258.—Skull of *Diprotodon, Australis*, greatly reduced. Post-Pliocene, Australia.

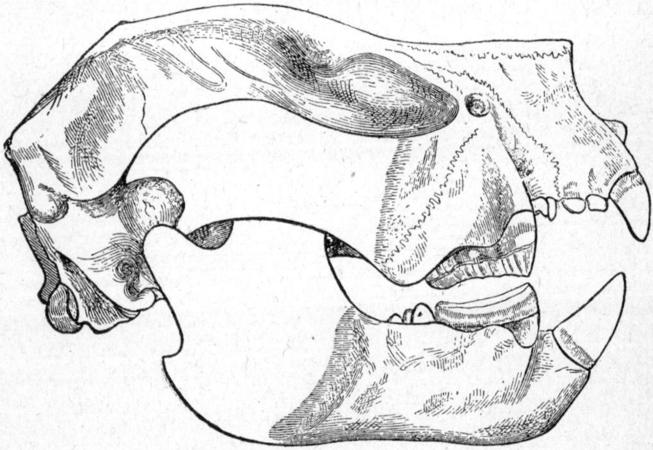

Fig. 259.—Skull of *Thylacoleo*. Post-Pliocene, Australia. Greatly reduced. (After Flower.)

living, grass-eating Kangaroos; but the hind-limbs were not so disproportionately long. In size, also, *Diprotodon* must have many times exceeded the dimensions of the largest of its living successors, since the skull measures no less than three feet in length. The other form in question is *Thylacoleo* (fig. 259), which is believed by Professor Owen to belong to the same group as the existing "Native Devil" (*Dasyurus*) of

Van Diemen's Land, and therefore to have been flesh-eating and rapacious in its habits, through this view is not accepted by others. The principal feature in the skull of *Thylacoleo* is the presence, on each side of each jaw, of a single huge tooth, which is greatly compressed, and has a cutting edge. This tooth is regarded by Owen as corresponding to the great cutting tooth of the jaw of the typical Carnivores, but Professor Flower considers that *Thylacoleo* is rather related to the Kan-

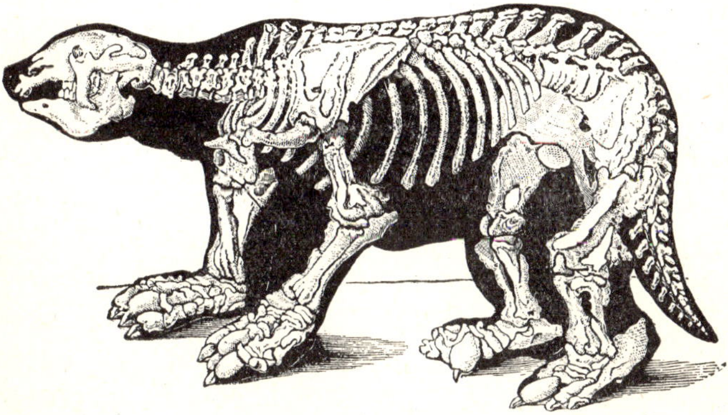

Fig. 260.—*Megatherium Cuvieri*. Post-Pliocene, South America.

garoo-rats. The size of the crown of the tooth in question is not less than two inches and a quarter; and whether carnivorous or not, it indicates an animal of a size exceeding that of the largest of existing Lions.

The order of the *Edentates,* comprising the existing Sloths, Ant-eaters, and Armadillos, and entirely restricted at the present day to South America, Southern Asia, and Africa, is one alike singular for the limited geographical range of its members, their curious habits of life, and the well-marked peculiarities of their anatomical structure. South America is the metropolis of the existing forms; and it is an interesting fact that there flourished within Post-Pliocene times in this continent, and to some extent in North America also, a marvelous group of extinct Edentates, representing the living Sloths and Armadillos, but of gigantic size. The most celebrated of these is the huge *Megatherium Cuvieri* (fig. 260) of the South American Pampas. The Megathere was a colossal Sloth-like animal which attained a length of from twelve to eighteen feet, with bones more massive than those of the Elephant. Thus the thigh-bone is nearly thrice the thickness of the same bone in the largest of existing Elephants, its circumference at its narrowest point nearly equalling its total length; the massive bones of the shank (tibia and fibula) are amalgamated at their extremities; the heel-bone (calcaneum) is nearly half a yard in length; the haunch-bones (ilia) are from four to five feet across at their crests; and the bodies of the vertebræ at the root of the tail are from five to seven inches in diameter, from which it has been computed that the circumference of the tail at this part might have been from five to six feet. The length of the fore-foot is about a yard, and the toes are armed with powerful curved claws. It is known now that the Megathere, in spite of its enormous weight and ponderous construction, walked, like the existing Ant-eaters and Sloths, upon the outside edge of the fore-feet, with the claws more or less bent inwards towards the palm of the hand. As in the great majority of the Edentate order, incisor and canine teeth are entirely wanting, the front of the jaws being toothless. The jaws, however, are furnished with five upper and four lower molar teeth on each side. These grinding teeth are from seven to eight inches in length, in the form of four-sided prisms, the crowns of which are provided with well-marked transverse ridges; and they continue to grow during the whole life of the animal. There are indications that the snout was prolonged, and more or less flexible; and the tongue was probably prehensile. From the characters of the molar teeth it is certain that the Megathere was purely herbivorous in its habits; and from the enormous size and weight of the body, it is equally certain that it could not have imitated its modern allies, the Sloths, in the feat of climbing, back downwards, amongst the trees. It is clear, therefore, that the Megathere sought its sustenance upon the ground; and it was originally supposed to have lived upon roots. By a masterly piece of deductive reasoning, however, Professor Owen showed that this great "Ground-Sloth" must have truly lived upon the foliage of trees, like the existing Sloths—but with this difference, that instead of climbing amongst the branches, it actually uprooted the tree bodily. In this *tour de force,* the animal sat upon its huge haunches and mighty tail, as on a tripod, and then grasping the trunk with its powerful arms, either wrenched it up by the roots or broke it short off above the ground. Marvellous as this may seem, it can be shown that every detail in the skeleton of the Megathere accords with the supposition that it obtained its food in this way. Similar habits were followed by the allied *Mylodon* (fig. 261), another of the great "Ground-Sloths," which inhabited South America during the Post-Pliocene period. In most respects, the *Mylo-*

Fig. 261.—Skeleton of *Mylodon robustus*. Post-Pliocene, South America.

don is very like the Megathere; but the crowns of the molar teeth are flat instead of being ridged. The nearly-related genus *Megalonyx*, unlike the Megathere, but like the Mylodon, extended its range northwards as far as the United States.

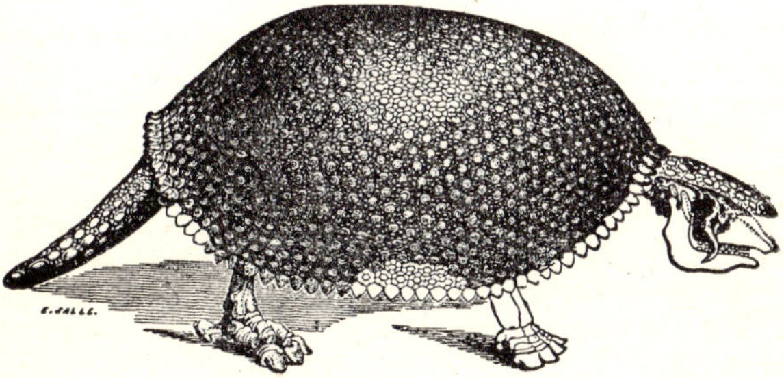

Fig. 262.—Skeleton of *Glyptodon clavipes*. Post-Pliocene, South America.

Just as the Sloths of the present day were formerly represented in the same geographical area by the gigantic Megatheroids, so the little banded and cuirassed Armadillos of South America were formerly represented by gigantic species, constituting the genus *Glyptodon.* The *Glyptodons* (fig. 262) differed from the living Armadillos in having no bands in their armour, so that they must have been unable to roll themselves up. It is rare at the present day to meet with any Armadillo over two or three feet in length; but the length of the *Glyptodon clavipes,* from the tip of the snout to the end of the tail, was more than nine feet.

There are no canine or incisor teeth in the *Glyptodon,* but there are eight molars on each side of each jaw, and the crowns of these are fluted and almost trilobed. The head is covered by a helmet of bony plates, and the trunk was defended by an armour of almost hexagonal bony pieces united by sutures, and exhibiting special patterns of sculpturing in each species. The tail was almost defended by a similar armour, and the vertebræ were mostly fused together so as to form a cylindrical bony rod. In addition to the above-mentioned forms, a number of other Edentate animals have been discovered by the researches of M. Lund in the Post-Pliocene deposits of the Brazilian bone-caves. Amongst these are true Ant-eaters, Armadillos, and Sloths, many of them of gigantic size, and all specifically or generically distinct from existing forms.

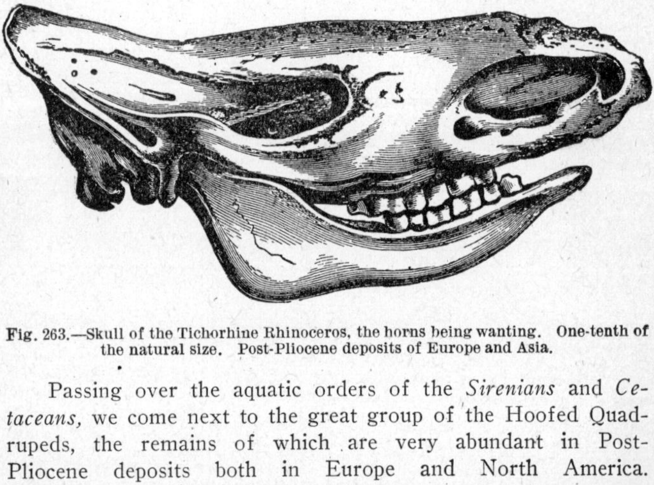

Fig. 263.—Skull of the Tichorhine Rhinoceros, the horns being wanting. One-tenth of the natural size. Post-Pliocene deposits of Europe and Asia.

Passing over the aquatic orders of the *Sirenians* and *Cetaceans,* we come next to the great group of the Hoofed Quadrupeds, the remains of which are very abundant in Post-Pliocene deposits both in Europe and North America. Amongst the Odd-toed Ungulates the most important are the Rhinoceroses, of which three species are known to have existed in Europe during the Post-Pliocene period. Two of these are the well-known Pliocene forms, the *Rhinoceros Etruscus* and the *R. megarhinus,* still surviving in diminished numbers; but the most famous is the *Rhinoceros tichorhinus* (fig. 263), or so-called "Woolly Rhinoceros." This species is known not only by innumerable bones, but also by a carcass, at the time of its discovery complete, which was found embedded in the frozen soil of Siberia towards the close of last century, and which was partly saved from destruction by the exertions of the naturalist Pallas. From this, we know that the Tichorhine Rhinoceros, like its associate the Mammoth, was provided with a coating of hair, and therefore was enabled to endure a more severe climate than any existing species. The skin was not thrown into the folds which characterize most of the existing forms; and the technical name of the species refers to the fact that the nostrils were completely separated by a bony partition. The head carried two horns, placed one behind the other, the front one being unusually large. As regards its geographical range, the Woolly Rhinoceros is found in Europe in vast numbers north of the Alps and Pyrenees, and it also abounded in Siberia; so that it would appear to be a distinctly northern form, and to have been adapted for a temperate climate. It is not known to occur in Pliocene deposits, but it makes its first appearance in the Pre-Glacial deposits, surviving the Glacial period, and being found in abundance in Post-Glacial accumulations. It was undoubtedly a contemporary of the earlier races of men in Western Europe; and it may perhaps be regarded as being the actual substantial kernel of some of the "Dragons" of fable.

The only other Odd-toed Ungulate which needs notice is the so-called *Equus fossilis* of the Post-Pliocene of Europe. This made its appearance before the Glacial period, and appears to be in reality identical with the existing Horse (*Equus caballus*). True Horses also occur in the Post-Pliocene of North America; but, from some cause or another, they must have been exterminated before historic times.

Amongst the Even-toed Ungulates, the great *Hippopotamus major* of the Pliocene still continued to exist in Post-Pliocene times in Western Europe; and the existing Wild Boar (*Sus scrofa*), the parent of our domestic breeds of Pigs, appeared for the first time. The Old World possessed extinct representatives of its existing Camels, and lost types of the living Llamas inhabited South America. Amongst the Deer, the Post-Pliocene accumulations have yielded the remains of various living species, such as the Red Deer (*Cervus elaphus*), the Reindeer (*Cervus tarandus*), the Moose or Elk (*Alces malchis*), and the Roebuck (*Cervus capreolus*), together with a number of extinct forms. Among the latter, the great "Irish Elk" (*Cervus megaceros*) is justly celebrated both for

Fig. 264.—Skeleton of the "Irish Elk" (*Cervus megaceros.*) Post-Pliocene, Britain.

its size and for the number and excellent preservation of its discovered remains. This extinct species (fig. 264) has been found principally in peat-mosses and Post-Pliocene lake-deposits, and is remarkable for the enormous size of the spreading antlers, which are widened out towards their extremities, and attain an expanse of over ten feet from tip to tip. It is not a genuine Elk, but is intermediate between the Reindeer and the Fallow-deer. Among the existing Deer

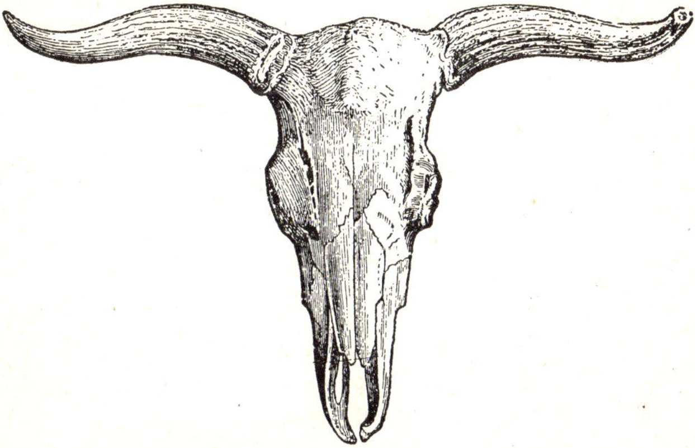

Fig. 265.—Skull of the Urus (*Bos primigenius*). Post-Pliocene and Recent. (After Owen.)

of the Post-Pliocene, the most noticeable is the Reindeer, an essentially northern type, existing at the present day in Northern Europe, and also (under the name of the "Caribou") in North America. When the cold of the Glacial period became established, this boreal species was enabled to invade Central and Western Europe in great herds, and its remains are found abundantly in cave-earths and other Post-Pliocene deposits as far south as the Pyrenees.

In addition to the above, the Post-Pliocene deposits of Europe and North America have yielded the remains of various Sheep and Oxen. One of the most interesting of the latter is the "Urus" or Wild Bull (*Bos primigenius*, fig. 265), which, though much larger than any of the existing forms, is believed to be specifically undistinguishable from the domestic Ox (*Bos taurus*), and to be possibly the ancestor of some of the larger European varieties of oxen. In the earlier part of its existence the Urus ranged over Europe and Britain in company with the Woolly Rhinoceros and the Mammoth; but it long survived these, and does not appear to have been finally exterminated till about the twelfth century. Another remarkable member of the Post-Pliocene Cattle, also to begin with an associate of the Mammoth and Rhinoceros, is the European Bison or "Aurochs" (*Bison priscus*). This "maned" ox formerly abounded in Europe in Post-Glacial times, and was not rare even in the later periods of the Roman empire, though much diminished in numbers, and driven back into the wilder and more inaccessible parts of the country. At present this fine species has been so nearly exterminated that it no longer exists in Europe save in Lithuania, where its preservation has been secured by rigid protective laws. Lastly, the Post-Pliocene deposits have yielded the remains of the singular living animals which is known as the Musk-ox or Musk-sheep (*Ovibos moschatus*). At the present day, the Musk-ox is an inhabitant of the "barren grounds" of Arctic America, and it is remarkable for the great length of its hair. It is, like the Reindeer, a distinctively northern animal; but it enjoyed during the Glacial period a much wider range than it has at the present day, the conditions suitable for its existence being then extended over a considerable portion of the northern hemisphere. Thus remains of the Musk-Ox are found in greater or less abundance in Post-Pliocene deposits over a great part of Europe, extending even to the south of France; and closely-related forms are found in similar deposits in the United States.

Coming to the *Proboscideans,* we find that the *Mastodons* seem to have disappeared in Europe at the close of the Pliocene period, or at the very commencement of the Post-Pliocene. In the New World, on the other hand, a species of Mastodon (*M. Americanus* or *M. Ohioticus*) is found abundantly in deposits of Post-Pliocene age, from Canada to Texas. Very perfect skeletons of this species have been exhumed from morasses and swamps, and large individuals attained a length (exclusive of the tusks) of seventeen feet and a height of eleven feet, the tusks being twelve feet in length. Remains of *Elephants* are also abundant in the Post-pliocene deposits of both the Old and the New World. Amongst these, we find in Europe the two familiar Pliocene species, *E. meridionalis* and *E. antiquus,* still surviving, but in diminished numbers. With these are found in vast abundance the remains of the characteristic Elephant of the Post-Pliocene, the well-known "Mammoth" (*Elephas primigenius*), which is accompanied in North America by the nearly-allied, but more southern species, the *Elephas Americanus.* The Mammoth (fig. 266) is considerably larger than the largest of the living Ele-

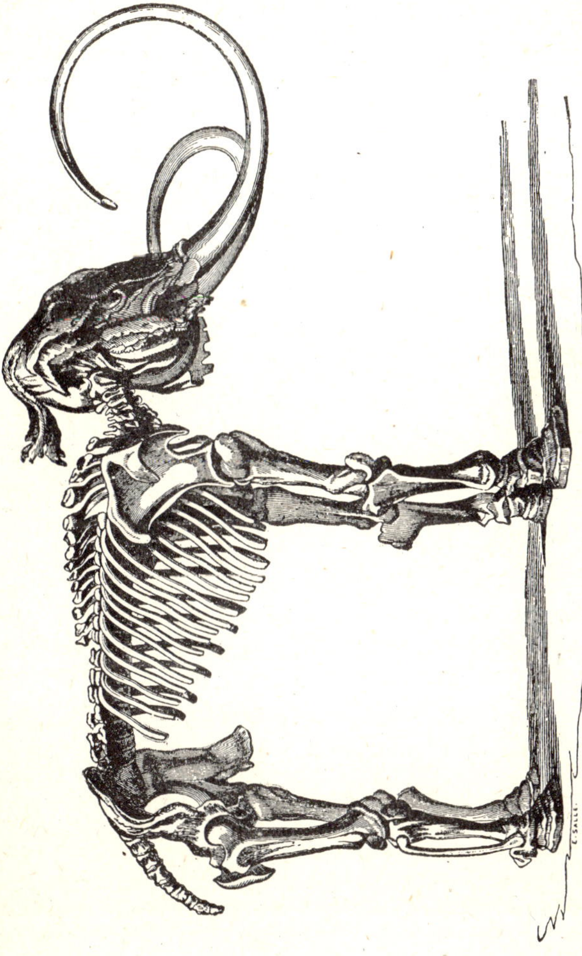

Fig. 266.—Skeleton of the Mammoth (*Elephas primigenius.*) Portions of the integument still adhere to the head, and the thick skin of the soles is still attached to the feet. Post-Pliocene.

phants, the skeleton being over sixteen feet in length, exclusive of the tusks, and over nine feet in height. The tusks are bent almost into a circle, and are sometimes twelve feet in length, measured along their curvature. In the frozen soil of Siberia several carcasses of the Mammoth have been discovered with the flesh and skin still attached to the bones, the most celebrated of these being a Mammoth which was discovered at the beginning of this century at the mouth of the Lena, on the borders of the Frozen Sea, and the skeleton of which is now preserved at St. Petersburg (fig. 266). From the occurrence of the remains of the Mammoth in vast numbers in Siberia, it might have been safely inferred that this ancient Elephant was able to endure a far more rigorous climate than its existing congeners. This inference has, however, been rendered a certainty by the specimens just referred to, which show that the Mammoth was protected

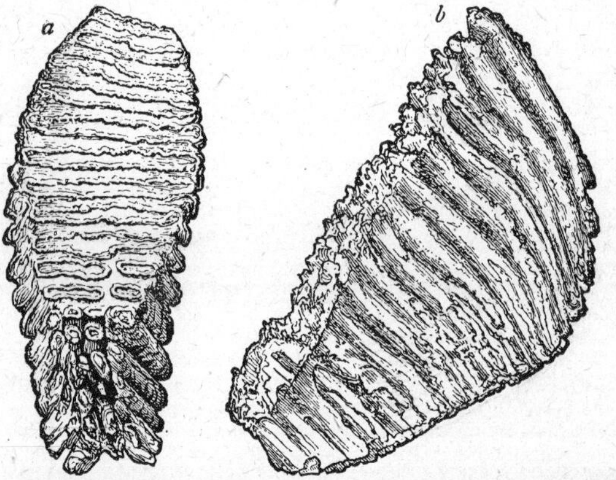

Fig. 267.—Molar tooth of the Mammoth (*Elephas primigenius*), upper jaw, right side, one-third of the natural size. *a*, Grinding surface ; *b*, Side view. Post-Pliocene.

against the cold by a thick coat of reddish-brown wool, some nine or ten inches long, interspersed with strong, coarse black hair more than a foot in length. The teeth of the Mammoth (fig. 267) are of the type of those of the existing Indian Elephant, and are found in immense numbers in certain localities. The Mammoth was essentially northern in its distribution, never passing south of a line drawn through the Pyrenees, the Alps, the northern shores of the Caspian, Lake Baikal, Kamschatka, and the Stanovi Mountains (Dawkins). It occurs in the Pre-Glacial forest-bed of Cromer in Norfolk, survived the Glacial period, and is found abundantly in Post-Glacial deposits in France, Germany, Britain, Russia in Europe, Asia, and North America, being often associated with the Reindeer, Lemming, and Musk-Ox. That it survived into the earlier portion of the human period is unquestionable, its remains having been found in a great number of instances associated with implements of human manufacture; whilst in one instance a recognizable portrait of it has been discovered, carved on bone.

Amongst other Elephants which occur in Post-Pliocene deposits may be mentioned, as of special interest, the pigmy Elephants of Malta. One of these—the *Elephas Melitensis,* or so-called "Donkey-Elephant"—was not more than four and a half feet in height. The other—the *Elephas Falconeri,* of Busk—was still smaller, its average height at the withers not exceeding two and a half to three feet.

Whilst herbivorous animals abounded during the Post-Pliocene, we have ample evidence of the coexistence with them of a number of Carnivorous forms, both in the New and the Old World. The Bears are represented in Europe by at least three species, two of which—namely, the great Grizzly Bear (*Ursus ferox*) and the smaller Brown Bear (*Ursus arctos*) —are in existence at the present day. The third species is the celebrated Cave-bear (*Ursus spelæus,* fig. 268), which is now

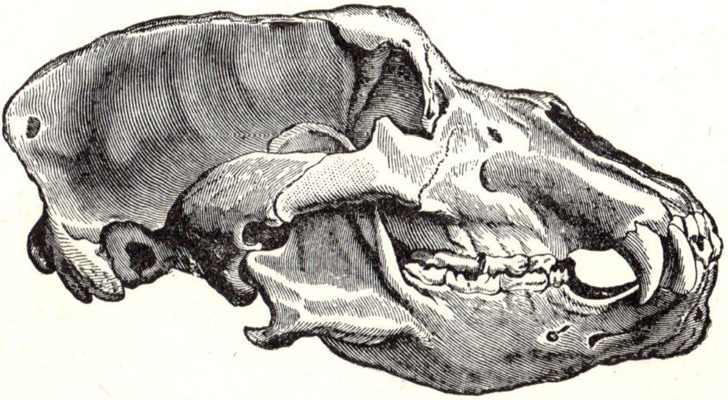

Fig. 268.—Skull of *Ursus spelæus.* Post-Pliocene, Europe. One-sixth of the natural size.

extinct. The Cave-bear exceeded in its dimensions the largest of modern Bears; and its remains, as its name implies, have been found mainly in cavern-deposits. Enormous numbers of this large and ferocious species must have lived in Europe in Post-Glacial times; and that they survived into the human period, is clearly shown by the common association of their bones with the implements of man. They are occasionally accompanied by the remains of a Glutton (the *Gulo spelæus*), which does not appear to be really separable from the existing Wolverine or Glutton of northern regions (the *Gulo luscus*). In addition, we meet with the bones of the Wolf, Fox, Weasel, Otter, Badger, Wild Cat, Panther, Hyæna, and Lion, &c., together with the extinct *Machairodus or* "Sabre-toothed Tiger." The only two of these that deserve further mention are the Hyæna and the Lion. The Cave-hyæna (*Hyæna spelæa*, fig. 269) is regarded by high authorities as nothing more than a variety of the living Spotted Hyæna (*H. crocuta*) of South Africa. This well-known species inhabited Britain and a considerable portion of Europe during a large part of the Post-Pliocene period; and its remains often occur in great abundance. Indeed, some caves, such as the Kirkdale Cavern in Yorkshire, were dens inhabited during long periods by these animals, and thus contain the remains of numerous individuals and of successive generations of Hyænas, together with innumerable gnawed and bitten bones of their prey. That the

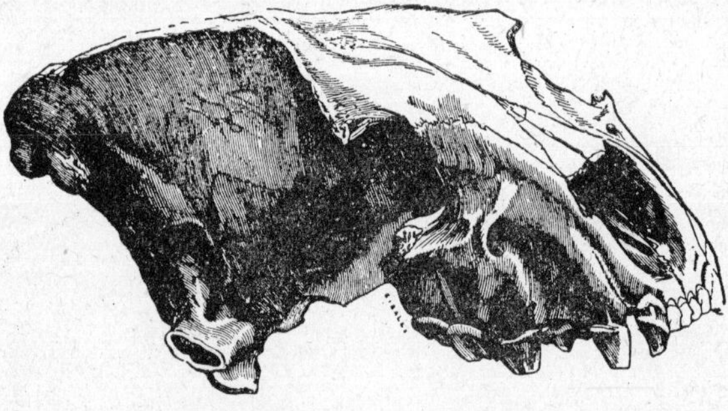

Fig. 269.—Skull of *Hyæna spelæa*, one-fourth of the natural size. Post-Pliocene, Europe.

Cave-hyæna was a contemporary with Man in Western Europe during Post-Glacial times is shown beyond a doubt by the common association of its bones with human implements.

Lastly, the so-called Cave-lion (*Felis Spelæa*), long supposed to be a distinct species, has been shown to be nothing more than a large variety of the existing Lion (*Felis leo*). This animal inhabited Britain and Western Europe in times posterior to the Glacial period, and was a contemporary of the Cave-hyæna, Cave-bear, Woolly Rhinoceros, and Mammoth. The Cave-lion also unquestionably survived into the earlier portion of the human period in Europe.

The Post-Pliocene deposits of Europe have further yielded the remains of numerous *Rodents*—such as the Beaver, the Northern Lemming, Marmots, Mice, Voles, Rabbits, &c.—together with the gigantic extinct Beaver known as the *Trogontherium Cuvieri* (fig. 270). The great *Castoroides Ohioensis* of the Post-Pliocene of North America is also a great extinct Beaver, which reached a length of about five feet. Lastly, the Brazilian bone-caves have yielded the remains of numerous Rodents of types now characteristic of South America, such as Guinea-pigs, Capybaras, tree-inhabiting Porcupines, and Coypus.

Fig. 270.—Lower jaw of *Trogontherium Cuvieri*, one-fourth of the natural size. Post-Pliocene, Britain.

The deposits just alluded to have further yielded the remains of various Monkeys, such as Howling Monkeys, Squirrel Monkeys, and Marmosets, all of which belong to the group of *Quadrumana* which is now exclusively confined to the South American continent—namely, the "Platyrhine" Monkeys.

We still have very briefly to consider the occurrence of Man in Post-Pliocene deposits; but before doing so, it will be well to draw attention to the evidence afforded by the Post-Pliocene Mammals as to the climate of Western Europe at this period. The chief point which we have to notice is, that a considerable revolution of opinion has taken place on this point. It was originally believed that the presence of such animals as Elephants, Lions, the Rhinoceros, and the Hippopotamus afforded an irrefragable proof that the climate of Europe must have been a warm one, at any rate during Post-Glacial times. The existence, also, of numbers of Mammoths in Siberia, was further supposed to indicate that this high temperature extended itself very far north. Upon the whole, however, the evidence is against this view. Not only is there great difficulty in supposing that the Arctic conditions of the Glacial period were immediately followed by anything warmer than a cold-temperate climate; but there is nothing in the nature of the Mammals themselves which would absolutely forbid their living in a temperate climate. The *Hippopotamus major,* though probably clad in hair, offers some difficulty—since, as pointed out by Professor Busk, it must have required a climate sufficiently warm to insure that the rivers were not frozen over in the winter; but it was probably a migratory animal, and its occurrence may be accounted for by this. The Wooly Rhinoceros and the Mammoth are known with certainty to have been protected with a thick covering of wool and hair; and their extension northwards need not necessarily have been limited by anything except the absence of a sufficiently luxuriant vegetation to afford them food. The great American Mastodon, though not certainly known to have possessed a hairy covering, has been shown to have lived upon the shoots of Spruce and Firs, trees characteristic of temperate regions—as shown by the undigested food which has been found with its skeleton, occupying the place of the stomach. The Lions and Hyænas, again, as shown by Professor Boyd Dawkins, do not indicate necessarily a warm climate. Wherever a sufficiency of herbivorous animals to supply them with food can live, there they can live also; and they have therefore no special bearing upon the question of climate. After a review of the whole evidence, Professor Dawkins concludes that the nearest approach at the present day to the Post-Pliocene climate of Western Europe is to be found in the climate of the great Siberian plains which stretch from the Altai Mountains to the Frozen Sea. "Covered by impenetrable forests, for the most part of Birch, Poplar, Larch, and Pines, and low creeping dwarf Cedars, they present every gradation in climate from the temperate to that in which the cold is too severe to admit of the growth of trees, which decrease in size as the traveler advances northwards, and are replaced by the grey mosses and lichens that cover the low marshy 'tundras.' The maximum winter cold, registered by Admiral Von Wrangel at Nishne Kolymsk, on the banks of the Kolyma, is—65° in January. 'Then breathing becomes difficult; the Reindeer, that citizen of the Polar region, withdraws to the deepest thicket of the forest, and stands there motionless as if deprived of life;' and trees burst asunder with the cold. Throughout this area roam Elks, Black Bears, Foxes, Sables, and Wolves, that afford subsistence to the

Jakutian and Tungusian fur-hunters. In the northern part countless herds of Reindeer, Elks, Foxes, and Wolverines make up for the poverty of vegetation by the rich abundance of animal life. 'Enormous flights of Swans, Geese, and Ducks arrive in the spring, and seek deserts where they may moult and build their nests in safety. Ptarmigans run in troops amongst the bushes; little Snipes are busy along the brooks and in the morasses; the social Crows seek the neighborhood of new habitations; and when the sun shines in spring, one may even sometimes hear the cheerful note of the Finch, and in autumn that of the Thrush.' Throughout this region of woods, a hardy, middle-sized breed of horses lives under the mastership and care of man, and is eminently adapted to bear the severity of the climate. . . . The only limit to their northern range is the difficulty of obtaining food. The severity of the winter through the southern portion of this vast wooded area is almost compensated for by the summer heat and its marvellous effect on vegetation."—(Dawkins, 'Monograph of Pleistocene Mammalia.')

Finally, a few words must be said as to the occurrence of the remains of Man in Post-Pliocene deposits. That Man existed in Western Europe and in Britain during the Post-pliocene period, is placed beyond a doubt by the occurrence of his bones in deposits of this age, along with the much more frequent occurrence of implements of human manufacture. At what precise point of time during the Post-Pliocene period he first made his appearance is still a matter of conjecture. Recent researches would render it probable that the early inhabitants of Britain and Western Europe were witnesses of the stupendous phenomena of the Glacial period; but this cannot be said to have been demonstrated. That Man existed in these regions during the Post-Glacial division of Post-Pliocene time cannot be doubted for a moment. As to the physical peculiarities of the ancient races that lived with the Mammoth and the Woolly Rhinoceros, little is known compared with what we may some day hope to know. Such information as we have, however, based principally on the skulls of the Engis, Neanderthal, Cro-Magnon, and Bruniquel caverns, would lead to the conclusion that Post-Pliocene Man was in no respect inferior in his organization to, or less highly developed than, many existing races. All the known skulls of this period, with the single exception of the Neanderthal cranium, are in all respects average and normal in their characters; and even the

Neanderthal skull possessed a cubic capacity at least equal to that of some existing races. The implements of Post-Pliocene Man are exclusively of stone or bone; and the former are invariably of rude shape and *undressed.* These "palæolithic" tools (Gr. *palaios,* ancient; *lithos,* stone) point to a very early condition of the arts; since the men of the earlier portion of the Recent period, though likewise unacquainted with the metals, were in the habit of polishing or dressing the stone implements which they fabricated.

It is impossible here to enter further into this subject; and it would be useless to do so without entering as well into a consideration of the human remains of the Recent period—a period which lies outside of the province of the present work. So far as Post-Pliocene Man is concerned, the chief points which the palæontological student has to remember have been elsewhere summarized by the author as follows:—

1. Man unquestionably existed during the later portion of what Sir Charles Lyell has termed the "Post-Pliocene" period. In other words, Man's existence dates back to a time when several remarkable Mammals, previously mentioned, had not yet become extinct; but he does not date back to a time anterior to the present *Molluscan* fauna.

2. The antiquity of the so-called Post-Pliocene period is a matter which must be mainly settled by the evidence of Geology proper, and need not be discussed here.

3. The extinct Mammals with which man coexisted in Western Europe are mostly of large size, the most important being the Mammoth (*Elephas primigenius*), the Woolly Rhinoceros (*Rhinoceros tichorhinus*), the Cave-lion (*Felis spelæa*), the Cave-hyæna (*Hyæna spelæa*), and the Cave-bear (*Ursus spelæus*). We do not know the cause which led to the extinction of these Mammals; but we know that hardly any Mammalian species has become extinct during the historical period.

4. The extinct Mammals with which man coexisted are referable in many cases to species which presumably required a very different climate to that now prevailing in Western Europe. How long a period, however, has been consumed in the bringing about of the climatic changes thus indicated, we have no means of calculating with any approach to accuracy.

5. Some of the deposits in which the remains of man have been found associated with the bones of extinct Mammals, are such as to show incontestably that great changes in the physical geography and surface-configuration of Western Europe

have taken place since the period of their accumulation. We have, however, no means at present of judging of the lapse of time thus indicated except by analogies and comparisons which may be disputed.

6. The human implements which are associated with the remains of extinct Mammals, themselves bear evidence of an exceedingly barbarous condition of the human species. Post-Pliocene or "Palæolithic" Man was clearly unacquainted with the use of any of the metals. Not only so, but the workmanship of these ancient races was much inferior to that of the later tribes, who were also ignorant of the metals, and who also used nothing but weapons and tools of stone, bone, &c.

7. Lastly, it is only with the human remains of the Post-Pliocene period that the palæontologist proper has to deal. When we enter the "Recent" period, in which the remains of Man are associated with those of *existing species of Mammals*, we pass out of the region of pure palæontology into the domain of the Archæologist and the Ethnologist.

LITERATURE.

The following are some of the principal works and memoirs to which the student may refer for information as to the Post-Pliocene deposits and the remains which they contain, as well as the primitive races of mankind:—

(1) 'Elements of Geology.' Lyell.
(2) 'Antiquity of Man.' Lyell.
(3) 'Palæontological Memoirs.' Falconer.
(4) 'The Great Ice-age.' James Geikie.
(5) 'Manual of Palæontology.' Owen.
(6) 'British Fossil Mammals and Birds.' Owen.
(7) 'Cave-Hunting.' Boyd Dawkins.
(8) 'Prehistoric Times.' Lubbock.
(9) 'Ancient Stone Implements.' Evans.
(10) 'Prehistoric Man.' Daniel Wilson.
(11) 'Prehistoric Races of the United States.' Foster.
(12) 'Manual of Geology.' Dana.
(13) 'Monograph of Pleistocene Mammalia.' (Palæontographical Society). Boyd Dawkins and Sanford.
(14) 'Monograph of the Post-Tertiary Entomostraca of Scotland, &c., with an Introduction on the Post-Tertiary Deposits of Scotland' (Ibid.) G. S. Brady, H. W. Crosskey, and D. Robertson.
(15) "Reports on Kent's Cavern"—'British Association Reports.' Pengelly.
(16) "Reports on the Victoria Cavern, Settle"—'British Association Reports.' Tiddeman.
(17) 'Ossemens Fossiles.' Cuvier.

(18) 'Reliquiæ Diluvianæ.' Buckland.
(19) "Fossil Mammalia"—'Zoology of the Voyage of the Beagle.' Owen.
(20) 'Description of the Tooth and Part of the Skeleton of the *Glyptodon*.' Owen.
(21) "Memoir on the Extinct Sloth Tribe of North America"—'Smithsonian Contributions to Knowledge.' Leidy.
(22) "Report on Extinct Mammals of Australia"—'British Association,' 1844. Owen.
(23) 'Description of the Skeleton of an Extinct Gigantic Sloth (*Mylodon robustus*).' Owen.
(24) "Affinities and Probable Habits of Thylacoleo"—'Quart. Journ. Geol. Soc.,' vol. xxiv. Flower.
(25) 'Prodromus of the Palæontology of Victoria.' M'Coy.
(26) 'Les Ossemens Fossiles des Cavernes de Liège.' Schmerling.
(27) 'Die Fauna der Pfahlbauten in der Schweiz.' Rütimeyer.
(28) "Extinct and Existing Bovine Animals of Scandinavia"—'Annals of Natural History,' ser. 2, vol. iv., 1849. Nilsson.
(29) 'Man's Place in Nature.' Huxley.
(30) 'Les Temps Antéhistoriques en Belgique.' Dupont.
(31) "Classification of the Pleistocene Strata of Britain and the Continent"—'Quart. Journ. Geol. Soc.,' vol. xxviii. Boyd Dawkins.
(32) 'Distribution of the Post-Glacial Mammalia' (Ibid.) vol. xxv. Boyd Dawkins.
(33) 'On British Fossil Oxen' (Ibid.), vols. xxii. and xxiii. Boyd Dawkins.
(34) 'British Prehistoric Mammals' (Congress of Prehistoric Archæology, 1868). Boyd Dawkins.
(35) 'Reliquiæ Aquitanicæ.' Lartet and Christy.
(36) 'Zoologie et Paléontologie Françaises.' Gervais.
(37) 'Notes on the Post-Pliocene Geology of Canada.' Dawson.
(38) "On the Connection between the existing Fauna and Flora of Great Britain and certain Geological Changes"—'Mem. Geol. Survey.' Edward Forbes.
(39) 'Cavern-Researches.' M'Enery. Edited by Vivian.
(40) "Quaternary Gravels"—'Quart. Journ. Geol. Soc.,' vol. xxv. Tylor.

CHAPTER XXIII.

THE SUCCESSION OF LIFE UPON THE GLOBE.

In conclusion, it may not be out of place if we attempt to summarize, in the briefest possible manner, some of the principal results which may be deduced as to the succession of life upon the earth from the facts which have in the preceding portion of this work been passed in review. That there was a time when the earth was void of life is universally admitted, though it may be that the geological record gives us no direct evidence of this. That the globe of to-day is peopled with innumerable forms of life whose term of existence has been, for the most part, but as it were of yesterday, is likewise an assertion beyond dispute. Can we in any way connect the present with the remote past, and can we indicate even imperfectly the conditions and laws under which the existing order was brought about? The long series of fossiliferous deposits, with their almost countless organic remains, is the link between what has been and what is; and if any answer to the above question can be arrived at, it will be by the careful and conscientious study of the facts of Palæontology. In the present state of our knowledge, it may be safely said that anything like a dogmatic or positive opinion as to the precise sequence of living forms upon the globe, and still more as to the manner in which this sequence may have been brought about, is incapable of scientific proof. There are, however, certain general deductions from the known facts which may be regarded as certainly established.

In the first place, it is certain that there has been a *succession* of life upon the earth, different specific and generic types succeeding one another in successive periods. It follows from this, that the animals and plants with which we are familiar as living, were not always upon the earth, but that they have been preceded by numerous races more or less differing from them. What is true of the species of animals and plants, is true also of the higher zoological divisions; and it is, in the second place, quite certain that there has been a similar *succession* in the order of appearance of the primary groups ("sub-kingdoms," "classes," &c.) of animals and vegetables. These great groups did not all come into existence at once, but they made their appearance successively. It is true that we cannot be said to be certainly acquainted with the first *absolute* appearance of any great group of animals. No one dare assert positively that the apparent first appearance of Fishes in the Upper Silurian is really their first introduction upon the earth: indeed, there is a strong probability against any such supposition. To whatever extent, however, future discoveries may push back the first advent of any or all of the great groups of life, there is no likelihood that anything will be found out which will materially alter the *relative* succession of these groups as at present known to us. It is not likely, for example, that the future has in store for us any discovery by which it would be shown that Fishes were in existence before Molluscs, or that Mammals made their appearance before Fishes. The sub-kingdoms of Invertebrate animals were all represented in Cambrian times—and it might therefore be inferred that *these* had all come simultaneously into existence; but it is clear that this inference, though incapable of actual disproof, is in the last degree improbable. Anterior to the Cambrian is the great series of the Laurentian, which, owing to the metamorphism to which it has been subjected, has so far yielded but the singular *Eozoön.* We may be certain, however, that others of the Invertebrate sub-kingdoms besides the Protozoa were in existence in the Laurentian period; and we may infer from known analogies that they appeared successively, and not simultaneously.

When we come to smaller divisions than the sub-kingdoms—such as classes, orders, and families—a similar succession of groups is observable. The different classes of any given sub-kingdom, or the different orders of any given class, do not make their appearance together and all at once, but they are introduced upon the earth in *succession.* More than this, the different classes of a sub-kingdom, or the different orders of a class, *in the main succeed one another in the relative order of their zoological rank—the lower groups appearing first and the higher groups last.* It is true that in the Cambrian formation—the earliest series of sediments in which fossils are abundant—we find numerous groups, some very low, others very high, in the zoological scale, which *appear* to have simultaneously flashed into existence. For reasons stated above, however, we cannot accept this appearance as real; and we must believe that many of the Cambrian groups of animals really came into being long before the commencement of the Cambrian period. At any rate, in the long series of fossiliferous deposits of later date than the Cambrian the above-stated rule holds good as a broad generalization—that the lower groups, namely, precede the higher in point of time; and though there are apparent exceptions to the rule, there are none of such a nature as not to admit of explanation. Some of the leading facts upon which this generalization is founded will be enumerated immediately; but it will be well, in the first place, to consider briefly what we precisely mean when we speak of "higher" and "lower" groups.

It is well known that naturalists are in the habit of "classifying" the innumerable animals which now exist upon the globe; or, in other words, of systematically arranging them into groups. The precise arrangement adopted by one naturalist may differ in minor details from that adopted by another; but all are agreed as to the fundamental points of classification, and all, therefore, agree in placing certain groups in a certain sequence. What, then, is the principle upon which this sequence is based? Why, for example, are the Sponges placed below the Corals; these below the Sea-Urchins; and these, again, below the Shell-fish? Without entering into a discussion of the principles of zoological classification, which would here be out of place, it must be sufficient to say that the sequence in question is based upon the *relative type of organization* of the groups of animals classified. The Corals are placed above the Sponges upon the ground that, regarded as a whole, the *plan or type of structure* of a Coral is more complex than that of a Sponge. It is not in the slightest degree that the Sponge is in any respect less highly organized or less perfect, *as a Sponge,* than is the Coral as a Coral? Each is equally perfect in its own way; but the structural pattern of the Coral is the highest, and therefore it occupies a higher place in the zoological scale. It is upon this principal, then, that the primary subdivisions of the animal kingdom (the so-called "sub-kingdoms") are arranged in a certain order. Coming, again, to the minor subdivisions (classes, orders, &c.) of each sub-kingdom, we find a different but entirely analogous principal employed as a means of classification. The numerous animals belonging to any given sub-kingdom are formed upon the same fundamental plan of structure; but they nevertheless admit of being arranged in a regular series of groups. All the Shell-fish, for example, are built upon a common plan, this plan representing the ideal Mollusc; but there are at the same time various groups of the *Mollusca,* and these groups admit of an arrangement in a given sequence. The principle adopted in this case is simply of *the relative elaboration of the common type.* The Oyster is built upon the same ground-plan as the Cuttle-fish; but this plan is carried out with much greater elaboration, and with many more complexities, in the latter than in the former: and in accordance with this, the *Cephalopoda* consistitute a *higher* group than the Bivalve Shell-fish. As in the case of superiority of structural type, so in this case also, it is not in the least that the Oyster is an *imperfect* animal. On the contrary, it is just as perfectly adapted by its organization to fill its own sphere and to meet the exigencies of its own existence as is the Cuttle-fish; but the latter lives a life which is, physiologically, higher than the former, and its organization is correspondingly increased in complexity.

This being understood, it may be repeated that, in the main, the succession of life upon the globe in point of *time* has corresponded with the relative order of succession of the great groups of animals in *zoological rank;* and some of the more striking examples of this may be here alluded to. Amongst the *Echinoderms,* for instance, the two orders generally admitted to be the "lowest" in the zoological scale—namely, the *Crinoids* and the *Cystoids*—are likewise the oldest, both appearing in the Cambrian, the former slowly dying out as we approach the Recent period, and the latter disappearing wholly before the close of the Palæozoic period. Amongst the *Crustaceans,* the ancient groups of the Trilobites, Ostracodes, Phyllopods, Eurypterids, and Limuloids, some of which exist at the present day, are all "low" types; whereas the highly-organized Decapods do not make their appearance till the near the close of the Palæozoic epoch, and they do not become abundant till we reach Mesozoic times. Amongst the *Mollusca,* those Bivalves which possess breathing-tubes (the "siphonate" Bivalves) are generally admitted to be higher than those which are destitute of these organs (the "asiphonate" Bivalves); and the latter are especially characteristic of the Palæozoic period, whilst the former abound in Mesozoic and Kainozoic formations. Similarly, the Univalves with breathing-tubes and a corresponding notch in the mouth of the shell ("siphonostomatous" Univalves) are regarded as higher in the scale than the round-mouthed vegetable-eating Sea-snails, in which no respiratory siphons exist ("holostomatous" Univalves); but the latter abound in the Palæozoic rocks—whereas the former do not make their appearance till the Jurassic period, and their higher groups do not seem to have existed till the close of the Cretaceous. The *Cephalopods,* again—the highest of all the groups of Mollusca—are represented in the Palæozoic rocks exclusively by Tetrabranchiate forms, which constitute the lowest of the two orders of this class; whereas the more highly specialized Dibranchiates do not make their appearance till the commencement of the Mesozoic. The Palæozoic Tetrabranchiates, also are of a much simpler type than the highly complex *Ammonitidæ* of the Mesozoic.

Similar facts are observable amongst the *Vertebrate animals.* The *Fishes* are the lowest class of Vertebrates, and they are the first to appear, their first certain occurrence being in the Upper Silurian; whilst, even if the Lower Silurian and Upper Cambrian "Conodonts" were shown to be the teeth of Fishes, there would still remain the enormously long periods of the Laurentian and Lower Cambrian, during which there were Invertebrates, but no Vertebrates. The *Amphibians,* the next class in zoological order, appears later than the Fishes, and is not represented till the Carboniferous; whilst its highest group (that of the Frogs and Toads) does not make its entrance upon the scene till Tertiary times are reached. The class of the *Reptiles,* again, the next in order, does not appear till the Permian, and therefore not till after Amphibians of very varied forms had been in existence for a protracted period. The *Birds* seem to be undoubtedly later than the Reptiles; but, owing to the uncertainty as to the exact point of their first appearance, it cannot be positively asserted that they preceded Mammals, as they should have done. Finally, the Mesozoic types of *Mammals* are mainly, if not exclusively, referable to the *Marsupials,* one of the lowest orders of the class; whilst the higher orders of the "Placental" Quadrupeds are not with certainty known to have existed prior to the commencement of the Tertiary period.

Facts of a very similar nature are offered by the succession of Plants upon the globe. Thus the vegetation of the Palæozoic period consisted principally of the lowly-organized groups of the Cryptogamous or Flowerless plants. The Mesozoic formations, up to the Chalk, are especially characterized by the naked-seeded Flowering plants—the Conifers and the Cycads; whilst the higher groups of the Angiospermous Exogens and Monocotyledons characterize the Upper Cretaceous and Tertiary rocks.

Facts of the above nature—and they could be greatly multiplied—seem to point clearly to the existence of some law of progression, though we certainly are not yet in a position to formulate this law, or to indicate the precise manner in which it has operated. Two considerations, also, must not be overlooked. In the first place, there are various groups, some of them highly organized, which make their appearance at an extremely ancient date, but which continue throughout geological time almost unchanged, and certainly unprogressive. Many of these "persistent types" are known—such as various of the *Foraminifera,* the *Lingulæ,* the *Nautili,* &c.; and they indicate that under given conditions, at present unknown to us, it is possible for a life-form to subsist for an almost indefinite period without any important modification of its structure. In the second place, whilst the facts above mentioned point to some general law of progression of the great zoological groups, it cannot be asserted that the primeval types *of any given group* are necessarily "lower," zoologically speaking, than their modern representatives. Nor does this seem to be at all necessary for the establishment of the law in question. It cannot be asserted, for example, that the Ganoid and Placoid Fishes of the Upper Silurian are in themselves less highly organized than their existing representatives; nor can it even be asserted that the Ganoid and Placoid orders are low *groups* of the class *Pisces.* On the contrary, they are high groups; but then it must be remembered that these are probably not really the first Fishes, and that if we meet with Fishes at some future time in the Lower Silurian or Cambrian, these may easily prove to be representatives of the lower orders of the class. This question cannot be further entered into here, as its discussion could be carried out to an almost unlimited length; but whilst there are facts pointing both ways, it appears that at present we are not justified in asserting that the earlier types of each group—so far as these are known to us, or really are without predecessors—are *necessarily* or *invariably* more "degraded" or "embryonic" in their structure than their more modern representatives.

It remains to consider very briefly how far Palæontology supports the doctrine of "Evolution," as it is called; and this, too, is a question of almost infinite dimensions, which can but be glanced at here. Does Palæontology teach us that the almost innumerable kinds of animals and plants which we known to have successively flourished upon the earth in past times were produced separately and wholly independently of each other, at successive periods? or does it point to the theory that a large number of these supposed distinct forms have been in reality produced by the slow modification of a comparatively small number of primitive types? Upon the whole, it must be unhesitatingly replied that the evidence of Palæontology is in favor of the view that the succession of life-forms upon the globe has been to a large extent regulated by some orderly and constantly-acting law of modification and evolution. Upon no other theory can we comprehend how the fauna of any given formation is more closely related to that of the formation next below in the series, and to that of the formation next above, than to that of any other series of deposits. Upon no other view can we comprehend why the Post-Tertiary Mammals of South America should consist principally of Edentates, Llamas, Tapirs, Peccaries, Platyrhine Monkeys, and other forms now characterizing this continent; whilst those of Australia should be wholly referable to the order of Marsupials. On no other view can we explain the common occurrence of "intermediate" or "transitional" forms of life, filling in the gaps between groups now widely distinct.

On the other hand, there are facts which point clearly to the existence of some law other than that of evolution, and probably of a deeper and more far-reaching character. Upon no theory of evolution can we find a satisfactory explanation for the constant introduction throughout geological time of new forms of life, which do not appear to have been preceded by pre-existent allied types. The Graptolites and Trilobites have no known predecessors, and leave no known successors. The Insects appear suddenly in the Devonian, and the Arachnides and Myriapods in the Carboniferous, under well-differentiated and highly-specialized types. The Dibranchiate Cephalopods appear with equal apparent suddenness in the older Mesozoic deposits, and no known type of the Palæozoic period can be pointed to as a possible ancestor. The *Hippuritidæ* of the Cretaceous burst into a varied life to all appearance almost immediately after their first introduction into existence. The wonderful Dicotyledonous flora of the Upper Cretaceous period similarly surprises us without any prophetic annunciation from the older Jurassic.

Many other instances could be given; but enough has been said to show that there is a good deal to be said on both sides, and that the problem is one environed with profound difficulties. One point only seems now to be universally conceded, and that is, that the record of life in past time is not interrupted by gaps other than those due to the necessary imperfections of the fossiliferous series, to the fact that many animals are incapable of preservation in a fossil condition, or to other causes of a like nature. All those who are entitled to speak on this head are agreed that the introduction of new and the destruction of old species have been slow and gradual processes, in no sense of the term "catastrophistic." Most are also willing to admit that "Evolution" has taken place in the past, to a greater or less extent, and that a greater or less number of so-called species of fossil animals are really the modified descendants of pre-existent forms. *How* this process of evolution has been effected, to what extent it has taken place, under what conditions and laws it has been carried out, and how far it may be regarded as merely auxiliary and supplemental to some deeper law of change and progress, are questions to which, in spite of the brilliant generalizations of Darwin, no satisfactory answer can as yet be given. In the successful solution of this problem—if soluble with the materials available to our hands—will lie the greatest triumph that Palæontology can hope to attain; and there is reason to think that, thanks to the guiding-clue afforded by the genius of the author of the 'Origin of Species,' we are at least on the road to a sure, though it may be a far-distant, victory.

APPENDIX.

TABULAR VIEW OF THE CHIEF DIVISIONS OF THE ANIMAL KINGDOM.

(Extinct groups are marked with an asterisk. Groups not represented at all as fossils are marked with two asterisks.)

INVERTEBRATE ANIMALS.

SUB-KINGDOM I.—PROTOZOA.

Animal simple or compound; body composed of "sarcode," not definitely segmented; no nervous system; and no digestive apparatus, beyond occasionally a mouth and gullet.

CLASS I. GREGARINIDÆ. **
CLASS II. RHIZOPODA.
Order 1. *Monera.* **
Order 2. *Amœbea.* **
Order 3. *Foraminifera.*
Order 4. *Radiolaria* (Polycystines, &c.)
Order 5. *Spongida* (Sponges).
CLASS III. INFUSORIA. **

SUB-KINGDOM II.—CŒLENTERATA.

Animal simple or compound; body-wall composed of two principal layers; digestive canal freely communicating with the general cavity of the body; no circulating organs, and no nervous system or a rudimentary one; mouth surrounded by tentacles, arranged, like the internal organs, in a "radiate" or star-like manner.

CLASS I. HYDROZOA.

Sub-class 1. *Hydroida* ("Hydroid Zoophytes"). *Ex.* Freshwater Polypes,** Pipe-corallines (*Tubularia*), Sea-Firs (*Sertularia*).

Sub-class 2. *Siphonophora*** ("Oceanic Hydrozoa"). *Ex.* Portuguese Man-of-war (*Physalia*).

Sub-class 3. *Discophora* ("Jelly-fishes"). Only known as fossils by impressions of their stranded carcasses.

Sub-class 4. *Lucernarida* ("Sea-blubbers"). Also only known as fossils by impressions left in fine-grained strata.

Sub-class 5. *Graptolitidæ** ("Graptolites").

CLASS II. ACTINOZOA.

Order 1. *Zoantharia. Ex.* Sea-anemones** (*Actinidæ*), Star-corals (*Astræidæ*).

Order 2. *Alcyonaria. Ex.* Sea-pens (*Pennatula*), Organ-pipe Coral (*Tubipora*), Red Coral (*Corallium*).

Order 3. *Rugosa* ("Rugose Corals").

Order 4. *Ctenophora.*** *Ex.* Venus's Girdle (*Cestum*).

SUB-KINGDOM III.—ANNULOIDA.

Animals in which the digestive canal is completely shut off from the cavity of the body; a distinct nervous system; a system of branched "water-vessels," which usually communicate with the exterior. Body of the adult often "radiate," and never composed of a succession of definite rings.

CLASS I. ECHINODERMATA.

Order 1. *Crinoidea* ("Sea-lilies"). *Ex.* Feather-star (*Comatula*), Stone-lily (*Encrinus**).

Order 2. *Blastoidea** ("Pentremites").

Order 3. *Cystoidea** ("Globe-lilies").

Order 4. *Ophiuroidea* ("Brittle-stars"). *Ex.* Sand-stars (*Ophiura*), Brittle-stars (*Ophiocoma*).

Order 5. *Asteroidea* ("Star-fishes"). *Ex.* Cross-fish (*Uraster*), Sun-star (*Solaster*).

Order 6. *Echinoidea* ("Sea-urchins"). *Ex.* Sea-eggs (*Echinus*), Heart-urchins (*Spatangus*).

Order 7. *Holothuroidea* ("Sea-cucumbers"). *Ex.* Trepangs (*Holothuria*).

CLASS II. SCOLECIDA** (Intestinal Worms, Wheel Animalcules, &c.)

SUB-KINGDOM IV.—ANNULOSA.

Animal composed of numerous definite segments placed one behind the other; nervous system forming a knotted cord placed along the lower (ventral) surface of the body.

Division A. Anarthropoda. No jointed limbs.

CLASS I. GEPHYREA ** ("Spoon-worms").

CLASS II. ANNELIDA ("Ringed-worms"). *Ex.* Leeches ** (*Hirudinea*), Earthworms ** *Oligochæta*), Tube-worms (*Tubicola*), Sea-worms and Sea-centipedes (*Errantia*).

CLASS III. CHÆTOGNATHA ** ("Arrow-worms").

Division B. Arthropoda or Articulata. Limbs jointed to the body.

CLASS I. CRUSTACEA ("Crustaceans"). *Ex.* Barnacles and Acorn-shells (*Cirripedia*), Water-fleas (*Ostracoda*), Brine-shrimps and Fairy-shrimps (*Phyllopoda*), Trilobites * (*Trilobita*), King-crabs and Eurypterids * (*Merostomata*), Wood-lice and Slaters (*Isopoda*), Sand-hoppers (*Amphipoda*), Lobsters, Shrimps, Hermit-crabs, and Crabs (*Decapoda*).

CLASS II. ARACHNIDA. *Ex.* Mites (*Acarina*), Scorpions (*Pedipalpi*), Spiders (*Araneida*).

CLASS III. MYRIAPODA. *Ex.* Centipedes (*Chilopoda*), Millipedes and Galley-worms (*Chilognatha*).

CLASS IV. INSECTA ("Insects"). *Ex.* Field-bugs (*Hemiptera*); Crickets, Grasshoppers, &c. (*Orthoptera*); Dragon-flies and May-flies (*Neuroptera*); Gnats and House-flies (*Diptera*); Butterflies and Moths (*Lepidoptera*); Bees, Wasps and Ants (*Hymenoptera*); Beetles (*Coleoptera*).

SUB-KINGDOM V.—MOLLUSCA.

Animal soft-bodied, generally with a hard covering or shell; no distinct segmentation of the body; nervous system of scattered masses.

CLASS I. POLYZOA ("Sea-Mosses"). *Ex.* Sea-mats (*Flustra*), Lace-corals (*Fenestellidæ* *).

CLASS II. TUNICATA ** ("Tunicaries") *Ex.* Sea-squirts (*Ascidia*).

CLASS III. BRACHIOPODA ("Lamp-shells"). *Ex.* Goose-bill Lamp-shell (*Lingula*).

CLASS IV. LAMELLIBRANCHIATA ("Bivalves"). *Ex.* Oyster (*Ostrea*), Mussel (*Mytilus*), Scallop (*Pecten*), Cockle (*Cardium*).

CLASS V. GASTEROPODA ("Univalves"). *Ex.* Whelks (*Buccinum*), Limpets (*Patella*), Sea-slugs ** (*Doris*), Land-snails (*Helix*).

CLASS VI. PTEROPODA ("Winged Snails"). Ex *Hyalea, Cleodora.*

CLASS VII. CEPHALOPODA ("Cuttle-fishes"). *Ex.* Calamary (*Loligo*), Poulpe (*Octopus*), Paper Nautilus (*Argonauta*), Pearly Nautilus (*Nautilus*), Belemnites,* Orthoceratites,* Ammonites.*

VERTEBRATE ANIMALS.

SUB-KINGDOM VI.—VERTEBRATA.

Body composed of definite segments arranged longitudinally one behind the other; main masses of the nervous system placed dorsally; a backbone or "vertebral column" in the majority.

CLASS I. PISCES ("Fishes"). *Ex.* Lancelet** (*Amphioxus*); Lampreys and Hag-fishes (*Marsipobranchii***); Herring, Salmon, Perch, &c. (*Teleostei* or "Bony Fishes"); Gar-pike, Sturgeon, &c. (*Ganoidei*); Sharks, Dog-fishes, Rays, &c. (*Elasmobranchii* or "Placoids").

CLASS II. AMPHIBIA ("Amphibians"). Ex. *Labyrinthodontia,** Cæcilians,** Newts and Salamanders (*Urodela*), Frogs and Toads (*Anoura*).

CLASS III. REPTILIA ("Reptiles"). Ex. *Deinosauria,* Pterosauria,* Anomodontia,** Plesiosaurs (*Sauropterygia**), Ichthyosaurs (*Ichthyopterygia**), Tortoises and Turtles (*Chelonia*), Snakes (*Ophidia*), Lizards (*Lacertilia*), Crocodiles (*Crocodilia*).

CLASS IV. AVES ("Birds"). *Ex.* Toothed Birds (*Odontornithes**); Lizard-tailed Birds (*Archæopteryx**); Ducks, Geese, Gulls, &c. (*Natatores*); Storks, Herons, Snipes, Plovers, &c. (*Grallatores*); Ostrich, Emeu, Cassowary, Dinornis,* Æpiornis,* &c. (*Cursores*); Fowls, Game Birds and Doves (*Rasores*); Cuckoos, Woodpeckers, Parrots, &c. (*Scansores*); Crows, Starlings, Finches, Humming-birds, Swallows, &c. (*Insessores*); Owls, Hawks, Eagles, Vultures (*Raptores*).

CLASS V. MAMMALIA ("Quadrupeds"). *Ex.* Duck-mole and Spiny Ant-eater (*Monotremata***); Kangaroos, Phalangers, Opossums, Tasmanian Devil, &c. (*Marsupialia*); Sloths, Ant-eaters, Armadillos (*Edentata*); Manatees and Dugongs (*Sirenia*); Whales, Dolphins, Porpoises (*Cetacea*); Rhinoceros, Tapir, Horses, Hippopotamus, Pigs, Camels and Llamas, Giraffes, Deer, Antelopes, Sheep, Goats, Oxen (*Ungulata*); Hyrax (*Hyracoidea***); Elephants, Mastodon,* Deinotherium* (*Proboscidea*); Seals, Walrus, Bears, Dogs, Wolves, Cats, Lions, Tigers, &c. (*Carnivora*); Hares, Rabbits, Porcupines, Beavers, Rats, Mice, Lemmings Squirrels, Marmots, &c. (*Rodentia*); *Bats* (*Cheiroptera*); Moles, Shrew-mice, Hedgehogs (*Insectivora*); Lemurs, Spider-monkeys, Macaques, Baboons, Apes (*Quadrumana*); Man (*Bimana*).

GLOSSARY

ABDOMEN (Lat. *abdo,* I conceal). The posterior cavity of the body, containing the intestines and others of the viscera. In many Invertebrates there is no separation of the body-cavity into thorax and abdomen, and it is only in the higher *Annulosa* that a distinct abdomen can be said to exist.

ABERRANT (Lat. *aberro,* I wander away). Departing from the regular type.

ABNORMAL (Lat. *ab,* from; *norma,* a rule). Irregular; deviating from the ordinary standard.

ACRODUS (Gr. *akros,* high; *odous,* tooth). A genus of the Cestraciont fishes; so-called from the elevated teeth.

ACROGENS (Gr. *akros,* high; *gennao,* I produce). Plants which increase in height by additions made to the summit of the stem by the union of the bases of the leaves.

ACROTRETA (Gr. *akros,* high; *trētos,* pierced). A genus of Brachiopods, so-called from the presence of a foramen at the summit of the shell.

ACTINOCRINUS (Gr. *aktin,* a ray; *krinon,* a lily). A genus of Crinoids.

ACTINOZOA (Gr. *aktin,* a ray; and *zoön,* an animal). That division of the *Cœlenterata* of which the Sea-anemones may be taken as the type.

ÆGLINA (*Æglé,* a sea-nymph). A genus of Trilobites.

ÆPIORNIS (Gr. *aipus,* huge; *ornis,* bird). A genus of gigantic Cursorial birds.

AGNOSTUS (Gr. *a,* not; *gignosko,* I know). A genus of Trilobites.

ALCES (Lat. *alces,* elk). The European Elk or Moose.

ALECTO (the proper name of one of the Furies). A genus of *Polyzoa.*

ALETHOPTERIS (Gr. *alēthēs,* true; *pteris,* fern). A genus of Ferns.

ALGÆ (Lat. *alga,* a marine plant). The order of plants comprising the Sea-weeds and many fresh-water plants.

ALVEOLUS (Lat. *alvus,* belly). Applied to the sockets of the teeth.

AMBLYPTERUS (Gr. *amblus,* blunt; *pteron,* fin). An order of Ganoid Fishes.

AMBŌNYCHIA (Gr. *ambon,* a boss; *onux,* claw). A genus of Palæozoic Bivalves.

AMBULACRA (Lat. *ambulacrum,* a place for walking). The perforated spaces or "avenues" through which are protruded the tube-feet, by means of which locomotion is effected in the *Echinodermata.*

AMMONITIDÆ. A family of Tetrabranchiate Cephalopods, so-called from the resemblance of the shell of the type-genus, *Ammonites,* to the horns of the Egyptian God, Jupiter-Ammon.

AMORPHOZOA (Gr. *a,* without; *morphe,* shape; *zoön,* animal). A name sometimes used to designate the *Sponges.*

AMPHIBIA (Gr. *amphi,* both; *bios,* life). The Frogs, Newts, and the like, which have gills when young, but can always breathe air directly when adult.

AMPHICYON (Gr. *amphi,* both—implying doubt; *kuōn,* dog). An extinct genus of *Carnivora.*

AMPHILESTES (Gr. *amphi,* both; *lēstēs,* a thief). A genus of Jurassic Mammals.

AMPHISPONGIA (Gr. *amphi,* both; *spoggos,* sponge). A genus of Silurian sponges.

AMPHISTEGINA (Gr. *amphi,* both; *stegé,* roof). A genus of *Foraminifera.*

AMPHITHERIUM (Gr. *amphi,* both; *thērion,* beast). A genus of Jurassic Mammals.

AMPHITRAGULUS (Gr. *amphi,* both; dim. of *tragos,* goat). An extinct genus related to the living Musk-deer.

AMPLEXUS (Lat. an embrace). A genus of Rugose Corals.

AMPYX (Gr. *ampux,* a wreath or wheel). A genus of Trilobites.

ANARTHROPODA (Gr. *a,* without; *arthros,* a joint; *pous,* foot). That division of *Annulose* animals in which there are no articulated appendages.

ANCHITHERIUM (Gr. *agchi,* near; *thērion,* beast). An extinct genus of Mammals.

ANCYLOCERAS (Gr. *agkulos,* crooked; *ceras,* horn). A genus of *Ammonitidæ.*

ANCYLOTHERIUM (Gr. *agkulos,* crooked; *thērion,* beast). An extinct genus of Edentate Mammals.

ANDRIAS (Gr. *andrias,* image of man). An extinct genus of tailed Amphibians.

ANGIOSPERMS (Gr. *angeion,* a vessel; *sperma,* seed). Plants which have their seeds enclosed in a seed-vessel.

ANNELIDA (a Gallicised form of *Annulata*). The Ringed Worms, which form one of the divisions of the *Anarthropoda.*

ANNULARIA (Lat. *annulus,* a ring). A genus of Palæozoic plants, with leaves in whorls.

ANNULOSA (Lat. *annulus*). The sub-kingdom comprising the *Anarthropoda* and the *Arthropoda* or *Articulata,* in all of which the body is more or less evidently composed of a succession of rings.

ANOMODONTIA (Gr. *anomos,* irregular; *odous,* tooth). An extinct order of Reptiles, often called *Dicynodontia.*

ANOMURA (Gr. *anomos,* irregular; *oura,* tail). A tribe of Decapod *Crustacea,* of which the Hermit-crab is the type.

ANOPLOTHERIDÆ (Gr. *anoplos,* unarmed; *ther,* beast). A family of Tertiary Ungulates.

ANOURA (Gr. *a,* without; *oura,* tail). The order of *Amphibia* comprising the Frogs and Toads, in which the adult is destitute of a tail. Often called *Batrachia.*

ANTENNÆ (Lat. *antenna,* a yard-arm). The jointed horns or feelers possessed by the majority of the *Articulata.*

ANTENNULES (dim. of *Antennæ*). Applied to the smaller pair of antennæ in the *Crustacea.*

ANTHRACOSAURUS (Gr. *anthrax,* coal; *saura,* lizard). A genus of Labyrinthodont Amphibians.

ANTHRAPALÆMON (Gr. *anthrax,* coal; *palæmōn,* a prawn—originally a proper name). A genus of long-tailed Crustaceans from the Coal-measures.

ANTLERS. Properly the branches of the horns of the Deer tribe (*Cervidæ*), but generally applied to the entire horns.

APIOCRINIDÆ (Gr. *apion,* a pear; *krinon,* lily). A family of Crinoids—the "Pear-encrinites."

APTERYX (Gr. *a,* without; *pterux,* a wing). A wingless bird of New Zealand, belonging to the order *Cursores.*

AQUEOUS (Lat. *aqua,* water). Formed in or by water.

ARACHNIDA (Gr. *arachne,* a spider). A class of the *Articulata,* comprising Spiders, Scorpions, and allied animals.

ARBORESCENT. Branched like a tree.

ARCHÆOCIDARIS (Gr. *archaios,* ancient; Lat. *cidaris,* a diadem). A Palæozoic genus of Sea-urchins, related to the existing *Cidaris.*

ARCHÆOCYATHUS (Gr. *archaios,* ancient; *kuathos,* cup). A genus of Palæozoic fossils allied to the Sponges.

ARCHÆOPTERYX (Gr. *archaios,* ancient; *pterux,* a wing). The singular fossil bird which alone constitutes the order of the *Saururæ.*

ARCTOCYON (Gr. *arctos,* bear; *kuon,* dog). An extinct genus of Carnivora.

ARENACEOUS. Sandy, or composed of grains of sand.

ARENICOLITES (Lat. *arena,* sand; *colo,* I inhabit). A genus founded on burrows supposed to be formed by worms resembling the living Lobworms (*Arenicola*).

ARTICULATA (Lat. *articulus,* a joint). A division of the animal kingdom, comprising Insects, Centipedes, Spiders and Crustaceans, characterized by the possession of jointed bodies or jointed limbs. The term *Arthropoda* is now more usually employed.

ARTIODACTYLA (Gr. *artios,* even; *daktulos,* a finger or toe). A division of the hoofed quadrupeds (*Ungulata*) in which each foot has an even number of toes (two or four).

ASAPHUS (Gr. *asaphēs,* obscure). A genus of Trilobites.

ASCOCERAS (Gr. *askos,* a leather bottle; *keras,* horn). A genus of Tetrabranchiate Cephalopods.

ASIPHONATE. Not possessing a respiratory tube or siphon. (Applied to a division of the *Lamellibranchiate* Molluscs.)

ASTEROID (Gr. *aster,* a star; and *eidos,* form). Star-shaped, or possessing radiating lobes or rays like a star-fish.

ASTEROIDEA. An order of *Echinodermata,* comprising the Star-fishes, characterized by their rayed form.

ASTEROPHYLLITES (Gr. *aster,* a star; *phullon,* leaf). A genus of Palæozoic plants, with leaves in whorls.

ASTRÆIDÆ (Gr. *Astræ,* a proper name). The family of the Star-corals.

ASTYLOSPONGIA (Gr. *a,* without; *stulos,* a column; *spoggos,* a sponge). A genus of Silurian Sponges.

ATHYRIS (Gr. *a,* without; *thura,* door). A genus of Brachiopods.

ATRYPA (Gr. *a,* without; *trupa,* a hole). A genus of Brachiopods.

AVES (Lat. *avis,* a bird). The class of the Birds.

AVICULA (Lat. a little bird). The genus of Bivalve Molluscs comprising the Pearl-oysters.

AXOPHYLLUM (Gr. *axon,* a pivot; *phullon,* a leaf). A genus of Rugose Corals.

AZOIC (Gr. *a,* without; *zoé,* life). Destitute of traces of living beings.

B

BACULITES (Lat. *baculum,* a staff). A genus of the *Ammonitidæ.*

BALÆNA (Lat. a whale). The genus of the Whalebone Whales.

BALANIDÆ (Gr. *balanos,* an acorn). A family of sessile *Cirripedes,* commonly called "Acorn-shells."

BATRACHIA (Gr. *batrachos,* a frog). Often loosely applied to any of the *Amphibia,* but sometimes restricted to the Amphibians as a class, or to the single order of the *Anoura.*

BELEMNITIDÆ (Gr. *belemnon,* a dart). An extinct group of Dibranchiate Cephalopods, comprising the Belemnites and their allies.

BELEMNOTEUTHIS (Gr. *belemnon,* a dart; *teuthis,* a cuttle-fish). A genus allied to the Belemnites proper.

BELINURUS (Gr. *belos,* a dart; *oura,* tail). A genus of fossil King-crabs.

BELLEROPHON (Gr. proper name). A genus of oceanic Univalves (*Heteropoda*).

BELOTEUTHIS (Gr. *belos,* a dart; *teuthis,* a cuttle-fish). An extinct genus of Dibranchiate Cephalopods.

BEYRICHIA (named after Prof. Beyrich). A genus of Ostracode Crustaceans.

BILATERAL. Having two symmetrical sides.

BIMANA (Lat. *bis,* twice; *manus,* a hand). The order of *Mammalia* comprising man alone.

BIPEDAL (Lat. *bis,* twice; *pes,* foot). Walking upon two legs.

BIVALVE (Lat. *bis,* twice *valvæ,* folding-doors). Composed of two plates or valves; applied to the shell of the *Lamellibranchiata* and *Brachiopoda,* and to the carapace of certain *Crustacea.*

BLASTOIDEA (Gr. *blastos,* a bud; and *eidos,* form). An extinct order of *Echinodermata,* often called *Pentremites.*

BRACHIOPODA (Gr. *brachion*, an arm; *pous*, the foot). A class of the *Molluscoida*, often called "Lamp-shells," characterized by possessing two fleshy arms continued from the sides of the mouth.

BRACHYURA (Gr. *brachus*, short; *oura*, tail). A tribe of the Decapod *Crustaceans* with short tails (*i. e.*, the Crabs.)

BRADYPODIDÆ (Gr. *bradus*, slow; *podes*, feet). The family of *Edentata* comprising the Sloths.

BRANCHIA (Gr. *bragchia*, the gill of a fish). A respiratory organ adapted to breathe air dissolved in water.

BRANCHIATE. Possessing gills or branchiæ.

BRONTEUS (Gr. *brontē*, thunder—an epithet of Jupiter the Thunderer). A genus of Trilobites.

BRONTOTHERIUM (Gr. *brontē*, thunder; *therion*, beast). An extinct genus of Ungulate Quadrupeds.

BRONTOZOUM (Gr. *brontē*, thunder; *zoön*, animal). A genus founded on the largest footprints of the Triassic Sandstones of Connecticut.

BUCCINUM (Lat. *buccinum*, a trumpet). The genus of Univalves comprising the Whelks.

C

CAINOZOIC (*See* Kainozoic.)

CALAMITES (Lat. *calamus*, a reed). Extinct plants with reed-like stems, believed to be gigantic representatives of the *Equisetaceæ*.

CALCAREOUS (Lat *calx*, lime). Composed of carbonate of lime.

CALICE. The little cup in which the polype of a coralligenous Zoophyte (*Actinozoön*) is contained.

CALYMENE (Gr. *kalumenē*, concealed). A genus of Trilobites.

CALYX (Lat. a cup). Applied to the cup-shaped body of a *Crinoid* (*Echinodermata*).

CAMAROPHORIA (Gr. *kamara*, a chamber; *phero*, I carry). A genus of Brachiopods.

CAMELOPARDALIDÆ (Lat. *camelus*, a camel; *pardalis*, a panther). The family of the Giraffes.

CANINE (Lat. *canis*, a dog). The eye-tooth of Mammals, or the tooth which is placed at or close to the præmaxillary suture in the upper jaw, and the corresponding tooth in the lower jaw.

CARAPACE. A protective shield. Applied to the upper shell of Crabs, Lobsters, and many other *Crustacea*. Also the upper half of the immovable case in which the body of a Chelonian is protected.

CARCHARODON (Gr. *karcharos*, rough; *odous*, tooth). A genus of Sharks.

CARDIOCARPON (Gr. *kardia*, the heart; *karpos*, fruit). A genus of fossil fruit from the Coal-measures.

CARDIUM (Gr. *kardia*, the heart). The genus of Bivalve Molluscs comprising the Cockles. *Cardinia, Cardiola*, and *Cardita* have the same derivation.

CARNIVORA (Lat. *caro*, flesh; *voro*, I devour). An order of the *Mammalia*. The "Beasts of Prey."

CARNIVOROUS (Lat. *caro*, flesh; *voro*, I devour). Feeding upon flesh.

CARYOCARIS (Gr. *karua*, a nut; *karis*, a shrimp). A genus of Phyllopod Crustaceans.

CARYOCRINUS (Gr. *karua*, a nut; *krinon*, a lily). A genus of Cystideans.

CAUDAL (Lat. *cauda*, the tail). Belonging to the tail.

CAVICORNIA (Lat. *cavus*, hollow; *cornu*, a horn). The "hollow-horned" Ruminants, in which the horn consists of a central bony "horn-core" surrounded by a horny sheath.

CENTRUM (Gr. *kentron*, the point round which a circle is described by a pair of compasses). The central portion or "body" of a vertebra.

CEPHALASPIDÆ (Gr. *kephale*, head; *aspis*, shield). A family of fossil fishes.

CEPHALIC (Gr. *kephale*, head). Belonging to the head.

CEPHALOPODA (Gr. *kephale*; and *podes*, feet). A class of the *Mollusca*, comprising the Cuttle-fishes and their allies, in which there is a series of arms ranged round the head.

CERATIOCARIS (Gr. *keras*, a horn; *karis*, a shrimp). A genus of Phyllopod Crustaceans.

CERATITES (Gr. *keras*, a horn). A genus of *Ammonitidæ*.

CERATODUS (Gr. *keras*, a horn; *odous*, tooth). A genus of Dipnoous fishes.

CERVICAL (Lat. *cervix*, the neck). Connected with or belonging to the region of the neck.

CERVIDÆ (Lat. *cervus*, a stag). The family of the deer.

CESTRAPHORI (Gr. *kestra*, a weapon; *phero*, I carry). The group of the "Cestraciont Fishes," represented at the present day by the Port-Jackson Shark; so-called from their defensive spines.

CETACEA (Gr. *kētos*, a whale). The order of Mammals comprising the Whales and the Dolphins.

CETIOSAURUS (Gr. *kētos*, whale; *saura*, lizard). A genus of Deinosaurian Reptiles.

CHEIROPTERA (Gr. *cheir*, hand; *pteron*, wing). The Mammalian order of the Bats.

CHEIROTHERIUM (Gr. *cheir*, hand; *thērion*, beast). The generic name applied originally to the hand-shaped footprints of Labyrinthodonts.

CHEIRURUS (Gr. *cheir*, hand; *oura*, tail). A genus of Trilobites.

CHELONIA (Gr. *chelonē*, a tortoise). The Reptilian order of the Tortoises and Turtles.

CHONETES (Gr. *chōnē* or *chōanē*, a chamber or box). A genus of Brachiopods.

CIDARIS (Lat. a diadem). A genus of Sea-urchins.

CLADODUS (Gr. *klados*, branch; *odous*, tooth). A genus of Fishes.

CLATHROPORA (Lat. *clathri*, a trellis; *porus*, a pore). A genus of Lace-corals (*Polyzoa*).

CLISIOPHYLLUM (Gr. *klision*, a hut; *phullon*, leaf). A genus of Rugose Corals.

CLYMENIA (*Clumene,* a proper name). A genus of Tetrabranchiate Cephalopods.

COCCOSTEUS (Gr. *kokkos,* berry; *osteon,* bone). A genus of Ganoid Fishes.

COCHLIODUS (Gr. *kochlion,* a snail-shell; *odous,* tooth). A genus of Cestraciont Fishes.

CŒLENTERATA (Gr. *koilos,* hollow; *enteron,* the bowel). The sub-kingdom which comprises the *Hydrozoa* and *Actinozoa.* Proposed by Frey and Leuckhart in place of the old term *Radiata,* which included other animals as well.

COLEOPTERA (Gr. *koleos,* a sheath; *pteron,* wing). The order of Insects (Beetles) in which the anterior pair of wings are hardened, and serve as protective cases for the posterior pair of membranous wings.

COLOSSOCHELYS (Gr. *kolossos,* a gigantic statue; *chelus,* a tortoise). A huge extinct Land-tortoise.

COMATULA (Gr. *koma,* the hair). The Feather-star, so-called in allusion to its tress-like arms.

CONDYLE (Gr. *kondulos,* a knuckle). The surface by which one bone articulates with another. Applied especially to the articular surface or surfaces by which the skull articulates with the vertebral column.

CONIFERÆ (Lat. *conus,* a cone; *fero,* I carry). The order of the Firs, Pines and their allies, in which the fruit is generally a "cone" or "fir-apple."

CONULARIA (Lat. *conulus,* a little cone). An extinct genus of Pteropods.

COPROLITES (Gr. *kopros,* dung; *lithos,* stone). Properly applied to the fossilized excrements of animals; but often employed to designate phosphatic concretions which are not of this nature.

CORALLITE. The corallum secreted by an *Actinozoön* which consists of a single polype; or the portion of a composite corallum which belongs to, and is secreted by, an individual polype.

CORALLUM (from the Latin for Red Coral). The hard structures deposited in, or by, the tissues of an *Actinozoön*—commonly called a "coral."

CORIACEOUS (Lat. *corium,* hide). Leathery.

CORYPHODON (Gr. *korus,* helmet; *odous,* tooth). An extinct genus of Mammals, allied to the Tapirs.

CRANIUM (Gr. *kranion,* the skull). The bony or cartilaginous case in which the brain is contained.

CRETACEOUS (Lat. *creta,* chalk). The formation which in Europe contains white chalk as one of its most conspicuous members.

CRINOIDEA (Gr. *krinon,* a lily; *eidos,* form). An order of *Echinodermata,* comprising forms which are usually stalked, and sometimes resemble lilies in shape.

CRIOCERAS (Gr. *krios,* a ram; *keras,* a horn). A genus of *Ammonitidæ.*

CROCODILIA (Gr. *krokodeilos,* a crocodile). An order of Reptiles.

CROSSOPTERYGIDÆ (Gr. *krossotos,* a fringe; *pterux,* a fin). A sub-order of Ganoids in which the paired fins possess a central lobe.

CRUSTACEA (Lat. *crusta,* a crust). A class of Articulate animals, comprising Crabs, Lobsters, etc., characterized by the possession of a hard shell or crust, which they cast periodically.

CRYPTOGAMS (Gr. *kruptos,* concealed; *gamos,* marriage). A division of plants in which the organs of reproduction are obscure and there are no true flowers.

CTENACANTHUS (Gr. *kteis,* a comb; *akantha,* a thorn). A genus of fossil fishes, named from its fin-spines.

CTENOID (Gr. *kteis,* a comb; *eidos,* form). Applied to those scales of fishes the hinder margins of which are fringed with spines or comb-like projections.

CURSORES (Lat. *curro,* I run). An order of *Aves,* comprising birds destitute of the power of flight, but formed for running vigorously (*e. g.,* the Ostrich and Emeu).

CUSPIDATE. Furnished with small pointed eminences or "cusps."

CYATHOCRINUS (Gr. *kuathos,* a cup; *krinon,* a lily). A genus of Crinoids.

CYATHOPHYLLUM (Gr. *kuathos,* a cup; *phullon,* a leaf). A genus of Rugose Corals.

CYCLOID (Gr. *kuklos,* a circle; *eidos,* form). Applied to those scales of fishes which have a regularly circular or elliptical outline with an even margin.

CYCLOPHTHALMUS (Gr. *kuklos,* a circle; *opthalmos,* eye). A genus of fossil Scorpions.

CYCLOSTOMI (Gr. *kuklos,* and *stoma,* mouth). Sometimes used to designate the Hag-fishes and Lampreys, forming the order *Marsipobranchii.*

CYPRÆA (a name of Venus). The genus of Univalve Molluscs comprising the Cowries.

CYRTOCERAS (Gr. *kurtos,* crooked; *keras,* horn). A genus of Tetrabranchiate Cephalopods.

CYSTIPHYLLUM (Gr. *kustis,* a bladder; *phullon,* a leaf). A genus of Rugose Corals.

CYSTOIDEA (Gr. *kustis,* a bladder; *eidos,* form). The "Globe-crinoids," an extinct order of *Echinodermata.*

D

DADOXYLON (Gr. *dadion,* a torch; *xulon,* wood). An extinct genus of Coniferous trees.

DECAPODA (Gr. *deka,* ten; *podes,* feet). The division of *Crustacea* which have ten feet; also the family of Cuttle-fishes, in which there are ten arms or cephalic processes.

DECIDUOUS (Lat. *decido,* I fall off). Applied to parts which fall off or are shed during the life of the animal.

DEINOSAURIA (Gr. *deinos,* terrible; *saura,* lizard). An extinct order of Reptiles.

DEINOTHERIUM (Gr. *deinos,* terrible; *therion,* beast). An extinct genus of Proboscidean Mammals.

DENDROGRAPTUS (Gr. *dendron,* tree; *grapho,* I write). A genus of Graptolites.

DESMIDIÆ. Minute fresh-water plants, of a green color, without a siliceous epidermis.

DIATOMACEÆ (Gr. *diatemno,* I sever). An order of minute plants which are provided with siliceous envelopes.
DIBRANCHIATA (Gr. *dis,* twice; *bragchia,* gill). The order of *Cephalopoda* (comprising the Cuttle-fishes, etc.,) in which only two gills are present.
DICERAS (Gr. *dis,* twice; *keras,* horn). An extinct genus of Bivalve Molluscs.
DICTYONEMA (Gr. *diktuon,* a net; *nema,* thread). An extinct genus of *Polyzoa* (?).
DICYNODONTIA (Gr. *dis,* twice; *kuon,* dog; *odous,* tooth). An extinct order of Reptiles.
DIDYMOGRAPTUS (Gr. *didumos,* twin; *grapho,* I write). A genus of Graptolites.
DIMORPHODON (Gr. *dis,* twice; *morphé,* shape; *odous,* tooth). A genus of Pterosaurian Reptiles.
DINICHTHYS (Gr. *deinos,* terrible; *ichthus,* fish). An extinct genus of Fishes.
DINOCERAS (Gr. *deinos,* terrible; *keras,* horn). An extinct genus of Mammals.
DINOPHIS (Gr. *deinos,* terrible; *ophis,* snake). An extinct genus of Snakes.
DINORNIS (Gr. *deinos,* terrible; *ornis,* bird). An extinct genus of Birds.
DIPLOGRAPTUS (Gr. *diplos,* double; *grapho,* I write). A genus of Graptolites.
DIPNOI (Gr. *dis,* twice; *pnoé,* breath). An order of Fishes, comprising the Mud-fishes, so-called in allusion to their double mode of respiration.
DIPROTODON (Gr. *dis,* twice; *protos,* first; *odous,* tooth). A genus of extinct Marsupials.
DIPTERA (Gr. *dis,* twice; *pteron,* wing). An order of Insects characterized by the possession of two wings.
DISCOID (Gr. *diskos,* a quoit; *eidos,* form). Shaped like a round plate or quoit.
DOLOMITE (named after M. Dolomieu). Magnesian limestone.
DORSAL (Lat. *dorsum,* the back). Connected with or placed upon the back.
DROMATHERIUM (Gr. *dromaios,* nimble; *thērion,* beast). A genus of Triassic Mammals.
DRYOPITHECUS (Gr. *drus,* an oak; *pithekos,* an ape). An extinct genus of Monkeys.

E

ECHINODERMATA (Gr. *echinos*; and *derma,* skin). A class of animals comprising the Sea-urchins, Star-fishes and others, most of which have spiny skins.
ECHINOIDEA (Gr. *echinos*; and *eidos,* form). An order of *Echinodermata,* comprising the Sea-urchins.
EDENTATA (Lat. *e,* without; *dens,* tooth). An order of *Mammalia* often called *Bruta.*
EDENTULOUS. Toothless, without any dental apparatus. Applied to the mouth of any animal, or to the hinge of the Bivalve Molluscs.

ELASMOBRANCHII (Gr. *elasma,* a plate; *bragchia,* gill). An order of Fishes, including the Sharks and Rays.

ENALIOSAURIA (Gr. *enalios,* marine; *saura,* lizard). Sometimes employed as a common term to designate the extinct Reptilian orders of the *Ichthyosauria* and *Plesiosauria.*

EOCENE (Gr. *eos,* dawn; *kainos,* new or recent). The lowest division of the Tertiary rocks, in which species of existing shells are to a small extent represented.

EOPHYTON (Gr. *eos,* dawn; *phuton,* a plant). A genus of Cambrian fossils, supposed to be of a vegetable nature.

EOZOÖN (Gr. *eos,* dawn; *zoön,* animal). A genus of chambered calcareous organisms found in the Laurentian and Huronian formations.

EQUILATERAL (Lat. *æquus,* equal; *latus,* side). Having its sides equal. Usually applied to the shells of the *Brachiopoda.* When applied to the spiral shells of the *Foraminifera,* it means that all the convolutions of the shell lie in the same plane.

EQUISETACEÆ (Lat. *equus,* horse; *seta,* bristle). A group of Cryptogamous plants, commonly known as "Horse-tails."

EQUIVALVE (Lat. *æquus,* equal; *valvæ,* folding-doors). Applied to shells which are composed of two equal pieces or valves.

ERRANTIA (Lat. *erro,* I wander). An order of *Annelida,* often called *Nereidea,* distinguished by their great locomotive powers.

EUOMPHALUS (Gr. *eu,* well; *omphalos,* navel). An extinct genus of Univalve Molluscs.

EURYPTERIDA (Gr. *eurus,* broad; *pteron,* wing). An extinct sub-order of *Crustacea.*

EXOGYRA (Gr. *exo,* outside; *guros,* circle). An extinct genus of Oysters.

F

FAUNA (Lat. *Fauni,* the rural deities of the Romans). The general assemblage of the animals of any region or district.

FAVOSITES (Lat. *favos,* a honeycomb). A genus of Tabulate Corals.

FENESTELLIDÆ (Lat. *fenestella,* a little window). The "Lace-corals," a group of Palæozoic Polyzoans.

FILICES (Lat. *filix,* a fern). The order of Cryptogamic plants comprising the Ferns.

FILIFORM (Lat. *filum,* a thread; *forma,* shape). Thread-shaped.

FLORA (Lat. *Flora,* the goddess of flowers). The general assemblage of the plants of any region or district.

FORAMINIFERA (Lat. *foramen,* an aperture; *fero,* I carry). An order of *Protozoa,* usually characterized by the possession of a shell perforated by numerous pseudopodial apertures.

FRUGIVOROUS (Lat. *frux,* fruit; *voro,* I devour). Living upon fruits.

FUCOIDS (Lat. *fucus,* sea-weed; Gr. *eidos,* likeness). Fossils often of an obscure nature, believed to be the remains of sea-weeds.

FUSULINA (Lat. *fusus,* a spindle). An extinct genus of *Foraminifera.*

G

GANOID (Gr. *ganos,* splendour, brightness). Applied to those scales or plates which are composed of an inferior layer of true bone covered by a superior layer of polished enamel.

GANOIDEI. An order of Fishes.

GASTEROPODA (Gr. *gaster,* stomach; *pous,* foot). The class of the *Mollusca* comprising the ordinary Univalves, in which locomotion is usually effected by a muscular expansion of the under surface of the body (the "foot").

GLOBIGERINA (Lat. *globus,* a globe; *gero,* I carry). A genus of *Foraminifera.*

GLYPTODON (Gr. *glupho,* I engrave; *odous,* tooth). An extinct genus of Armadillos, so-named in allusion to the fluted teeth.

GONIATITES (Gr. *gonia,* angle). A genus of Tetrabanchiate Cephalopods.

GRALLATORES (Lat. *grallæ,* stilts). The order of the long-legged Wading Birds.

GRAPTOLITIDÆ (Gr. *grapho,* I write; *lithos,* stone). An extinct sub-class of the *Hydrozoa.*

GYMNOSPERMS (Gr. *gumnos,* naked; *sperma,* seed). The Conifers and Cycads, in which the seed is not protected within a seed-vessel.

H

HALITHERIUM (Gr. *hals,* sea; *thērion,* beast). An extinct genus of Sea-cows (*Sirenia*).

HAMITES (Lat. *hamus,* a hook). A genus of the *Ammonitidæ.*

HELIOPHYLLUM (Gr. *helios,* the sun; *phullon,* leaf). A genus of Rugose Corals.

HELLADOTHERIUM (Gr. *Hēllas,* Greece; *thērion,* beast). An extinct genus of Ungulate Mammals.

HEMIPTERA (Gr. *hemi*; and *pteron,* wing). An order of Insects in which the anterior wings are sometimes "hemelytra."

HESPERORNIS (Gr. *Hesperos,* the evening star; *ornis,* bird). An extinct genus of Birds.

HETEROCERCAL (Gr. *heteros,* diverse; *kerkos,* tail). Applied to the tail of Fishes when it is unsymmetrical, or composed of two unequal lobes.

HETEROPODA (Gr. *heteros,* diverse; *podes,* feet). An aberrant group of the Gasteropods, in which the foot is modified so as to form a swimming organ.

HIPPARION (Gr. *hipparion,* a little horse). An extinct genus of *Equidæ.*

HIPPOPOTAMUS (Gr. *hippos,* horse; *potamos,* river). A genus of hoofed Quadrupeds—the "River-horses."

HIPPURITIDÆ (Gr. *hippos,* horse; *oura,* tail). An extinct family of Bivalve Molluscs.

HOLOPTYCHIUS (Gr. *holos,* whole; *ptuchē,* wrinkle). An extinct genus of Ganoid Fishes.

HOLOSTOMATA (Gr. *holos,* whole; *stoma,* mouth). A division of *Gasteropodous Molluscs,* in which the aperture of the shell is rounded, or "entire."

HOLOTHUROIDEA (Gr. *holothourion,* a kind of zoöphyte; and *eidos,* form). An order of *Echinodermata* comprising the Trepangs.

HOMOCERCAL (Gr. *homos,* same; *kerkos,* tail). Applied to the tail of Fishes when it is symmetrical, or composed of two equal lobes.

HYBODONTS (Gr. *hubos,* curved; *odous,* tooth). A group of Fishes of which *Hybodus* is the type-genus.

HYDROIDA (Gr. *hudra,* water-serpent; and *eidos,* form). The sub-class of the *Hydrozoa,* which comprises the animals most nearly allied to the *Hydra.*

HYDROZOA (Gr. *hudra*; and *zoōn,* animal). The class of the *Cœlenterata* which comprises animals constructed after the type of the *Hydra.*

HYMENOPTERA (Gr. *humen,* a membrane; *pteron,* a wing). An order of Insects (comprising Bees, Ants, etc.) characterized by the possession of four membranous wings.

I

ICHTHYODORULITE (Gr. *ichthus,* fish; *dorus,* spear; *lithos,* stone). The fossil fin-spine of Fishes.

ICHTHYOPTERYGIA (Gr. *ichthus*; *pterux,* wing). An extinct order of Reptiles.

ICHTHYORNIS (Gr. *ichthus,* fish; *ornis,* bird). An extinct genus of Birds.

ICHTHYOSAURIA (Gr. *ichthus*; *saura,* lizard). Synonymous with *Ichthyopterygia.*

IGUANODON (*Iguana,* a living lizard; Gr. *odous,* tooth). A genus of Deinosaurian Reptiles.

INCISOR (Lat. *incido,* I cut). The cutting teeth fixed in the intermaxillary bones of the *Mammalia,* and the corresponding teeth in the lower jaw.

INEQUILATERAL. Having the two sides unequal, as in the case of the shells of the ordinary bivalves (*Lamellibranchiata*). When applied to the shells of the *Foraminifera,* it implies that the convolutions of the shell do not lie in the same plane, but are obliquely wound round an axis.

INEQUIVALVE. Composed of two unequal pieces or valves.

INOCERAMUS (Gr. *is,* a fibre; *keramos,* an earthen vessel). An extinct genus of Bivalve Molluscs.

INSECTA (Lat. *inseco,* I cut into). The class of articulate animals commonly known as Insects.

INSECTIVORA (Lat. *insectum,* an insect; *voro,* I devour). An order of Mammals.

INSECTIVOROUS. Living upon Insects.

INSESSORES (Lat. *insedeo,* I sit upon). The order of the Perching Birds, often called *Passeres.*

INTERAMBULACRA. The rows of plates in an *Echinoid* which are not perforated for the emission of the "tube-feet."

INTERMAXILLÆ or PRÆMAXILLÆ. The two bones which are situated between the two superior maxillæ in *Vertebrata.* In man, and some monkeys, the præmaxillæ anchylose with the maxillæ, so as to be irrecognizable in the adult.

INVERTEBRATA (Lat. *in,* without; *vertebra,* a bone of the back). Animals without a spinal column or backbone.

ISOPODA (Gr. *isos,* equal; *podes,* feet). An order of *Crustacea* in which the feet are like one another and equal.

K

KAINOZOIC (Gr. *kainos,* recent; *zoe,* life). The Tertiary period in Geology comprising those formations in which the organic remains approximate more or less closely to the existing fauna and flora.

L

LABYRINTHODONTIA (Gr. *laburinthos,* a labyrinth; *odous,* tooth). An extinct order of *Amphibia,* so-called from the complex microscopic structure of the teeth.

LACERTILIA (Lat. *lacerta,* a lizard). An order of *Reptilia* comprising the Lizards and Slow-worms.

LAMELLIBRANCHIATA (Lat. *lamella,* a plate; Gr. *bragchia,* gill). The class of *Mollusca* comprising the ordinary bivalves, characterized by the possession of lamellar gills.

LEPIDODENDRON (Gr. *lepis,* a scale; *dendron,* a tree). A genus of extinct plants, so-named from the scale-like scars upon the stem left by the falling off of the leaves.

LEPIDOPTERA (Gr. *lepis,* a scale; *pteron,* a wing). An order of Insects, comprising Butterflies and Moths, characterized by possessing four wings which are usually covered with minute scales.

LEPIDOSIREN (Gr. *lepis,* a scale; *seirēn,* a siren—the generic name of the Mud-eel or *Siren lacertina*). A genus of Dipnoous fishes, comprising the "Mud-fishes."

LEPIDOSTROBUS (Gr. *lepis,* a scale; *strobilos,* a fir-cone). A genus founded on the cones of *Lepidodendron.*

LEPTÆNA (Gr. *leptos,* slender). A genus of Brachiopods.

LINGULA (Lat. *lingula,* a little tongue). A genus of Brachiopods.

LYCOPODIACEÆ (Gr. *lupos, a* wolf; *pous,* foot). The group of Cryptogamic plants generally known as "Club-mosses."

M

MACHÆRACANTHUS (Gr. *machaira,* a sabre; *acantha,* thorn or spine). An extinct genus of Fishes.

MACHAIRODUS (Gr. *machaira,* a sabre; *odous,* tooth). An extinct genus of Carnivora.

MACROTHERIUM (Gr. *makros,* long; *therion,* beast). An extinct genus of Edentata.

MACRURA (Gr. *makros,* long; *oura,* tail). A tribe of Decapod *Crustaceans* with long tails (*e. g.,* the Lobster, Shrimp, etc.)

MAMMALIA (Lat. *mamma,* the breast). The class of Vertebrate animals which suckle their young.

MANDIBLE (Lat. *mandibulum,* a jaw). The upper pair of jaws in Insects; also applied to one of the pairs of jaws in *Crustacea* and Spiders, to the beak of Cephalopods, the lower jaw of Vertebrates, etc.

MANTLE. The external integument of most of the Mollusca, which is largely developed, and forms a cloak in which the viscera are protected. Technically called the "pallium."

MANUS (Lat. the hand). The hand of the higher Vertebrates.

MARSIPOBRANCHII (Gr. *marsipos,* a pouch; *bragchia,* gill). The order of Fishes comprising the Hag-fishes and Lampreys, with pouch-like gills.

MARSUPIALIA (Lat. *marsupium,* a pouch). An order of Mammals in which the females mostly have an abdominal pouch in which the young are carried.

MASTODON (Gr. *mastos,* nipple; *odous,* tooth). An extinct genus of Elephantine Mammals.

MEGALONYX (Gr. *megas,* great; *onux,* nail). An extinct genus of Edentate Mammals.

MEGALOSAURUS (Gr. *megas,* great; *saura,* lizard). A genus of Deinosaurian Reptiles.

MEGATHERIUM (Gr. *megas,* great; *thērion,* beast). An extinct genus of Edentata.

MESOZOIC (Gr. *mesos,* middle; and *zoe,* life). The Secondary period in Geology.

MICROLESTES (Gr. *mikros,* little; *lēstes,* thief). An extinct genus of Triassic Mammals.

MILLEPORA (Lat. *mille,* one thousand; *porus,* a pore). A genus of "Tabulate Corals."

MIOCENE (Gr. *meion,* less; *kainos,* new). The Middle Tertiary period.

MOLARS (Lat. *mola,* a mill). The "grinders" in man, or the teeth in diphyodont Mammals which are not preceded by milk-teeth.

MOLLUSCA (Lat. *mollis,* soft). The sub-kingdom which includes the Shell-fish proper, the *Polyzoa,* the *Tunicata,* and the Lamp-shells; so-called from the generally soft nature of their bodies.

MOLLUSCOIDA (*Mollusca*; Gr. *eidos,* form). The lower division of the *Mollusca,* comprising the *Polyzoa, Tunicata,* and *Brachiopoda.*

MONOGRAPTUS (Gr. *monos,* single; *grapho,* I write). A genus of Graptolites.

MYLODON (Gr. *mulos,* a mill; *odous,* tooth). An extinct genus of Edentate Mammals.

MYRIAPODA OR MYRIOPODA (Gr. *murios,* ten thousand; *podes,* feet). A class of *Arthropoda* comprising the Centipedes and their allies, characterized by their numerous feet.

N

NATATORES (Lat. *nare,* to swim). The order of the Swimming Birds.

NATATORY (Lat. *nare,* to swim). Formed for swimming.

NAUTILOID. Resembling the shell of the *Nautilus* in shape.

NERVURES (Lat. *nervus,* a sinew). The ribs which support the membranous wings of insects.

NEUROPTERA (Gr. *neuron,* a nerve; *pteron,* a wing). An order of Insects characterized by four membranous wings with numerous reticulated nervures (*e. g.,* Dragon-flies).

NEUROPTERIS (Gr. *neuron,* a nerve; *pteris,* a fern). An extinct genus of Ferns.

NOTHOSAURUS (Gr. *nothos,* spurious; *saura,* lizard). A genus of *Plesiosaurian* Reptiles.

NOTOCHORD (Gr. *notos,* back; *chorde,* string). A cellular rod which is developed in the embryo of Vertebrates immediately beneath the spinal cord, and which is usually replaced in the adult by the vertebral column. Often it is spoken of as the "chorda dorsalis."

NUDIBRANCHIATA (Lat. *nudus,* naked; and Gr. *bragchia,* gill). An order of the *Gasteropoda* in which the gills are naked.

NUMMULINA (Lat. *nummus,* a coin). A genus of *Foraminifera,* comprising the coin-shaped "Nummulites."

O

OBOLELLA (Lat. dim of *obolus,* a small coin). An extinct genus of Brachiopods.

OCCIPITAL. Connected with the *occiput,* or the back part of the head.

OCEANIC. Applied to animals which inhabit the open ocean (=pelagic).

ODONTOPTERYX (Gr. *odous,* tooth; *pterux,* wing). An extinct genus of Birds.

ODONTORNITHES (Gr. *odous,* tooth; *ornis,* bird). The extinct order of Birds, comprising forms with distinct teeth in sockets.

OLIGOCENE (Gr. *oligos,* few; *kainos,* new). A name used by many Continental geologists as synonymous with the Lower Miocene.

OPHIDIA (Gr. *ophis,* a serpent). The order of Reptiles comprising the Snakes.

OPHIUROIDEA (Gr. *ophis,* snake; *oura,* tail; *eidos,* form). An order of *Echinodermata,* comprising the Brittle-stars and Sand-stars.

ORNITHOSCELIDA (Gr. *ornis,* bird; *skelo,* leg). Applied by Huxley to the Deinosaurian Reptiles, together with the genus *Compsognathus,* on account of the bird-like character of their hind-limbs.

ORTHIS (Gr. *orthos,* straight). A genus of Brachiopods, named in allusion to the straight hinge-line.

ORTHOCERATIDÆ (Gr. *orthos,* straight; *keras,* horn). A family of the *Nautilidæ,* in which the shell is straight, or nearly so.

ORTHOPTERA (Gr. *orthos,* straight; *pteron,* wing). An order of Insects.

OSTEOLEPIS (Gr. *osteon,* bone; *lepis,* scale). An extinct genus of Ganoid Fishes.

OSTRACODA (Gr. *ostrakon,* a shell). An order of small Crustaceans which are enclosed in bivalve shells.

OTODUS (Gr. *ota,* ears; *odous,* tooth). An extinct genus of Sharks.

OUDENODON (Gr. *ouden,* none; *odous,* tooth). A genus of Dicynodont Reptiles.

OVIBOS (Lat. *ovis,* sheep; *bos,* ox). The genus comprising the Musk-ox.

P

PACHYDERMATA (Gr. *pachus,* thick; *derma,* skin). An old Mammalian order constituted by Cuvier for the reception of the Rhinoceros, Hippopotamus, Elephant, etc.

PALÆASTER (Gr. *palaios,* ancient; *aster,* star). An extinct genus of Star-fishes.

PALÆOCARIS (Gr. *palaios,* ancient; *karis,* shrimp). An extinct genus of Decapod Crustaceans.

PALÆOLITHIC (Gr. *palaios,* ancient; *lithos,* stone). Applied to the rude stone implements of the earliest known races of men, to the men who made these implements, or to the period at which they were made.

PALÆONTOLOGY (Gr. *palaios,* ancient; and *logos,* discourse). The science of fossil remains or of extinct organized beings.

PALÆOPHIS (Gr. *palaios,* ancient; *ophis,* serpent). An extinct genus of Snakes.

PALÆOSAURUS (Gr. *palaios,* ancient; *saura,* lizard). A genus of Thecodont Reptiles.

PALÆOTHERIDÆ (Gr. *palaios,* ancient; *therion,* beast). A group of Tertiary Ungulates.

PALÆOZOIC (Gr. *palaios,* ancient; and *zoe,* life). Applied to the oldest of the great geological epochs.

PARADOXIDES (Lat. *paradoxus,* marvellous). A genus of Trilobites.

PATAGIUM (Lat. the border of a dress). Applied to the expansion of the integument by which Bats, Flying Squirrels, and other animals support themselves in the air.

PECOPTERIS (Gr. *peko,* I comb; *pteris,* a fern). An extinct genus of Ferns.

PECTEN (Lat. a comb). The genus of Bivalve Molluscs comprising the Scallops.

PECTORAL (Lat. *pectus,* chest). Connected with, or placed upon, the chest.

PENTACRINUS (Gr. *penta,* five; *krinon,* lily). A genus of Crinoids in which the column is five-sided.

PENTAMERUS (Gr. *penta,* five; *meros,* part). An extinct genus of Brachiopods.

PENTREMITES (Gr. *penta,* five; *trema,* aperture). A genus of *Blastoidea,* so-named in allusion to the apertures at the summit of the calyx.

PERENNIBRANCHIATA (Lat. *perennis,* perpetual; Gr. *bragchia,* gill). Applied to those *Amphibia* in which the gills are permanently retained throughout life.

PERISSODACTYLA (Gr. *perissos,* uneven; *daktulos,* finger). Applied to those Hoofed Quadrupeds (*Ungulata*) in which the feet have an uneven number of toes.

PETALOID. Shaped like the petal of a flower.

PHACOPS (Gr. *phakē*, a lentil; *ops*, the eye). A genus of Trilobites.

PHALANGES (Gr. *phalanx*, a row). The small bones composing the digits of the higher *Vertebrata*. Normally, each digit has three phalanges.

PHANEROGAMS (Gr. *phaneros*, visible; *gamos*, marriage). Plants which have the organs of reproduction conspicuous, and which bear true flowers.

PHARYNGOBRANCHII (Gr. *pharugx*, pharynx; *bragchia*, gill). The order of Fishes comprising only the Lancelet.

PHASCOLOTHERIUM (Gr. *phaskolos*, a pouch; *thērion*, a beast). A genus of Oolitic Mammals.

PHRAGMACONE (Gr. *phragma*, a partition; and *konos*, a cone). The chambered portion of the internal shell of a *Belemnite*.

PHYLLOPODA (Gr. *phullon*, leaf; and *pous*, foot). An order of *Crustacea*.

PINNATE (Lat. *pinna*, a feather). Feather-shaped; or possessing lateral processes.

PINNIGRADA (Lat. *pinna*, a feather; *gradior*, I walk). The group of *Carnivora*, comprising the Seals and Walruses, adapted for an aquatic life. Often called *Pinnipedia*.

PINNULÆ (Lat. dim. of *pinna*). The lateral processes of the arms of *Crinoids*.

PISCES (Lat. *piscis*, a fish). The class of Vertebrates comprising the Fishes.

PLACOID (Gr. *plax*, a plate; *eidos*, form). Applied to the irregular bony plates, grains or spines which are found in the skin of various fishes (*Elasmobranchii*).

PLAGIOSTOMI (Gr. *plagios*, transverse; *stoma*, mouth). The Sharks and Rays, in which the mouth is transverse, and is placed on the under surface of the head.

PLATYCERAS (Gr. *platus*, broad; *keras*, horn). A genus of Univalve Molluscs.

PLATYCRINUS (Gr. *platus*, broad; *krinon*, lily). A genus of Crinoidea.

PLATYRHINA (Gr. *platus*, broad; *rhines*, nostrils). A group of the *Quadrumana*.

PLATYSOMUS (Gr. *platus*, wide; *soma*, body). A genus of Ganoid Fishes.

PLEISTOCENE (Gr. *pleistos*, most; *kainos*, new). Often used as synonymous with "Post-Pliocene."

PLEUROTOMARIA (Gr. *pleura*, the side; *tomē*, notch). A genus of Univalve shells.

PLIOCENE (Gr. *pleion*, more; *kainos*, new). The later Tertiary period.

PLIOPITHECUS (Gr. *pleion*, more; *pithekos*, ape). An extinct genus of Monkeys.

PLIOSAURUS (Gr. *pleion*, more; *saura*, lizard). A genus of Plesiosaurian Reptiles.

POLYCYSTINA (Gr. *polus*, many; and *kustis*, a cyst). An order of *Protozoa* with foraminated siliceous shells.

POLYPARY. The hard chitinous covering secreted by many of the *Hydrozoa*.

Polype (Gr. *polus,* many; *pous,* foot). Restricted to the single individual of a simple *Actinozoön,* such as Sea-anemone, or to the separate zooids of a compound *Actinozoön.* Often applied indiscriminately to any of the *Cœlenterata,* or even to the *Polyzoa.*

Polypora (Gr. *polus,* many; *poros,* a passage). A genus of Lace-corals (*Fenestellidæ.*)

Polythalamous (Gr. *polus*; and *thalamos,* chamber). Having many chambers; applied to the shells of *Foraminifera* and *Cephalopoda.*

Polyzoa (Gr. *polus*; and *zoön,* animal). A division of the *Molluscoida* comprising compound animals, such as the Sea-mat—sometimes called *Bryozoa.*

Porifera (Lat. *porus,* a pore; and *fero,* I carry). Sometimes used to designate the *Foraminifera,* or the *Sponges.*

Præmolars (Lat. *præ,* before; *molares,* the grinders). The molar teeth of Mammals which succeed the molars of the milk-set of teeth. In man, the bicuspid teeth.

Proboscidea (Lat. *proboscis,* the snout). The order of Mammals comprising the Elephants.

Procœlous (Gr. *pro,* before; *koilos,* hollow). Applied to vertebræ the bodies of which are hollow or concave in front.

Producta (Lat. *productus,* drawn out or extended). An extinct genus of Brachiopods, in which the shell is "eared," or has its lateral angles drawn out.

Protichnites (Gr. *protos,* first; *ichnos,* footprint). Applied to certain impressions in the Potsdam sandstone of North America, believed to have been produced by large Crustaceans.

Protophyta (Gr. *protos*; and *phuton,* plant). The lowest division of plants.

Protoplasm (Gr. *protos*; and *plasso,* I mould). The elementary basis of organized tissues. Sometimes used synonymously for the "sarcode" of the *Protozoa.*

Protorosaurus or Proterosaurus (Gr. *protos,* first; *orao,* I see or discover; *saura,* lizard; or *proteros,* earlier; *saura,* lizard). A genus of Permian lizards.

Protozoa (Gr. *protos*; and *zoön,* animal). The lowest division of the animal kingdom.

Psammodus (Gr. *psammos,* sand; *odous,* tooth). An extinct genus of Cestraciont Sharks.

Pseudopodia (Gr. *pseudos,* falsity; and *pous,* foot). The extensions of the body-substance which are put forth by the *Rhizopoda* at will, and which serve for locomotion and prehension.

Psilophyton (Gr. *psilos,* bare; *phuton,* plant). An extinct genus of Lycopodiaceous plants.

Pteranodon (Gr. *pteron,* wing; *a,* without; *odous,* tooth). A genus of Pterosaurian Reptiles.

Pteraspis (Gr. *pteron,* wing; *aspis,* shield). A genus of Ganoid Fishes.

Pterichthys (Gr. *pteron,* wing; *ichthus,* fish). A genus of Ganoid Fishes.

Pterodactylus (Gr. *pteron,* wing; *daktulos,* finger). A genus of Pterosaurian Reptiles.

PTEROPODA (Gr. *pteron,* wing; and *pous,* foot). A class of the *Mollusca* which swim by means of fins attached near the head.
PTEROSAURIA (Gr. *pteron,* wing; *saura,* lizard). An extinct order of Reptiles.
PTILODICTYA (Gr. *ptilon,* a feather; *diktuon,* a net). An extinct genus of *Polyzoa.*
PTYCHOCERAS (Gr. *ptuché,* a fold; *keras,* a horn). A genus of *Ammonitidæ*
PULMONATE. Possessing lungs.
PYRIFORM (Lat. *pyrus,* a pear; and *forma,* form). Pear-shaped.

Q

QUADRUMANA (Lat. *quatuor,* four; *manus,* hand). The order of Mammals comprising the Apes, Monkeys, Baboons, Lemurs, etc.

R

RADIATA (Lat. *radius,* a ray). Formerly applied to a large number of animals which are now placed in separate sub-kingdoms (*e. g.,* the *Cœlenterata,* the *Echinodermata,* the *Infusoria,* etc.)
RADIOLARIA (Lat. *radius,* a ray). A division of *Protozoa.*
RAMUS (Lat. a branch). Applied to each half or branch of the lower jaw, or mandible, of Vertebrates.
RAPTORES (Lat. *rapto,* I plunder). The order of the Birds of Prey.
RASORES (Lat. *rado,* I scratch). The order of the Scratching Birds (Fowls, Pigeons, &c.)
RECEPTACULITES (Lat. *receptaculum,* a storehouse). An extinct genus of *Protozoa.*
REPTILIA (Lat. *repto,* I crawl). The class of the *Vertebrata* comprising the Tortoises, Snakes, Lizards, Crocodiles, etc.
RETEPORA (Lat. *rete,* a net; *porus,* a pore). A genus of Lace-corals (*Polyzoa*).
RHAMPHORHYNCHUS (Gr. *rhamphos,* beak; *rhugchos,* nose). A genus of Pterosaurian Reptiles.
RHINOCEROS (Gr. *rhis,* the nose; *keras,* horn). A genus of Hoofed Quadrupeds.
RHIZOPODA (Gr. *rhiza,* a root; and *pous,* foot). The division of *Protozoa* comprising all those which are capable of emitting pseudopodia.
RHYNCHOLITES (Gr. *rhugchos,* beak; and *lithos,* stone). Beak-shaped fossils consisting of the mandibles of *Cephalopoda.*
RHYNCHONELLA (Gr. *rhugchos,* nose or beak). A genus of Brachiopods.
RODENTIA (Lat. *rodo,* I gnaw). An order of the Mammals; often called *Glires* (Lat. *glis,* a dormouse).
ROTALIA (Lat. *rota,* a wheel). A genus of *Foraminifera.*
RUGOSA (Lat. *rugosus,* wrinkled). An order of Corals.
RUMINANTIA (Lat. *ruminor,* I chew the cud). The group of Hoofed Quadrupeds (*Ungulata*) which "ruminate" or chew the cud.

S

SARCODE (Gr. *sarx,* flesh; *eidos,* form). The jelly-like substance of which the bodies of the *Protozoa* are composed. It is an albuminous body containing oil-granules, and is sometimes called "animal protoplasm."

SAURIA (Gr. *saura,* a lizard). Any lizard-like Reptile is often spoken of as a "Saurian;" but the term is sometimes restricted to the Crocodiles alone, or to the Crocodiles and Larcertilians.

SAUROPTERYGIA (Gr. *saura*; *pterux,* wing). An extinct order of Reptiles, called by Huxley *Plesiosauria,* from the typical genus *Plesiosaurus.*

SAURURÆ (Gr. *saura*; *oura,* tail). The extinct order of Birds comprising only the *Archœopteryx.*

SCANSORES (Lat. *scando,* I climb). The order of the Climbing Birds (Parrots, Woodpeckers, &c.)

SCAPHI (Lat. *scapha,* a boat). A genus of the *Ammonitidœ.*

SCOLITHUS (Gr. *skolex,* a worm; *lithos,* a stone). The vertical burrows of sea-worms in rocks.

SCUTA (Lat. *scutum,* a shield). Applied to any shield-like plates; especially to those which are developed in the integument of many Reptiles.

SELACHIA or SELACHII (Gr. *selachos,* a cartilaginous fish, probably a shark). The sub-order of *Elasmobranchii* comprising the Sharks and Dog-fishes.

SEPIOSTAIRE. The internal shell of the Sepia, commonly known as the "cuttle-bone."

SEPTA. Partitions.

SERPENTIFORM. Resembling a serpent in shape.

SERTULARIDA (Lat. *sertum,* a wreath). An order of *Hydrozoa.*

SESSILE (Lat. *sedo,* I sit). Not supported upon a stalk or peduncle; attached by a base.

SETÆ (Lat. bristles). Bristles or long stiff hairs.

SIGILLARIOIDS (Lat. *sigilla,* little images). A group of extinct plants of which *Sigillaria* is the type, so-called from the seal-like markings on the bark.

SILICEOUS (Lat. *silex,* flint). Composed of flint.

SINISTRAL (Lat. *sinistra,* the left hand). Left-handed; applied to the direction of the spiral in certain shells, which are said to be "reversed."

SIPHON (Gr. a tube). Applied to the respiratory tubes in the *Mollusca*; also to other tubes of different functions.

SIPHONIA (Gr. *siphon,* a tube). A genus of fossil Sponges.

SIPHONOSTOMATA (Gr. *siphon*; and *stoma,* mouth). The division of *Gasteropodous Molluscs* in which the aperture of the shell is not "entire," but possesses a notch or tube for the emission of the respiratory siphon.

SIPHUNCLE (Lat. *siphunculus,* a little tube). The tube which connects together the various chambers of the shell of certain *Cephalopoda* (*e. g.,* the Pearly Nautilus).

SIRENIA (Gr. *seiren,* a mermaid). The order of *Mammalia* comprising the Dugongs and Manatees.

SIVATHERIUM (*Siva*, a Hindoo deity; Gr. *therion*, beast). An extinct genus of Hoofed Quadrupeds.

SOLIDUNGULA (Lat. *solidus*, solid; *ungula*, a hoof). The group of Hoofed Quadrupeds comprising theHorse, Ass, and Zebra, in which each foot has only a single solid hoof. Often called *Solipedia*.

SPHENOPTERIS (Gr. *sphén*, a wedge; *pteris*, a fern). An extinct genus of ferns.

SPICULA (Lat. *spiculum*, a point). Pointed needle-shaped bodies.

SPIRIFERA (Lat. *spira*, a spire or coil; *fero*, I carry). An extinct genus of Brachiopods, with large spiral supports for the "arms."

SPIRORBIS (Lat. *spira*, a spire; *orbis*, a circle). A genus of tube-inhabiting Annelides, in which the shelly tube is coiled into a spiral disc.

SPONGIDA (Gr. *spoggos*, a sponge). The division of *Protozoa*, commonly known as sponges.

STALACTITES (Gr. *stalasso*, I drop). Icicle-like encrustations and deposits of lime, which hang from the roof of caverns in limestone.

STALAGMITE (Gr. *stalagma*, a drop). Encrustations of lime formed on the floor of caverns which are hollowed out of limestone.

STIGMARIA (Gr. *stigma*, a mark made with a pointed instrument). A genus founded on the roots of various species of *Sigillaria*.

STRATUM (Lat. *stratus*, spread out; or *stratum*, a thing spread out). A layer of rock.

STROMATOPORA (Gr. *stroma*, a thing spread out; *poros*, a passage or pore). A Palæzoic genus of *Protozoa*.

STROPHOMENA (Gr. *strophao*, I twist; *mēnē*, moon). An extinct genus of Brachiopods.

SUB-CALCAREOUS. Somewhat calcareous.

SUB-CENTRAL. Nearly central, but not quite.

SUTURE (Lat. *suo*, I sew). The line of junction of two parts which are immovably connected together. Applied to the line where the whorls of a univalve shell join one another; also to the lines made upon the exterior of the shell of a chambered *Cephalopod* by the margins of the septa.

SYRINGOPORA (Gr. *surigx*, a pipe; *poros*, a pore). A genus of Tabulate Corals.

T

TABULÆ (Lat. *tabula*, a tablet). Horizontal plates or floors found in some Corals, extending across the cavity of the "theca" from side to side.

TEGUMENTARY (Lat. *tegumentum*, a covering). Connected with the integument or skin.

TELEOSAURUS (Gr. *teleios*, perfect; *saura*, lizard). An extinct genus of Crocodilian Reptiles.

TELEOSTEI (Gr. *teleios*, perfect; *osteon*, bone.) The order of the "Bony Fishes."

TELSON (Gr. a limit). The last joint in the abdomen of *Crustacea;* variously regarded as a segment without appendages, or as an azygous appendage.

TENTACULITES (Lat. *tentaculum,* a feeler). A genus of *Pteropoda.*

TERERBRATULA (Lat. *terebratus,* bored or pierced.) A genus of *Brachiopoda,* so called in allusion to the perforated beak of the ventral valve.

TEST (Lat. *testa,* shell.) The shell of *Mollusca,* which are for this reason sometimes called "*Testacea,*" also, the calcareous case of *Echinoderms;* also, the thick leathery outer tunic in the *Tunicata.*

TESTACEOUS. Provided with a shell or hard covering.

TESTUDINIDÆ (Lat *testudo,* a tortoise). The family of the Tortoises.

TETRABRANCHIATA (Gr. *tetra,* four; *bragchia,* gill). The order of *Cephalopoda* characterized by the possession of four gills.

TEXULARIA (Lat. *texilis,* woven). A genus of *Foraminifera.*

THECA (Gr. *thēkē,* a sheath). A genus of Pteropods.

THECODONTOSAURUS (Gr. *theke,* a sheath; *odous,* tooth; *saura,* lizard). A genus of "Thecodont" Reptiles, so named in allusion to the fact that the teeth are sunk in distinct sockets.

THERIODONT (Gr. *thērion,* a beast; *odous,* tooth). A group of Reptiles so named by Owen in allusion to the Mammalian character of their teeth.

THORAX (Gr. a breastplate). The region of the chest.

THYLACOLEO (Gr. *thulakos,* a pouch; *leo,* a lion). An extinct genus of Marsupials.

TRIGONIA (Gr. *treis,* three; *gonia,* angle.) A genus of Bivalve Molluscs.

TRIGONOCARPON (Gr. *treis,* three; *gonia,* angle; *karpos,* fruit). A genus founded on fossil fruits of a three-angled form.

TRILOBITA (Gr. *treis,* three; *lobos,* a lobe). An extinct order of *Crustaceans.*

TRINUCLEUS (Lat. *tris,* three; *nucleus,* a kernel.) A genus of Trilobites.

TROGONTHERIUM (Gr. *trogo,* I gnaw; *thērion,* beast.) An extinct genus of Beavers.

TUBICOLA (Lat. *tuba,* a tube; and *colo,* I inhabit). The order of *Annelida* which construct a tubular case in which they protect themselves.

TUBICOLOUS. Inhabiting a tube.

TUNICATA (Lat. *tunica,* a cloak). A class of *Molluscoida* which are enveloped in a tough leathery case or "test."

TURBINATED (Lat. *turbo,* a top). Top-shaped; conical with a round base.

TURRILITES (Lat. *turris,* a tower). A genus of the *Ammonitidæ.*

U

UMBO (Lat. the boss of a shield). The beak of a bivalve shell.

UNGUICULATE (Lat. *unguis,* nail). Furnished with claws.

UNGULATA (Lat. *ungula,* hoof). The order of *Mammals* comprising the Hoofed Quadrupeds.

UNGULATE. Furnished with expanded nails constituting hoofs.

UNILOCULAR (Lat. *unus,* one; and *loculus,* a little purse. Possessing a single cavity or chamber. Applied to the shells of *Foraminifera* and *Mollusca.*

UNIVALVE (Lat. *unus,* one; *valvæ,* folding-doors). A shell composed of a single piece or valve.

URODELA (Gr. *oura,* tail; *delos,* visible). The order of the Tailed Amphibians (Newts, etc.).

VENTRAL (Lat. *venter,* the stomach). Relating to the inferior surface of the body.

VENTRICULITES (Lat. *ventriculum,* a little stomach). A genus of siliceous Sponges.

VERMIFORM (Lat. *vermis,* worm; and *forma,* form). Worm-like.

VERTEBRA (Lat. *verto,* I turn). One of the bony segments of the vertebral column or backbone.

VERTEBRATA (Lat. *vertebra,* a bone of the back, from *vertere,* to turn.) The division of the Animal Kingdom roughly characterized by the possession of a backbone.

VESICLE (Lat. *vesica,* a bladder). A little sac or cyst.

W

WHORL. The spiral turn of a univalve shell.

X

XIPHOSURA (Gr. *xiphos,* a sword; and *oura,* tail). An order of *Crustacea,* comprising the *Limuli* or King-Crabs, characterized by their long sword-like tails.

XYLOBIUS (Gr. *xulon,* wood; *bios,* life). An extinct genus of Myriapods, named in allusion to the fact that the animals lived on decaying wood.

Z

ZAPHRENTIS (proper name). A genus of Rugose Corals.

ZEUGLODONTIDÆ (Gr. *zeuglē,* a yolk; *odous,* a tooth). An extinct family of Cetaceans, in which the molar teeth are two-fanged, and look as if composed of two parts united by a neck.

ZOOPHYTE (Gr. *zoön,* animal; *phuton,* plant). Loosely applied to many plant-like animals, such as Sponges, Corals, Sea-anemones, Sea-mats, etc.).

INDEX.